BrainComix

T0327320

JEAN-FRANÇOIS MARMION & MONSIEUR B
COLOR BY MANOLO

BrainComix

graphic mundi

Table of Contents

Chapter 1

WHY A BRAIN
INSTEAD OF NOTHING?

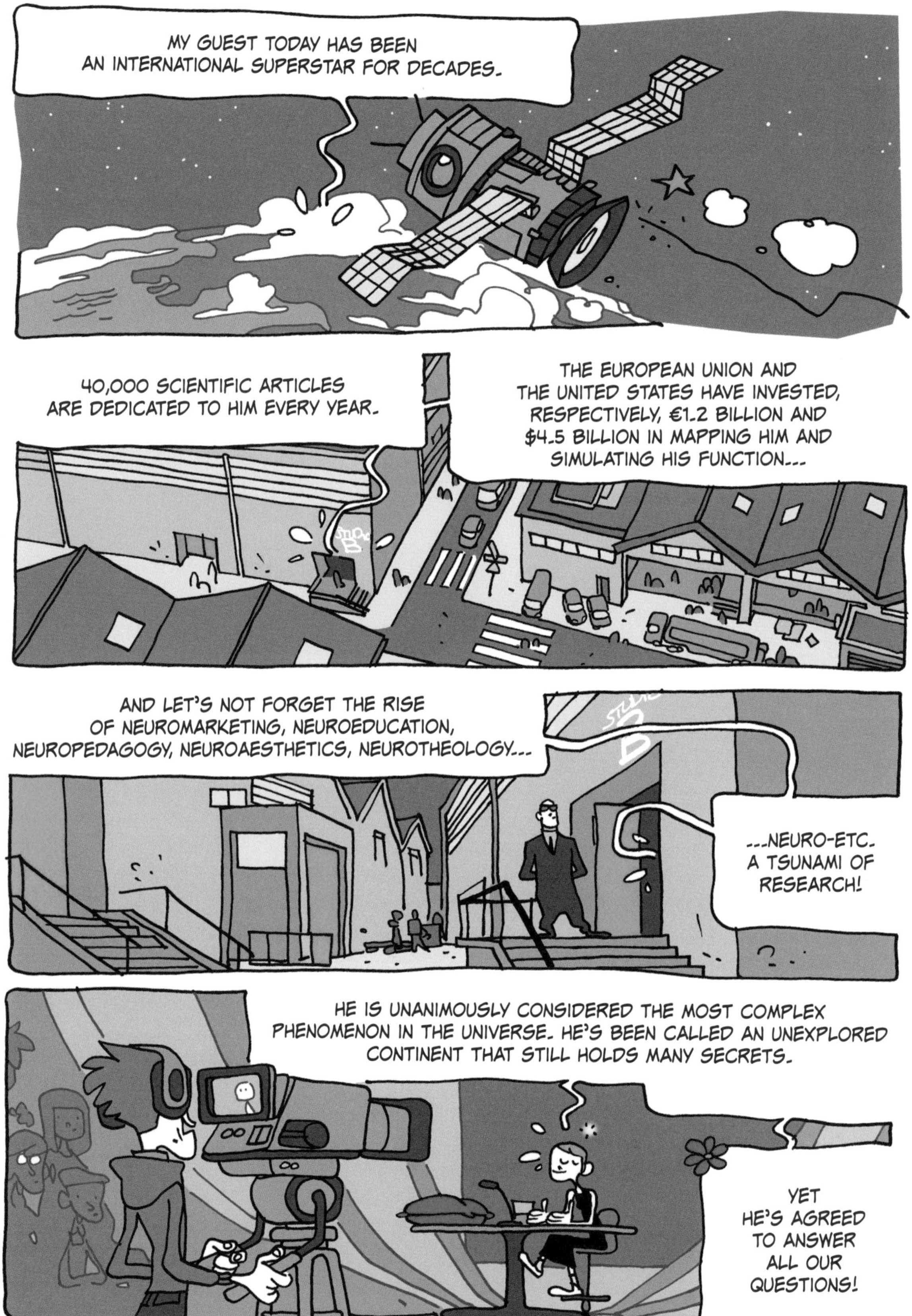
MY GUEST TODAY HAS BEEN AN INTERNATIONAL SUPERSTAR FOR DECADES.
40,000 SCIENTIFIC ARTICLES ARE DEDICATED TO HIM EVERY YEAR.
THE EUROPEAN UNION AND THE UNITED STATES HAVE INVESTED, RESPECTIVELY, €1.2 BILLION AND $4.5 BILLION IN MAPPING HIM AND SIMULATING HIS FUNCTION...
AND LET'S NOT FORGET THE RISE OF NEUROMARKETING, NEUROEDUCATION, NEUROPEDAGOGY, NEUROAESTHETICS, NEUROTHEOLOGY...
...NEURO-ETC. A TSUNAMI OF RESEARCH!
HE IS UNANIMOUSLY CONSIDERED THE MOST COMPLEX PHENOMENON IN THE UNIVERSE. HE'S BEEN CALLED AN UNEXPLORED CONTINENT THAT STILL HOLDS MANY SECRETS.
YET HE'S AGREED TO ANSWER ALL OUR QUESTIONS!

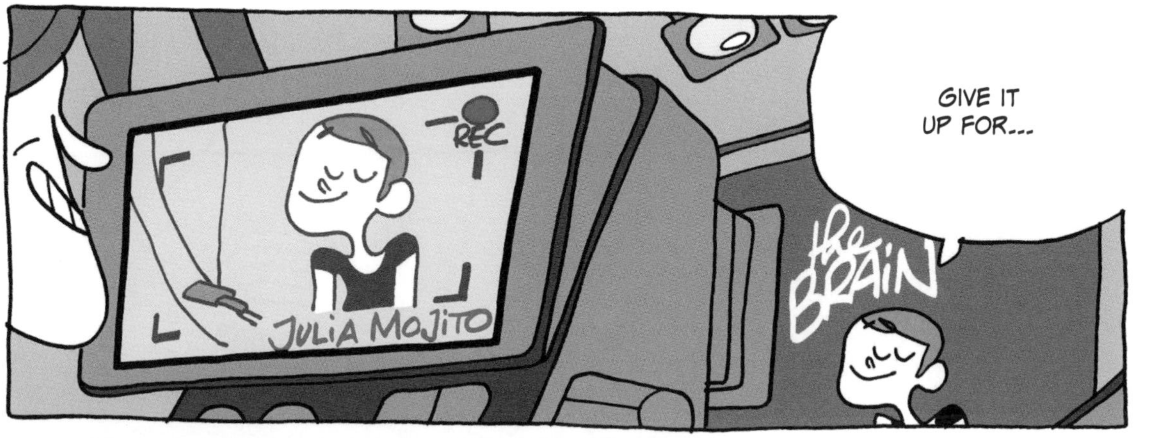
REC
JULIA MOJITO
GIVE IT UP FOR...
the BRAIN

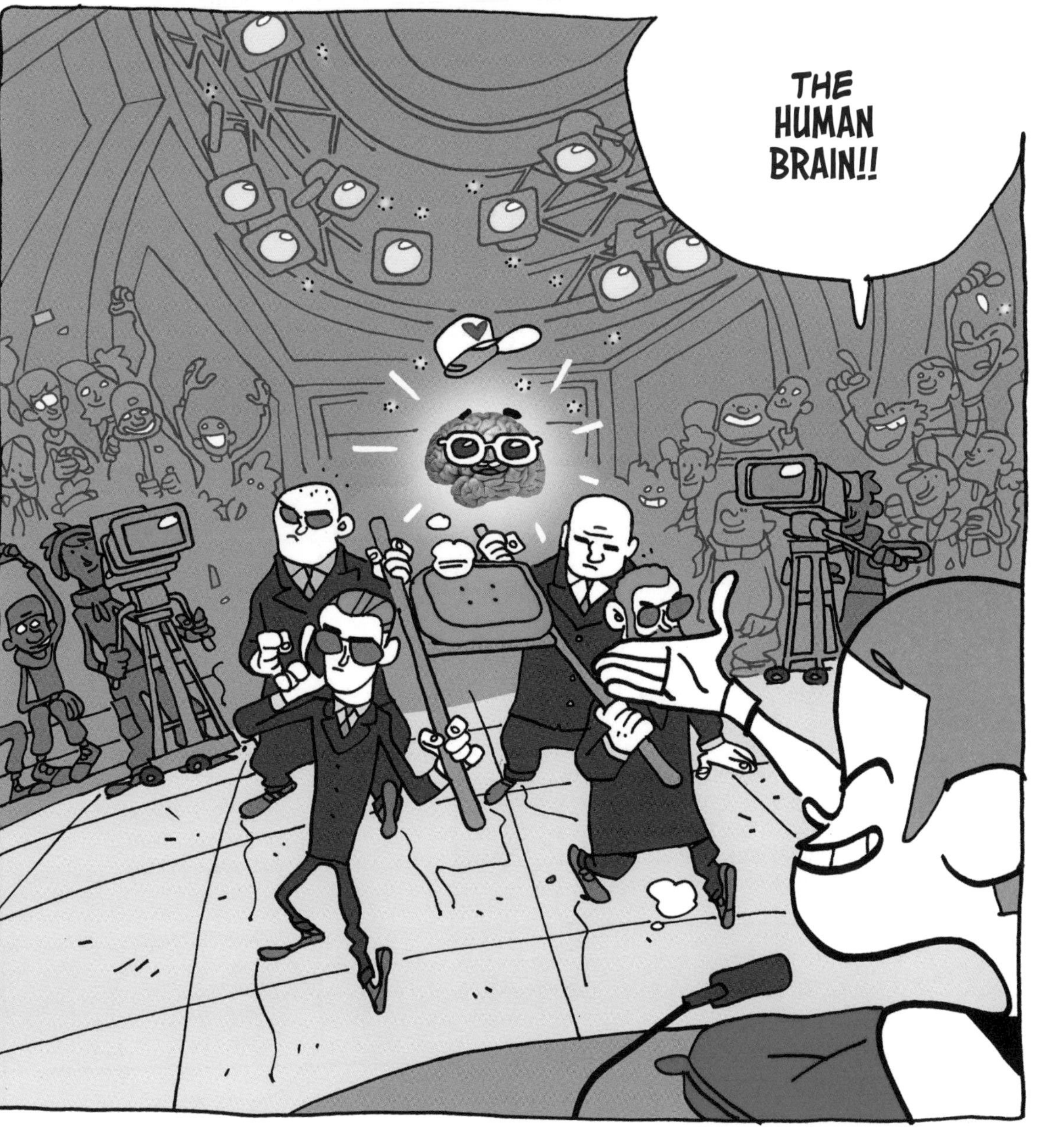
THE HUMAN BRAIN!!

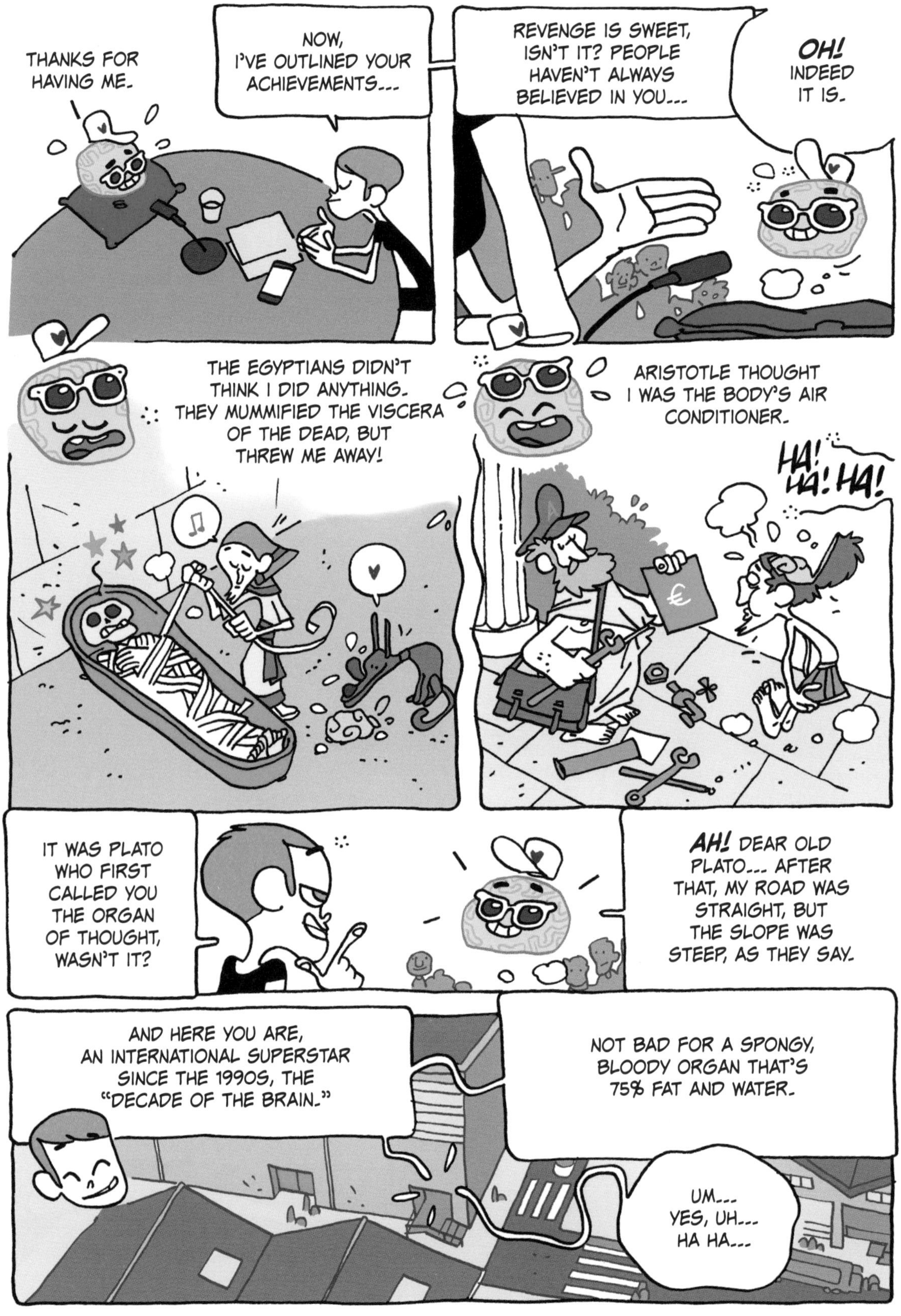
THANKS FOR HAVING ME.
NOW, I'VE OUTLINED YOUR ACHIEVEMENTS...
REVENGE IS SWEET, ISN'T IT? PEOPLE HAVEN'T ALWAYS BELIEVED IN YOU...
OH! INDEED IT IS.
THE EGYPTIANS DIDN'T THINK I DID ANYTHING. THEY MUMMIFIED THE VISCERA OF THE DEAD, BUT THREW ME AWAY!
ARISTOTLE THOUGHT I WAS THE BODY'S AIR CONDITIONER.
HA! HA! HA!
IT WAS PLATO WHO FIRST CALLED YOU THE ORGAN OF THOUGHT, WASN'T IT?
AH! DEAR OLD PLATO... AFTER THAT, MY ROAD WAS STRAIGHT, BUT THE SLOPE WAS STEEP, AS THEY SAY.
AND HERE YOU ARE, AN INTERNATIONAL SUPERSTAR SINCE THE 1990S, THE "DECADE OF THE BRAIN."
NOT BAD FOR A SPONGY, BLOODY ORGAN THAT'S 75% FAT AND WATER.
UM... YES, UH... HA HA...

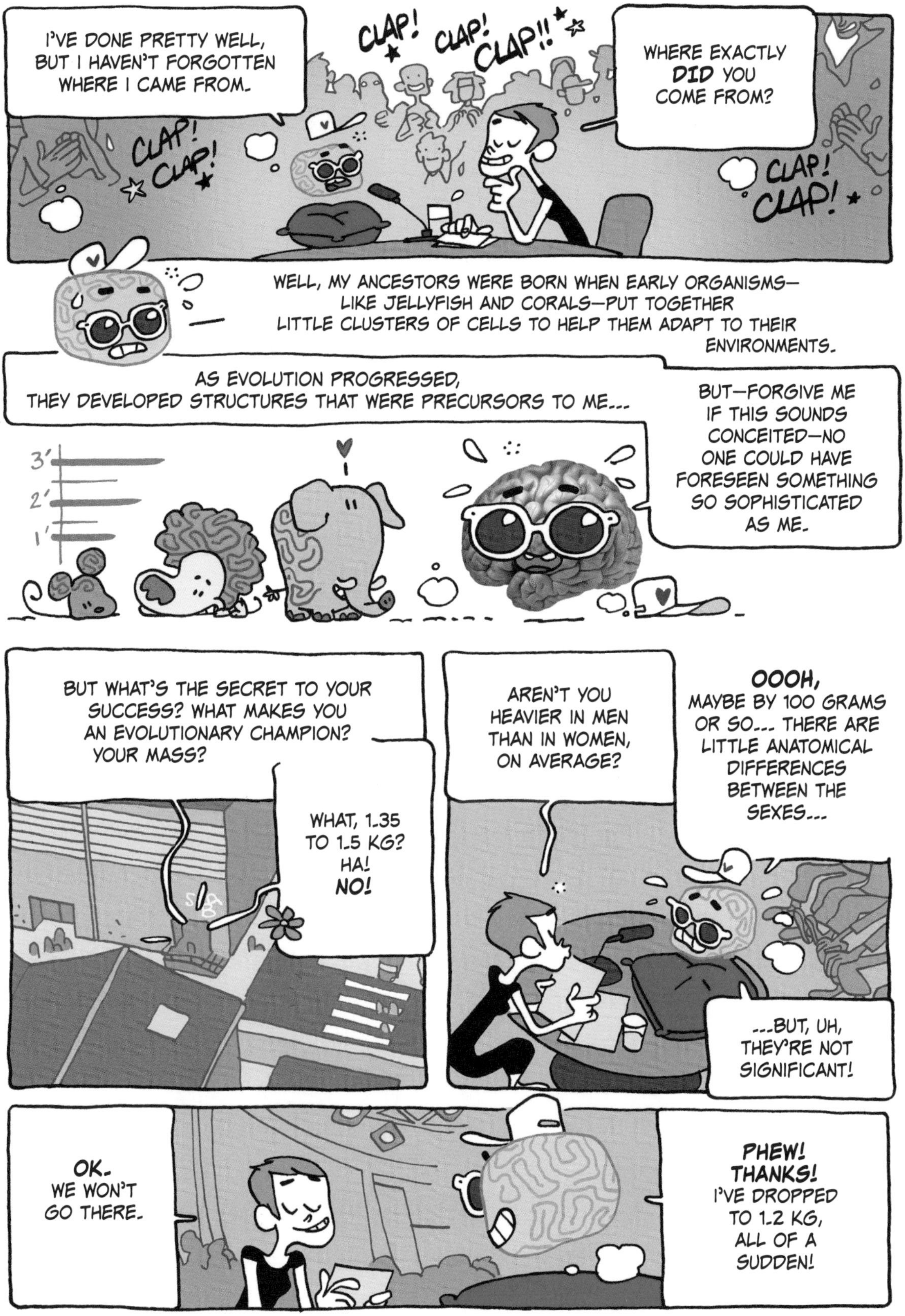
I'VE DONE PRETTY WELL, BUT I HAVEN'T FORGOTTEN WHERE I CAME FROM.
CLAP! CLAP! CLAP!!
WHERE EXACTLY DID YOU COME FROM?
CLAP! CLAP!
CLAP! CLAP!
WELL, MY ANCESTORS WERE BORN WHEN EARLY ORGANISMS—LIKE JELLYFISH AND CORALS—PUT TOGETHER LITTLE CLUSTERS OF CELLS TO HELP THEM ADAPT TO THEIR ENVIRONMENTS.
AS EVOLUTION PROGRESSED, THEY DEVELOPED STRUCTURES THAT WERE PRECURSORS TO ME...
BUT—FORGIVE ME IF THIS SOUNDS CONCEITED—NO ONE COULD HAVE FORESEEN SOMETHING SO SOPHISTICATED AS ME.
3'
2'
1'
BUT WHAT'S THE SECRET TO YOUR SUCCESS? WHAT MAKES YOU AN EVOLUTIONARY CHAMPION? YOUR MASS?
WHAT, 1.35 TO 1.5 KG? HA! NO!
AREN'T YOU HEAVIER IN MEN THAN IN WOMEN, ON AVERAGE?
OOOH, MAYBE BY 100 GRAMS OR SO... THERE ARE LITTLE ANATOMICAL DIFFERENCES BETWEEN THE SEXES...
...BUT, UH, THEY'RE NOT SIGNIFICANT!
OK. WE WON'T GO THERE.
PHEW! THANKS! I'VE DROPPED TO 1.2 KG, ALL OF A SUDDEN!

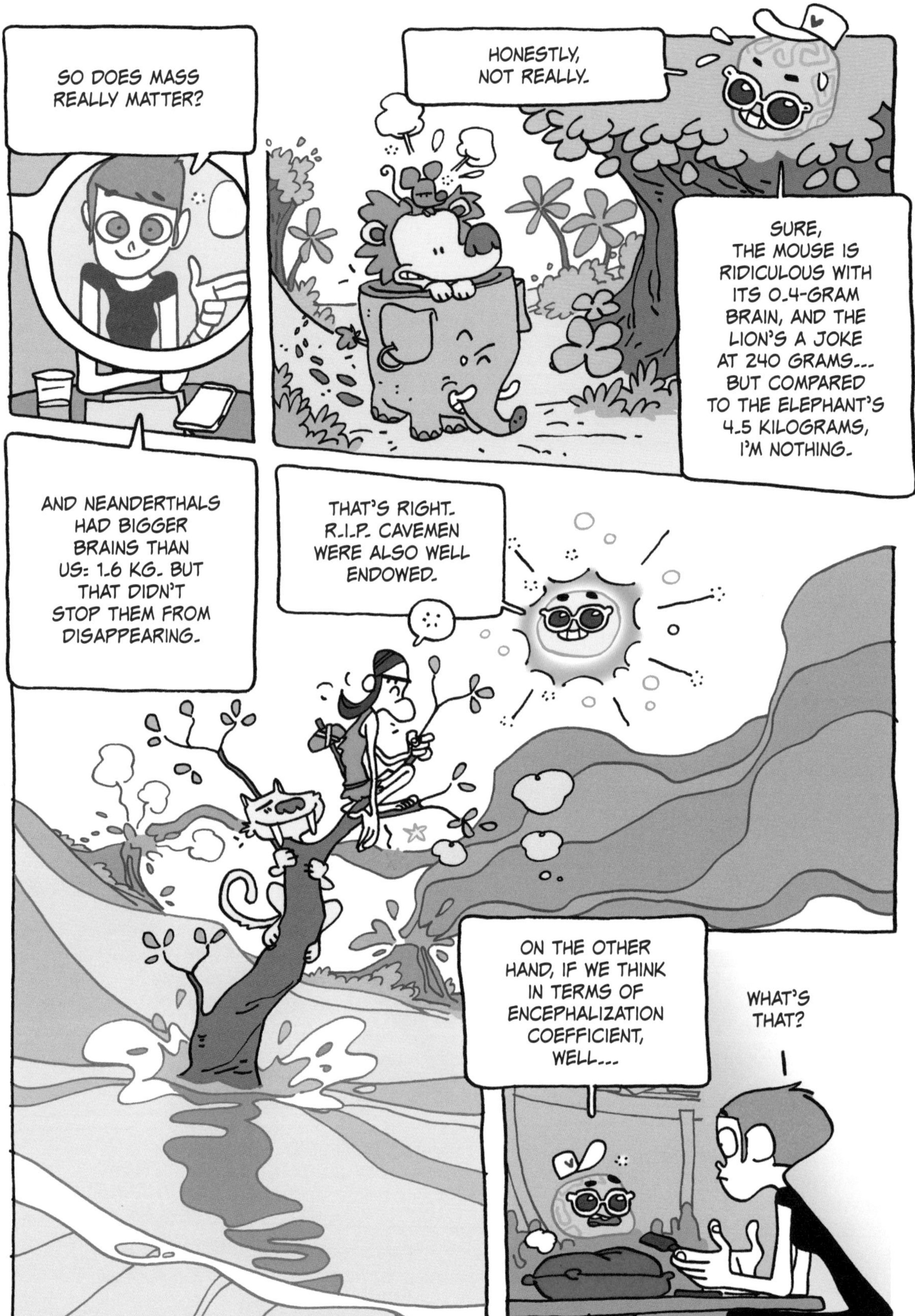
SO DOES MASS REALLY MATTER?
HONESTLY, NOT REALLY.
SURE, THE MOUSE IS RIDICULOUS WITH ITS 0.4-GRAM BRAIN, AND THE LION'S A JOKE AT 240 GRAMS... BUT COMPARED TO THE ELEPHANT'S 4.5 KILOGRAMS, I'M NOTHING.
AND NEANDERTHALS HAD BIGGER BRAINS THAN US: 1.6 KG. BUT THAT DIDN'T STOP THEM FROM DISAPPEARING.
THAT'S RIGHT. R.I.P. CAVEMEN WERE ALSO WELL ENDOWED.
ON THE OTHER HAND, IF WE THINK IN TERMS OF ENCEPHALIZATION COEFFICIENT, WELL...
WHAT'S THAT?

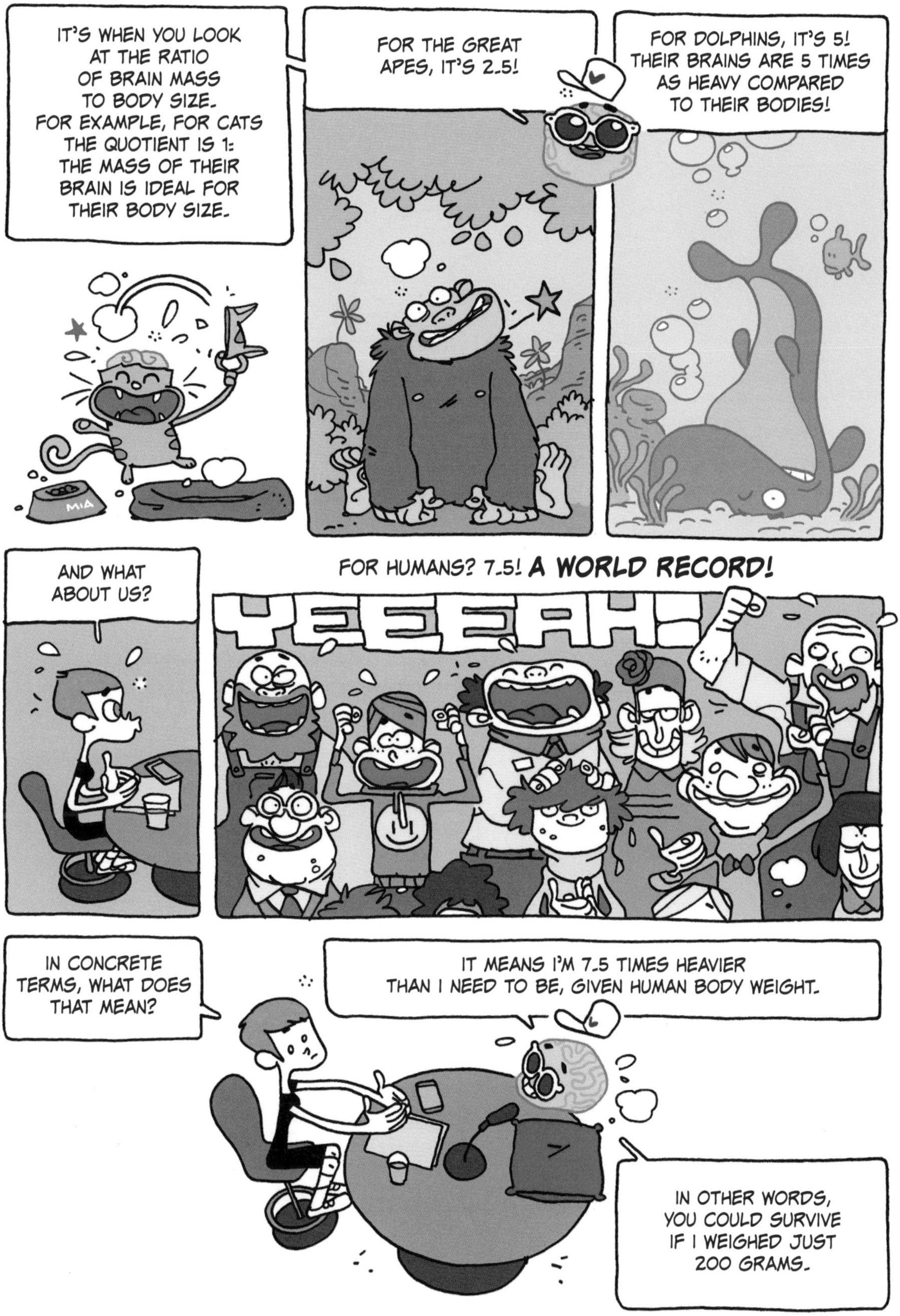
IT'S WHEN YOU LOOK AT THE RATIO OF BRAIN MASS TO BODY SIZE. FOR EXAMPLE, FOR CATS THE QUOTIENT IS 1: THE MASS OF THEIR BRAIN IS IDEAL FOR THEIR BODY SIZE.
MIA
FOR THE GREAT APES, IT'S 2.5!
FOR DOLPHINS, IT'S 5! THEIR BRAINS ARE 5 TIMES AS HEAVY COMPARED TO THEIR BODIES!
AND WHAT ABOUT US?
FOR HUMANS? 7.5! A WORLD RECORD!
YEEEAH!
IN CONCRETE TERMS, WHAT DOES THAT MEAN?
IT MEANS I'M 7.5 TIMES HEAVIER THAN I NEED TO BE, GIVEN HUMAN BODY WEIGHT.
IN OTHER WORDS, YOU COULD SURVIVE IF I WEIGHED JUST 200 GRAMS.

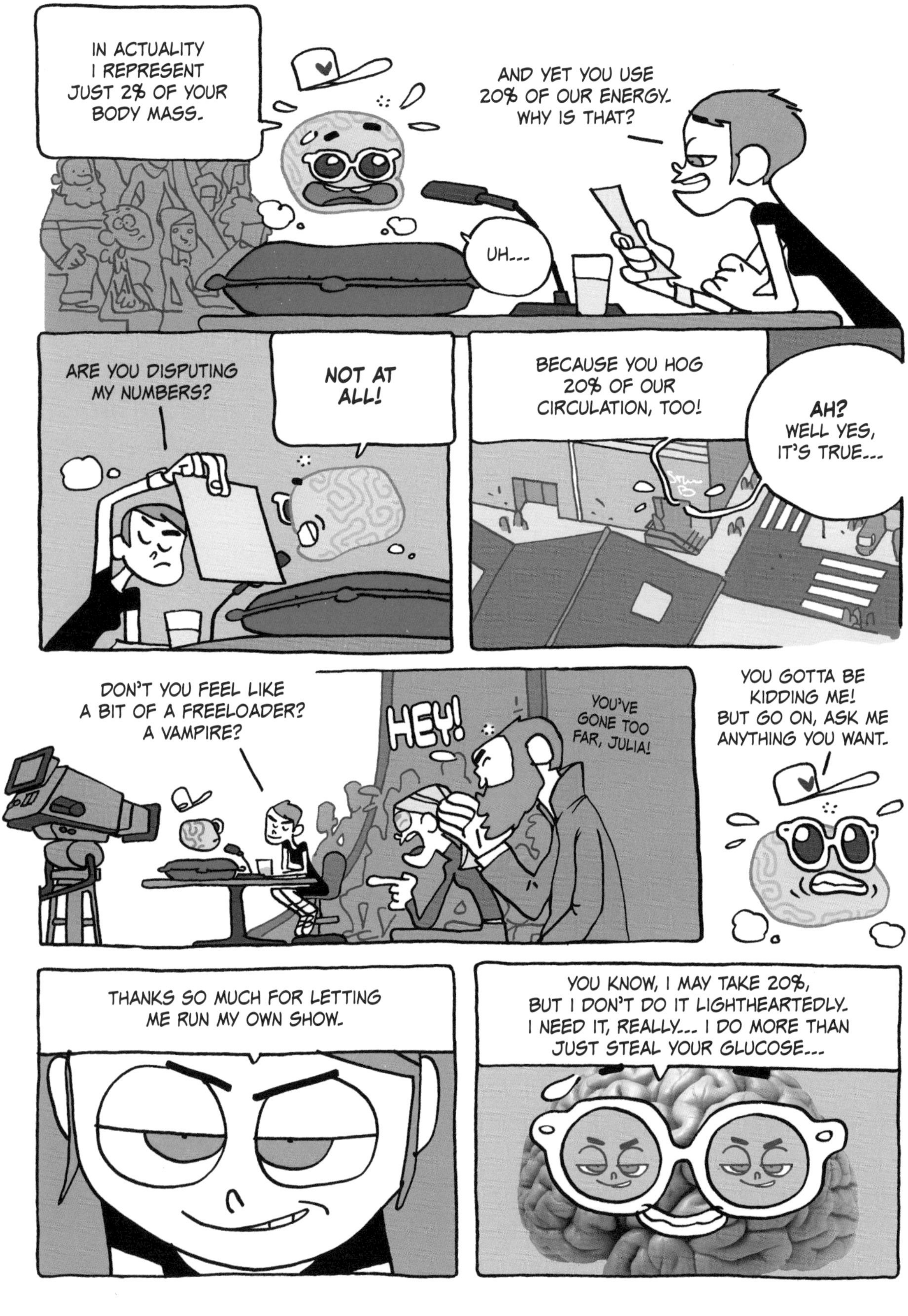
IN ACTUALITY I REPRESENT JUST 2% OF YOUR BODY MASS.
AND YET YOU USE 20% OF OUR ENERGY. WHY IS THAT?
UH...
ARE YOU DISPUTING MY NUMBERS?
NOT AT ALL!
BECAUSE YOU HOG 20% OF OUR CIRCULATION, TOO!
AH? WELL YES, IT'S TRUE...
DON'T YOU FEEL LIKE A BIT OF A FREELOADER? A VAMPIRE?
HEY!
YOU'VE GONE TOO FAR, JULIA!
YOU GOTTA BE KIDDING ME! BUT GO ON, ASK ME ANYTHING YOU WANT.
THANKS SO MUCH FOR LETTING ME RUN MY OWN SHOW.
YOU KNOW, I MAY TAKE 20%, BUT I DON'T DO IT LIGHTHEARTEDLY. I NEED IT, REALLY... I DO MORE THAN JUST STEAL YOUR GLUCOSE...

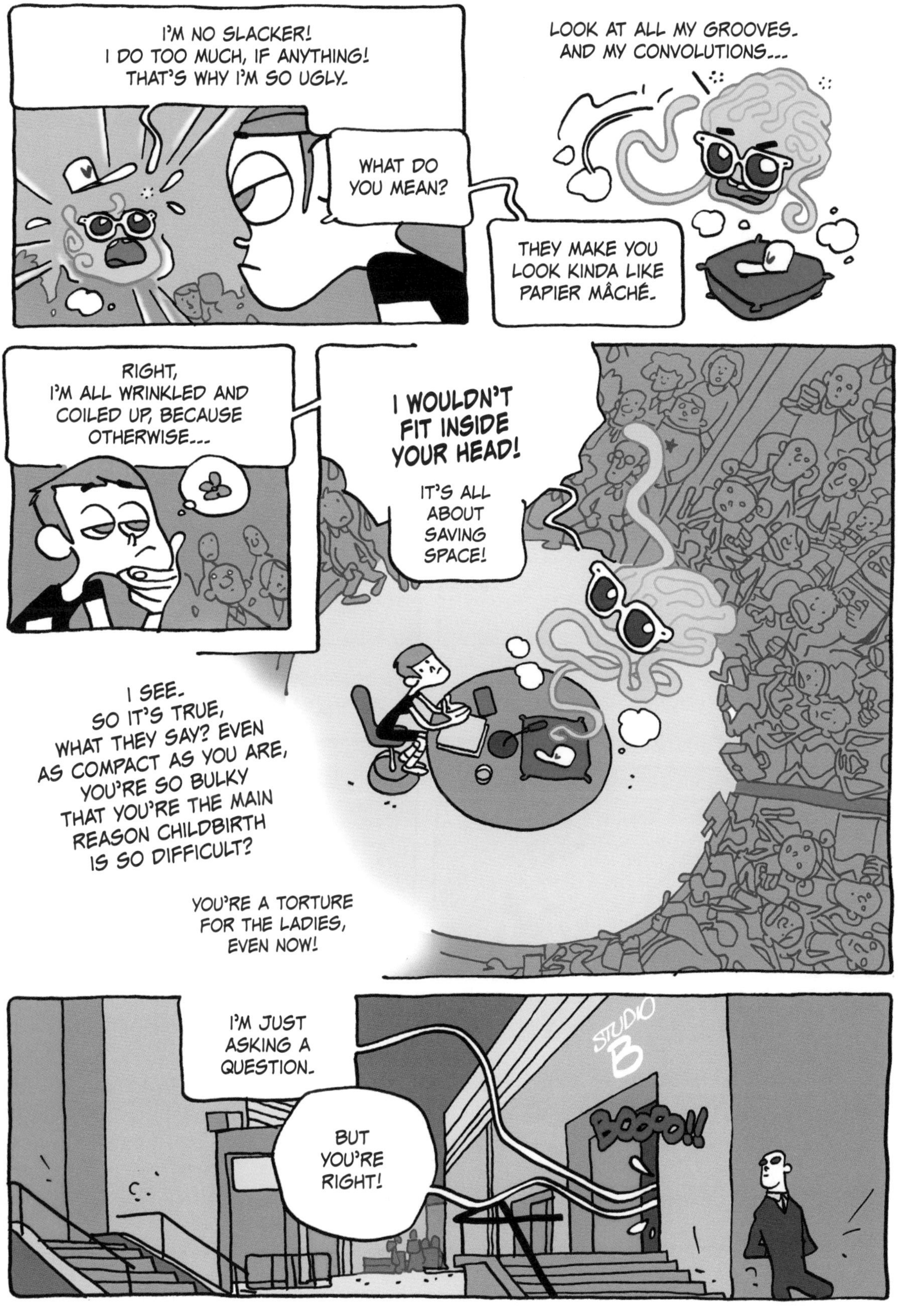
I'M NO SLACKER!
I DO TOO MUCH, IF ANYTHING!
THAT'S WHY I'M SO UGLY.
WHAT DO YOU MEAN?
LOOK AT ALL MY GROOVES.
AND MY CONVOLUTIONS...
THEY MAKE YOU LOOK KINDA LIKE PAPIER MÂCHÉ.
RIGHT, I'M ALL WRINKLED AND COILED UP, BECAUSE OTHERWISE...
I WOULDN'T FIT INSIDE YOUR HEAD!
IT'S ALL ABOUT SAVING SPACE!
I SEE.
SO IT'S TRUE, WHAT THEY SAY? EVEN AS COMPACT AS YOU ARE, YOU'RE SO BULKY THAT YOU'RE THE MAIN REASON CHILDBIRTH IS SO DIFFICULT?
YOU'RE A TORTURE FOR THE LADIES, EVEN NOW!
I'M JUST ASKING A QUESTION.
BUT YOU'RE RIGHT!
STUDIO B
BOOPO!!

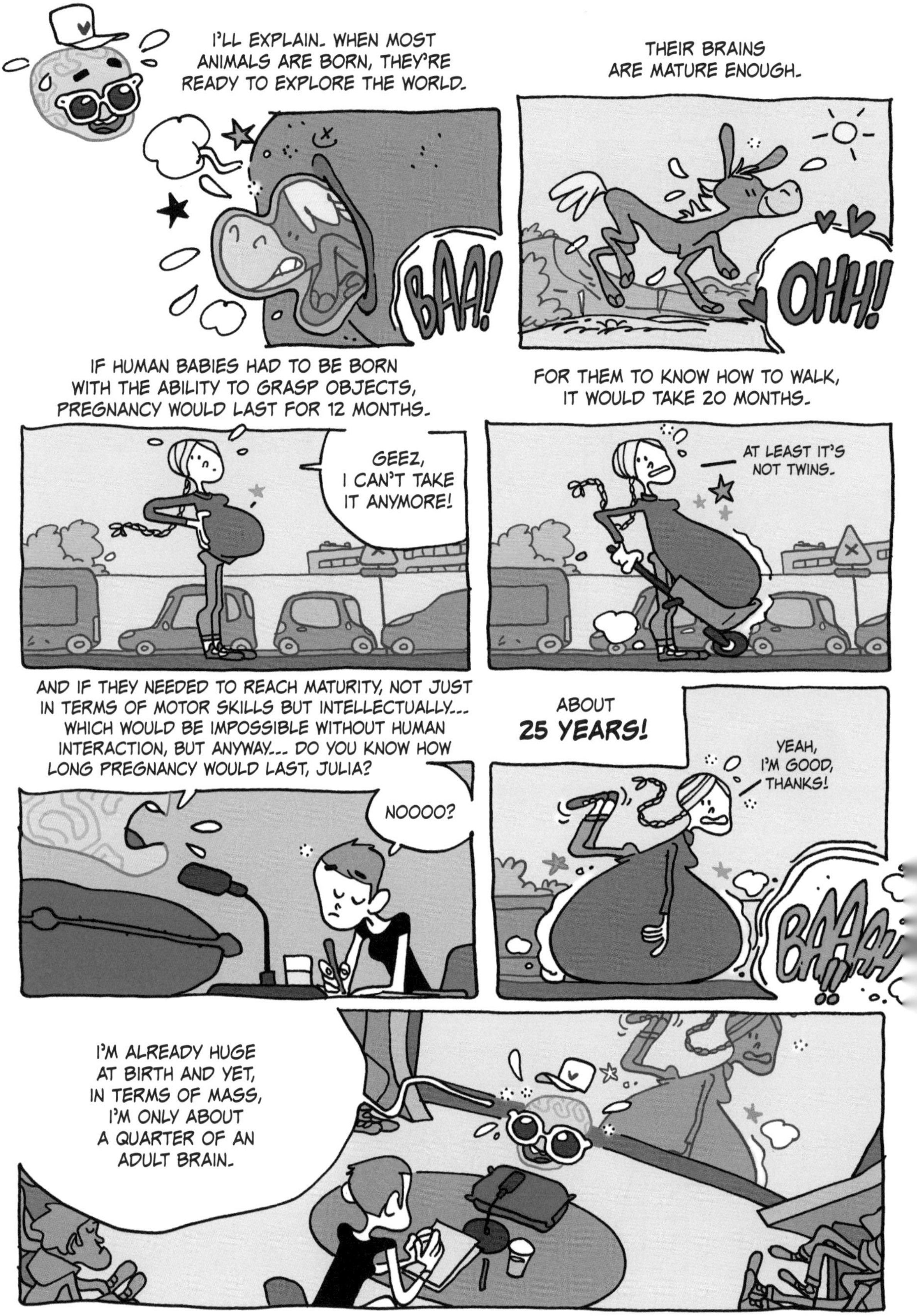
I'LL EXPLAIN. WHEN MOST ANIMALS ARE BORN, THEY'RE READY TO EXPLORE THE WORLD.
BAA!
THEIR BRAINS ARE MATURE ENOUGH.
OHH!
IF HUMAN BABIES HAD TO BE BORN WITH THE ABILITY TO GRASP OBJECTS, PREGNANCY WOULD LAST FOR 12 MONTHS.
GEEZ, I CAN'T TAKE IT ANYMORE!
FOR THEM TO KNOW HOW TO WALK, IT WOULD TAKE 20 MONTHS.
AT LEAST IT'S NOT TWINS.
AND IF THEY NEEDED TO REACH MATURITY, NOT JUST IN TERMS OF MOTOR SKILLS BUT INTELLECTUALLY... WHICH WOULD BE IMPOSSIBLE WITHOUT HUMAN INTERACTION, BUT ANYWAY... DO YOU KNOW HOW LONG PREGNANCY WOULD LAST, JULIA?
NOOOO?
ABOUT 25 YEARS!
YEAH, I'M GOOD, THANKS!
I'M ALREADY HUGE AT BIRTH AND YET, IN TERMS OF MASS, I'M ONLY ABOUT A QUARTER OF AN ADULT BRAIN.

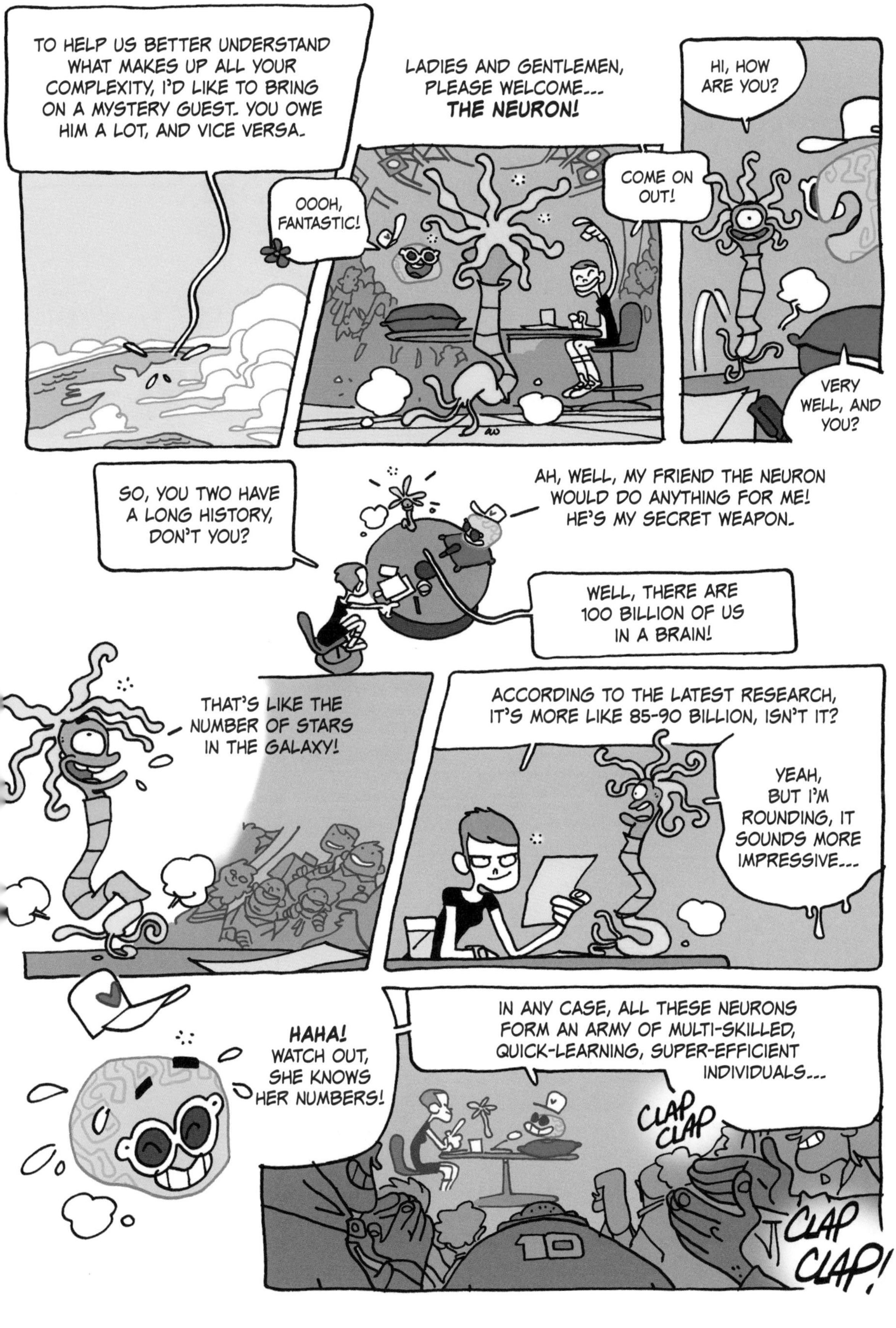
TO HELP US BETTER UNDERSTAND WHAT MAKES UP ALL YOUR COMPLEXITY, I'D LIKE TO BRING ON A MYSTERY GUEST. YOU OWE HIM A LOT, AND VICE VERSA.
LADIES AND GENTLEMEN, PLEASE WELCOME... **THE NEURON!**
OOOH, FANTASTIC!
COME ON OUT!
HI, HOW ARE YOU?
VERY WELL, AND YOU?
SO, YOU TWO HAVE A LONG HISTORY, DON'T YOU?
AH, WELL, MY FRIEND THE NEURON WOULD DO ANYTHING FOR ME! HE'S MY SECRET WEAPON.
WELL, THERE ARE 100 BILLION OF US IN A BRAIN!
THAT'S LIKE THE NUMBER OF STARS IN THE GALAXY!
ACCORDING TO THE LATEST RESEARCH, IT'S MORE LIKE 85-90 BILLION, ISN'T IT?
YEAH, BUT I'M ROUNDING, IT SOUNDS MORE IMPRESSIVE...
HAHA! WATCH OUT, SHE KNOWS HER NUMBERS!
IN ANY CASE, ALL THESE NEURONS FORM AN ARMY OF MULTI-SKILLED, QUICK-LEARNING, SUPER-EFFICIENT INDIVIDUALS...
CLAP CLAP
CLAP CLAP!

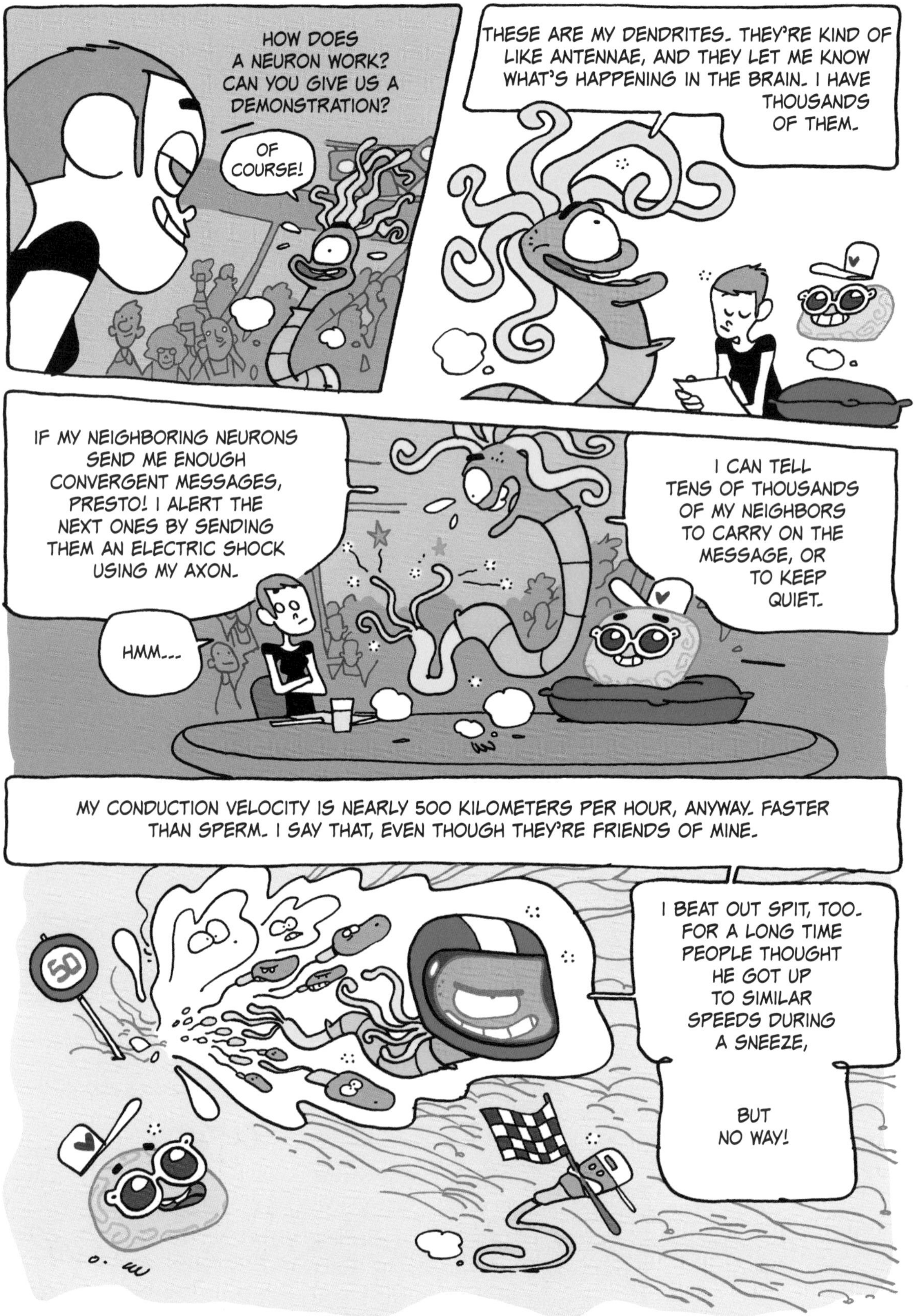

HOW DOES A NEURON WORK? CAN YOU GIVE US A DEMONSTRATION?
OF COURSE!
THESE ARE MY DENDRITES. THEY'RE KIND OF LIKE ANTENNAE, AND THEY LET ME KNOW WHAT'S HAPPENING IN THE BRAIN. I HAVE THOUSANDS OF THEM.
IF MY NEIGHBORING NEURONS SEND ME ENOUGH CONVERGENT MESSAGES, PRESTO! I ALERT THE NEXT ONES BY SENDING THEM AN ELECTRIC SHOCK USING MY AXON.
I CAN TELL TENS OF THOUSANDS OF MY NEIGHBORS TO CARRY ON THE MESSAGE, OR TO KEEP QUIET.
HMM...
MY CONDUCTION VELOCITY IS NEARLY 500 KILOMETERS PER HOUR, ANYWAY. FASTER THAN SPERM. I SAY THAT, EVEN THOUGH THEY'RE FRIENDS OF MINE.
I BEAT OUT SPIT, TOO. FOR A LONG TIME PEOPLE THOUGHT HE GOT UP TO SIMILAR SPEEDS DURING A SNEEZE,
BUT NO WAY!
50

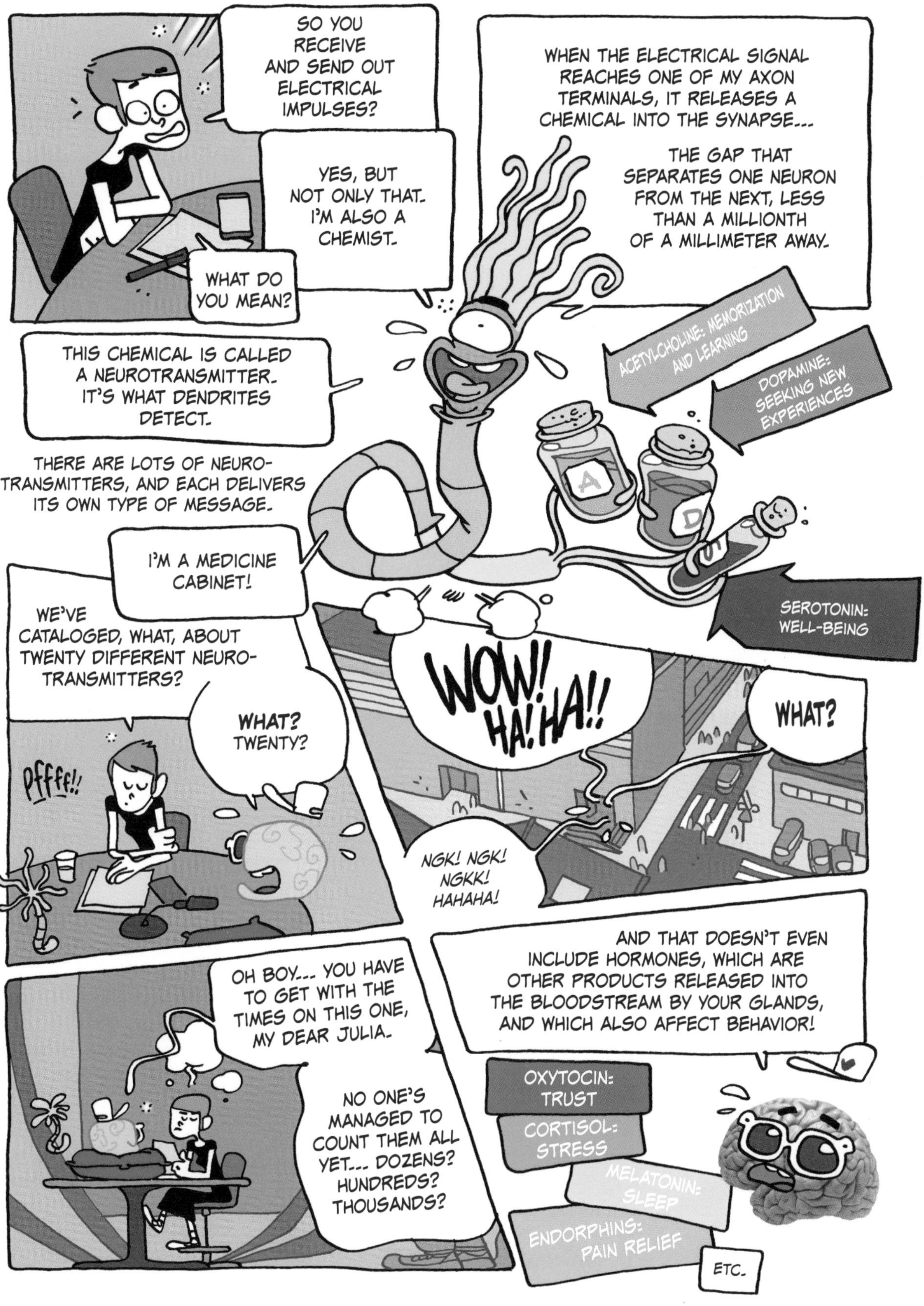
SO YOU RECEIVE AND SEND OUT ELECTRICAL IMPULSES?
YES, BUT NOT ONLY THAT. I'M ALSO A CHEMIST.
WHAT DO YOU MEAN?
WHEN THE ELECTRICAL SIGNAL REACHES ONE OF MY AXON TERMINALS, IT RELEASES A CHEMICAL INTO THE SYNAPSE...
THE GAP THAT SEPARATES ONE NEURON FROM THE NEXT, LESS THAN A MILLIONTH OF A MILLIMETER AWAY.
THIS CHEMICAL IS CALLED A NEUROTRANSMITTER. IT'S WHAT DENDRITES DETECT.
THERE ARE LOTS OF NEUROTRANSMITTERS, AND EACH DELIVERS ITS OWN TYPE OF MESSAGE.
ACETYLCHOLINE: MEMORIZATION AND LEARNING
DOPAMINE: SEEKING NEW EXPERIENCES
A
D
I'M A MEDICINE CABINET!
SEROTONIN: WELL-BEING
WE'VE CATALOGED, WHAT, ABOUT TWENTY DIFFERENT NEUROTRANSMITTERS?
WHAT? TWENTY?
PFFFF!!
WOW! HA! HA!!
WHAT?
NGK! NGK! NGKK! HAHAHA!
OH BOY... YOU HAVE TO GET WITH THE TIMES ON THIS ONE, MY DEAR JULIA.
NO ONE'S MANAGED TO COUNT THEM ALL YET... DOZENS? HUNDREDS? THOUSANDS?
AND THAT DOESN'T EVEN INCLUDE HORMONES, WHICH ARE OTHER PRODUCTS RELEASED INTO THE BLOODSTREAM BY YOUR GLANDS, AND WHICH ALSO AFFECT BEHAVIOR!
OXYTOCIN: TRUST
CORTISOL: STRESS
MELATONIN: SLEEP
ENDORPHINS: PAIN RELIEF
ETC.

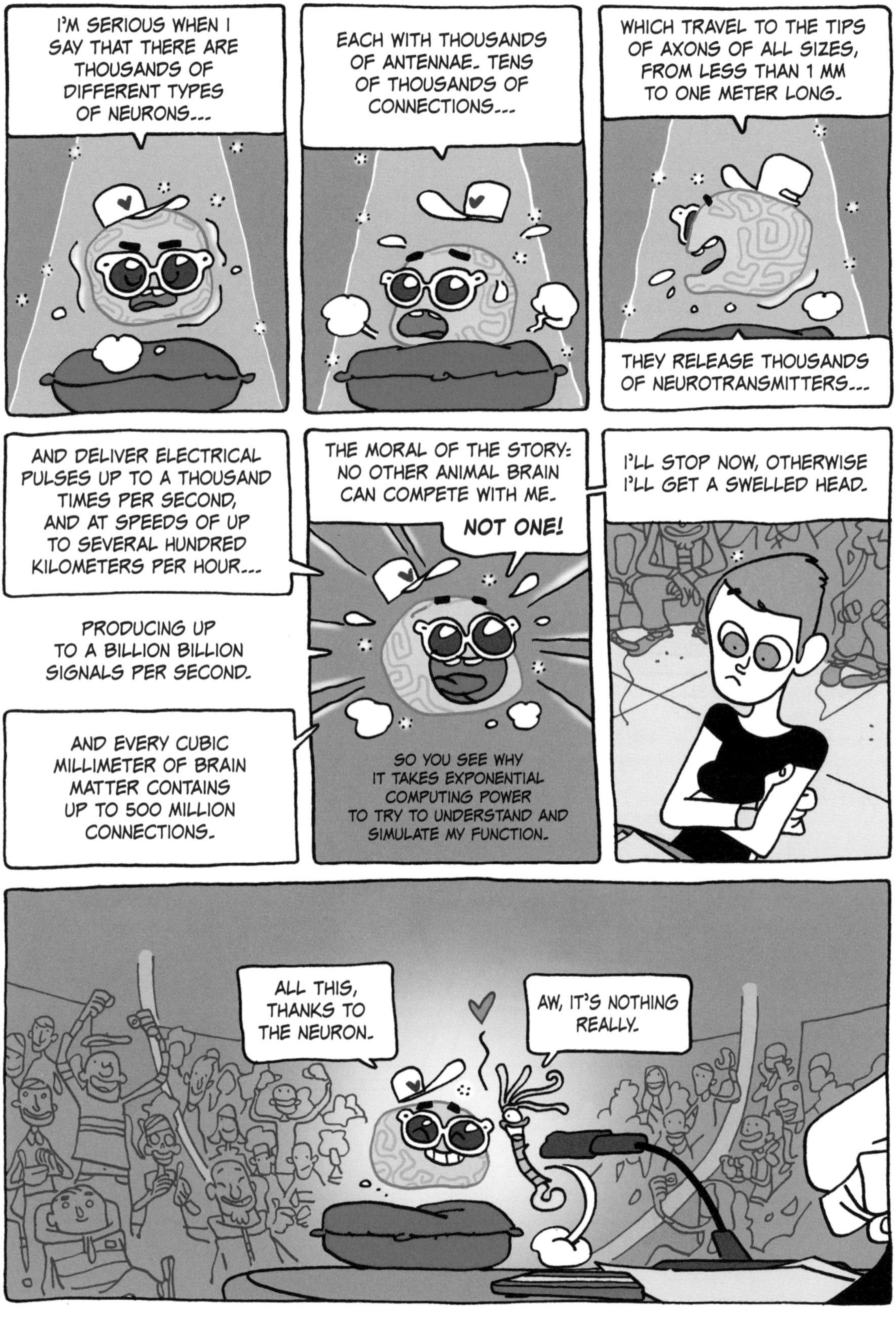
I'M SERIOUS WHEN I SAY THAT THERE ARE THOUSANDS OF DIFFERENT TYPES OF NEURONS...
EACH WITH THOUSANDS OF ANTENNAE. TENS OF THOUSANDS OF CONNECTIONS...
WHICH TRAVEL TO THE TIPS OF AXONS OF ALL SIZES, FROM LESS THAN 1 MM TO ONE METER LONG.
THEY RELEASE THOUSANDS OF NEUROTRANSMITTERS...
AND DELIVER ELECTRICAL PULSES UP TO A THOUSAND TIMES PER SECOND, AND AT SPEEDS OF UP TO SEVERAL HUNDRED KILOMETERS PER HOUR...
PRODUCING UP TO A BILLION BILLION SIGNALS PER SECOND.
AND EVERY CUBIC MILLIMETER OF BRAIN MATTER CONTAINS UP TO 500 MILLION CONNECTIONS.
THE MORAL OF THE STORY: NO OTHER ANIMAL BRAIN CAN COMPETE WITH ME.
NOT ONE!
SO YOU SEE WHY IT TAKES EXPONENTIAL COMPUTING POWER TO TRY TO UNDERSTAND AND SIMULATE MY FUNCTION.
I'LL STOP NOW, OTHERWISE I'LL GET A SWELLED HEAD.
ALL THIS, THANKS TO THE NEURON.
AW, IT'S NOTHING REALLY.

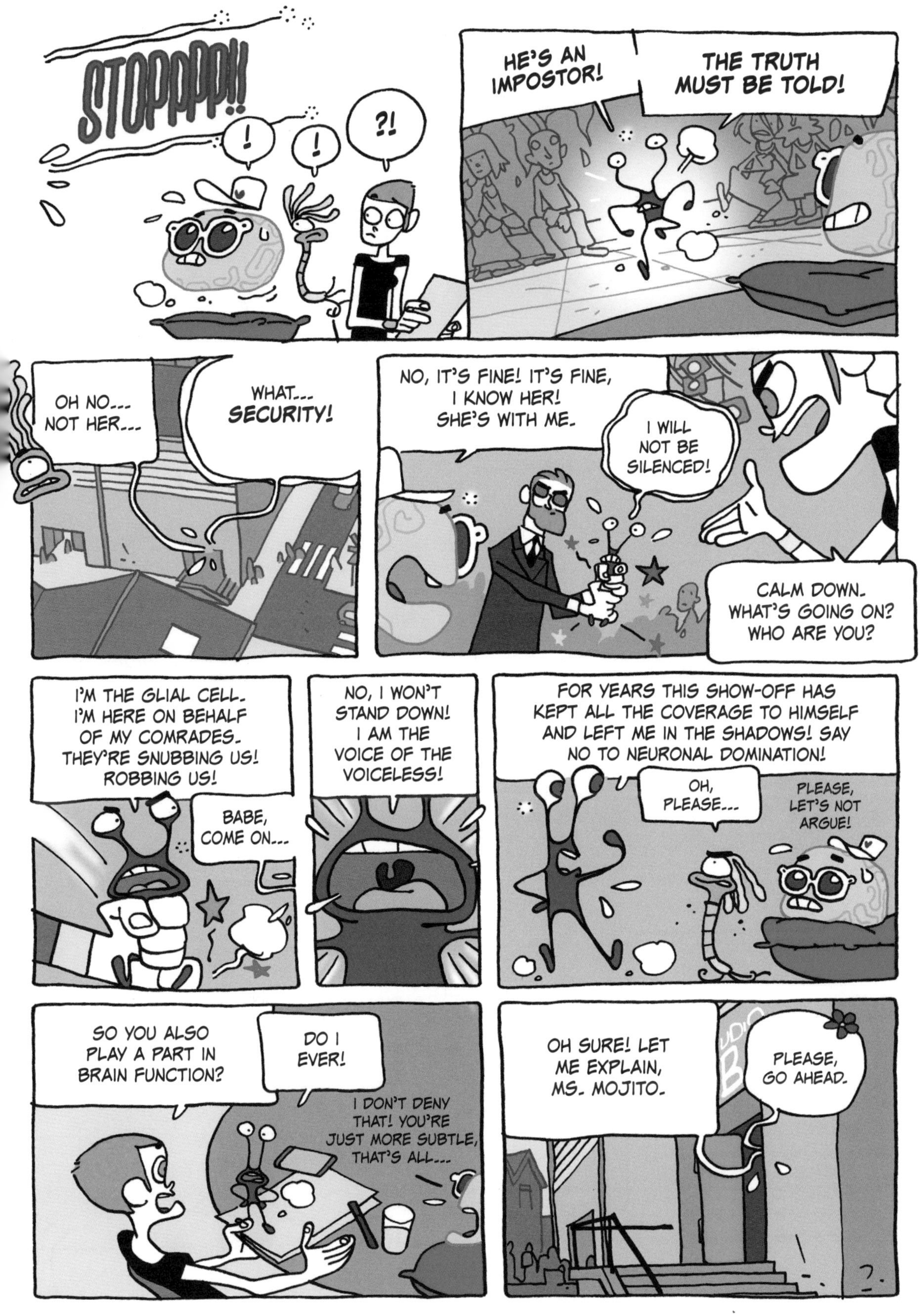
STOPPPP!!
!
!
?!
HE'S AN IMPOSTOR!
THE TRUTH MUST BE TOLD!
OH NO... NOT HER...
WHAT... SECURITY!
NO, IT'S FINE! IT'S FINE, I KNOW HER! SHE'S WITH ME.
I WILL NOT BE SILENCED!
CALM DOWN. WHAT'S GOING ON? WHO ARE YOU?
I'M THE GLIAL CELL. I'M HERE ON BEHALF OF MY COMRADES. THEY'RE SNUBBING US! ROBBING US!
BABE, COME ON...
NO, I WON'T STAND DOWN! I AM THE VOICE OF THE VOICELESS!
FOR YEARS THIS SHOW-OFF HAS KEPT ALL THE COVERAGE TO HIMSELF AND LEFT ME IN THE SHADOWS! SAY NO TO NEURONAL DOMINATION!
OH, PLEASE...
PLEASE, LET'S NOT ARGUE!
SO YOU ALSO PLAY A PART IN BRAIN FUNCTION?
DO I EVER!
I DON'T DENY THAT! YOU'RE JUST MORE SUBTLE, THAT'S ALL...
OH SURE! LET ME EXPLAIN, MS. MOJITO.
PLEASE, GO AHEAD.

THE MAJORITY OF BRAIN CELLS— AND I MEAN IT, THE MAJORITY— ARE NOT NEURONS BUT RATHER GLIAL CELLS.

"GLIAL"?

YES, "GLIAL" LIKE "GLUE"! WE'VE BEEN RELEGATED TO THE RANK OF CONNECTIVE TISSUE! SUPPORTIVE CEMENT!

FILLER!

AT BEST, WE'RE ACCUSED OF BEING LIKE MAIDS, THERE TO HELP THE SACRED NEURON AND CLEAR AWAY HIS WASTE! WE'RE STIGMATIZED!

BUT THAT'S NOT MY FAULT! FOR A LONG TIME, NEUROSCIENTISTS HAD THE TECHNOLOGY TO STUDY ME, BUT NOT YOU! THAT'S JUST BAD LUCK!

TECHNOLOGY EVOLVES... YOU'LL GET JUSTICE!

QUIET, YOU! WE ALSO PLAY A ROLE IN INFORMATION PROCESSING! AND EMOTIONAL REGULATION!

AND LEARNING!

AS MUCH AS THE NEURON?

BEHAVIORAL CONTROL!

BROADLY SPEAKING, YES, MS. MOJITO! BUT THE WORLD IGNORES US! PEOPLE ARE SPENDING BILLIONS TRYING TO SIMULATE BRAIN FUNCTION, AND THEY HARDLY TAKE ME INTO ACCOUNT! HOW DO YOU THINK THAT'LL WORK? THEY'LL BE MODELING HALF A BRAIN!

I ASSURE YOU, I'M GRATEFUL FOR YOUR CONTRIBUTIONS. I APPRECIATE THAT YOU AND THE NEURON WORK HAND IN HAND. I NEED YOU AS MUCH AS I NEED HIM.

ANYWAY WITHOUT ME, YOU'D KNOW NOTHING, MR. BRAIN! **NOTHING AT ALL!** YOUR NEURON BUDDY OVER THERE SENDS HIS LITTLE MESSAGES IN SEQUENCE, ONE BY ONE, WHILE I SEND THEM OUT IN WAVES, LIKE RINGS RIPPLING OUT OVER THE SURFACE OF WATER. I'M MORE EFFICIENT!

THEY WANT TO CALL ME A BUTTRESS? A WEDGE? A WINCH? WHEN ALL THE WHILE THE WORLD LETS ITSELF BE FOOLED BY THIS CHARLATAN OF AN ELECTRICIAN?

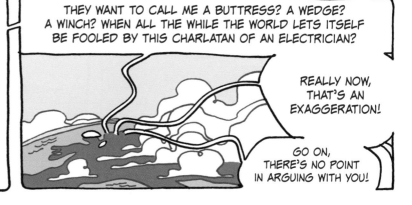

REALLY NOW, THAT'S AN EXAGGERATION!

GO ON, THERE'S NO POINT IN ARGUING WITH YOU!

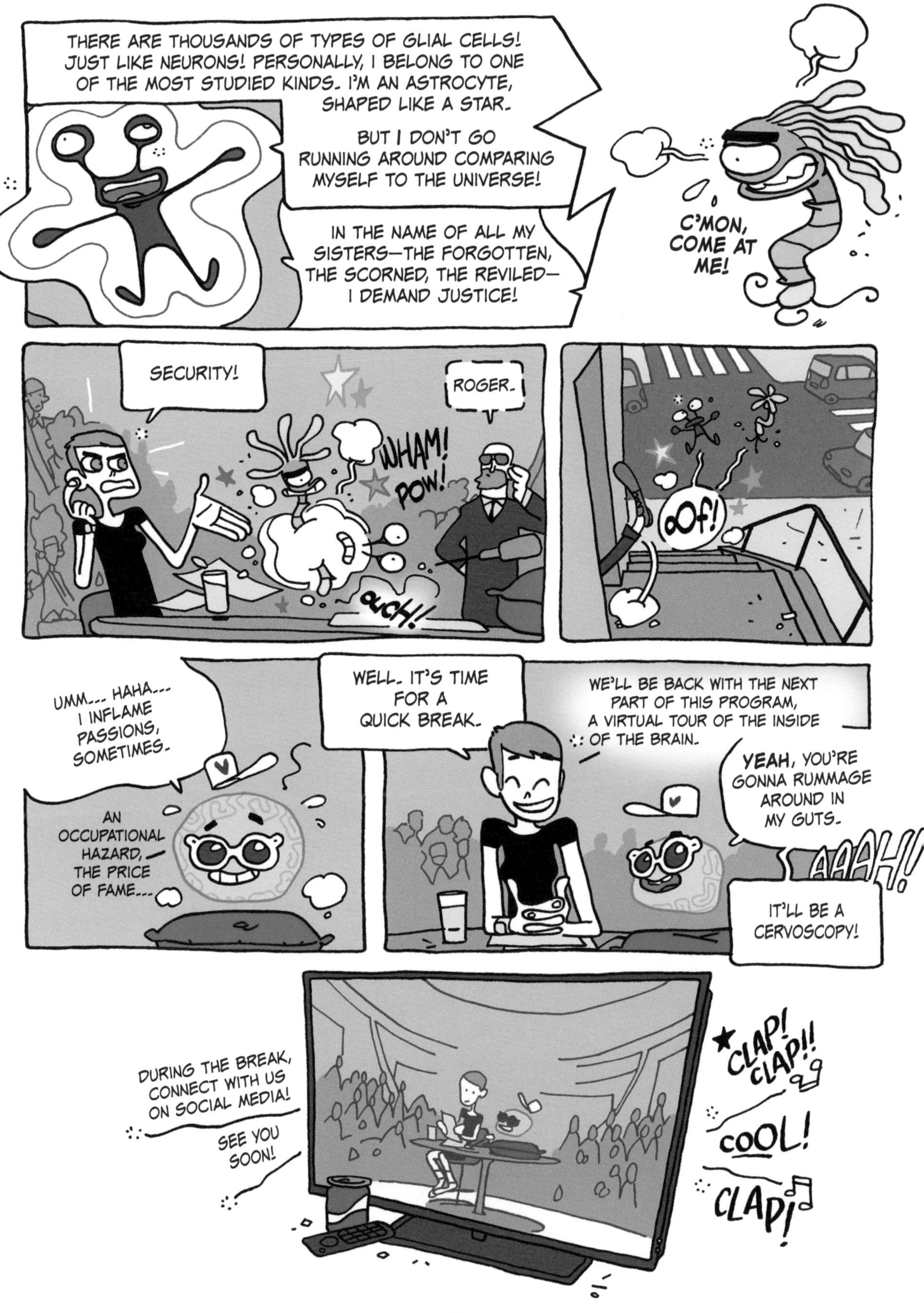

THERE ARE THOUSANDS OF TYPES OF GLIAL CELLS! JUST LIKE NEURONS! PERSONALLY, I BELONG TO ONE OF THE MOST STUDIED KINDS. I'M AN ASTROCYTE, SHAPED LIKE A STAR.
BUT I DON'T GO RUNNING AROUND COMPARING MYSELF TO THE UNIVERSE!
IN THE NAME OF ALL MY SISTERS—THE FORGOTTEN, THE SCORNED, THE REVILED—I DEMAND JUSTICE!
C'MON, COME AT ME!
SECURITY!
ROGER.
WHAM!
POW!
OUCH!
OOF!
UMM... HAHA... I INFLAME PASSIONS, SOMETIMES.
AN OCCUPATIONAL HAZARD, THE PRICE OF FAME...
WELL. IT'S TIME FOR A QUICK BREAK.
WE'LL BE BACK WITH THE NEXT PART OF THIS PROGRAM, A VIRTUAL TOUR OF THE INSIDE OF THE BRAIN.
YEAH, YOU'RE GONNA RUMMAGE AROUND IN MY GUTS.
AAAH!
IT'LL BE A CERVOSCOPY!
DURING THE BREAK, CONNECT WITH US ON SOCIAL MEDIA!
SEE YOU SOON!
CLAP! CLAP!!
COOL!
CLAP!

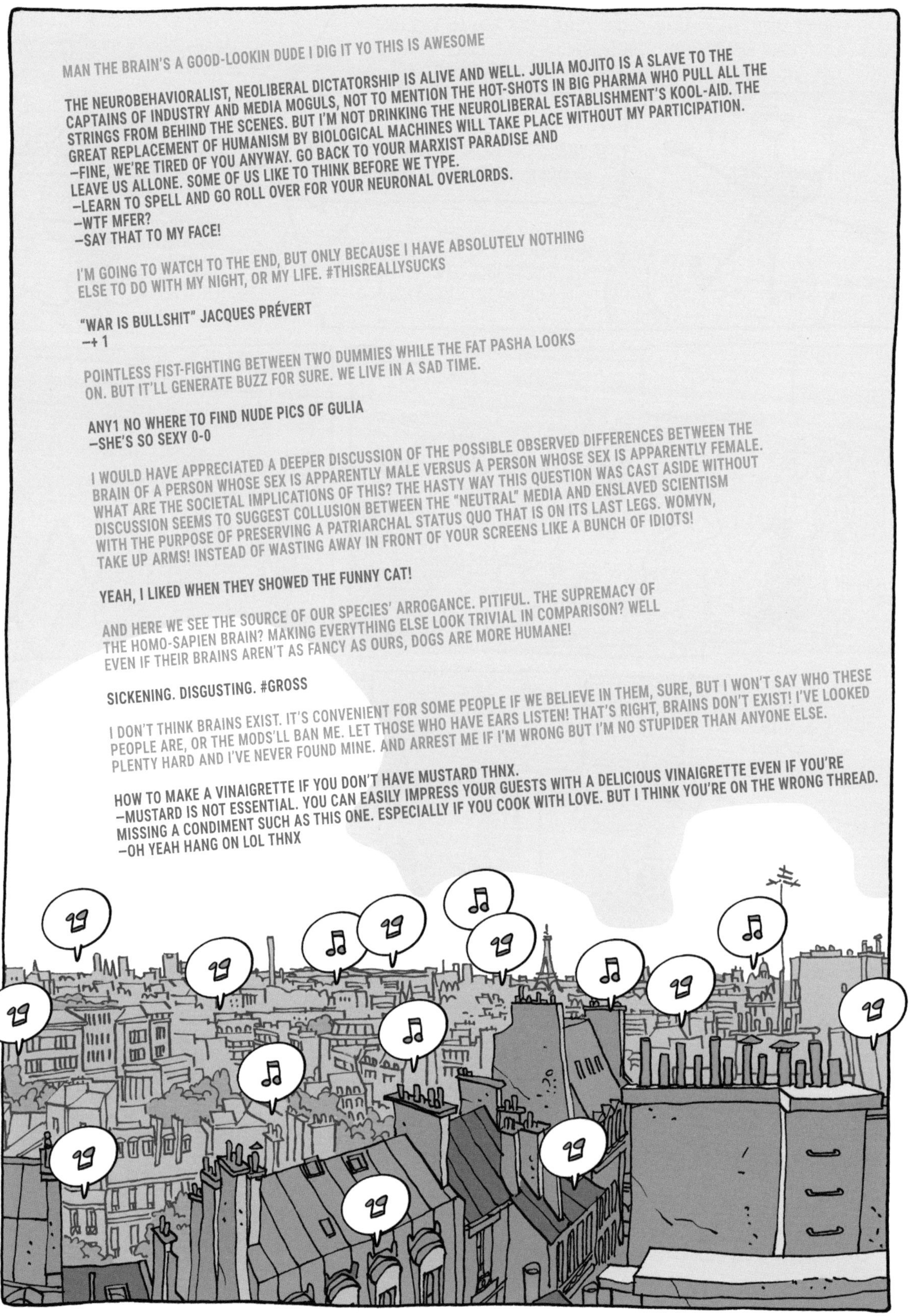
MAN THE BRAIN'S A GOOD-LOOKIN DUDE I DIG IT YO THIS IS AWESOME
THE NEUROBEHAVIORALIST, NEOLIBERAL DICTATORSHIP IS ALIVE AND WELL. JULIA MOJITO IS A SLAVE TO THE CAPTAINS OF INDUSTRY AND MEDIA MOGULS, NOT TO MENTION THE HOT-SHOTS IN BIG PHARMA WHO PULL ALL THE STRINGS FROM BEHIND THE SCENES. BUT I'M NOT DRINKING THE NEUROLIBERAL ESTABLISHMENT'S KOOL-AID. THE GREAT REPLACEMENT OF HUMANISM BY BIOLOGICAL MACHINES WILL TAKE PLACE WITHOUT MY PARTICIPATION.
–FINE, WE'RE TIRED OF YOU ANYWAY. GO BACK TO YOUR MARXIST PARADISE AND LEAVE US ALLONE. SOME OF US LIKE TO THINK BEFORE WE TYPE.
–LEARN TO SPELL AND GO ROLL OVER FOR YOUR NEURONAL OVERLORDS.
–WTF MFER?
–SAY THAT TO MY FACE!
I'M GOING TO WATCH TO THE END, BUT ONLY BECAUSE I HAVE ABSOLUTELY NOTHING ELSE TO DO WITH MY NIGHT, OR MY LIFE. #THISREALLYSUCKS
"WAR IS BULLSHIT" JACQUES PRÉVERT
–+ 1
POINTLESS FIST-FIGHTING BETWEEN TWO DUMMIES WHILE THE FAT PASHA LOOKS ON. BUT IT'LL GENERATE BUZZ FOR SURE. WE LIVE IN A SAD TIME.
ANY1 NO WHERE TO FIND NUDE PICS OF GULIA
–SHE'S SO SEXY 0-0
I WOULD HAVE APPRECIATED A DEEPER DISCUSSION OF THE POSSIBLE OBSERVED DIFFERENCES BETWEEN THE BRAIN OF A PERSON WHOSE SEX IS APPARENTLY MALE VERSUS A PERSON WHOSE SEX IS APPARENTLY FEMALE. WHAT ARE THE SOCIETAL IMPLICATIONS OF THIS? THE HASTY WAY THIS QUESTION WAS CAST ASIDE WITHOUT DISCUSSION SEEMS TO SUGGEST COLLUSION BETWEEN THE "NEUTRAL" MEDIA AND ENSLAVED SCIENTISM WITH THE PURPOSE OF PRESERVING A PATRIARCHAL STATUS QUO THAT IS ON ITS LAST LEGS. WOMYN, TAKE UP ARMS! INSTEAD OF WASTING AWAY IN FRONT OF YOUR SCREENS LIKE A BUNCH OF IDIOTS!
YEAH, I LIKED WHEN THEY SHOWED THE FUNNY CAT!
AND HERE WE SEE THE SOURCE OF OUR SPECIES' ARROGANCE. PITIFUL. THE SUPREMACY OF THE HOMO-SAPIEN BRAIN? MAKING EVERYTHING ELSE LOOK TRIVIAL IN COMPARISON? WELL EVEN IF THEIR BRAINS AREN'T AS FANCY AS OURS, DOGS ARE MORE HUMANE!
SICKENING. DISGUSTING. #GROSS
I DON'T THINK BRAINS EXIST. IT'S CONVENIENT FOR SOME PEOPLE IF WE BELIEVE IN THEM, SURE, BUT I WON'T SAY WHO THESE PEOPLE ARE, OR THE MODS'LL BAN ME. LET THOSE WHO HAVE EARS LISTEN! THAT'S RIGHT, BRAINS DON'T EXIST! I'VE LOOKED PLENTY HARD AND I'VE NEVER FOUND MINE. AND ARREST ME IF I'M WRONG BUT I'M NO STUPIDER THAN ANYONE ELSE.
HOW TO MAKE A VINAIGRETTE IF YOU DON'T HAVE MUSTARD THNX.
–MUSTARD IS NOT ESSENTIAL. YOU CAN EASILY IMPRESS YOUR GUESTS WITH A DELICIOUS VINAIGRETTE EVEN IF YOU'RE MISSING A CONDIMENT SUCH AS THIS ONE. ESPECIALLY IF YOU COOK WITH LOVE. BUT I THINK YOU'RE ON THE WRONG THREAD.
–OH YEAH HANG ON LOL THNX

Chapter 2

A PATCHWORK CATHEDRAL

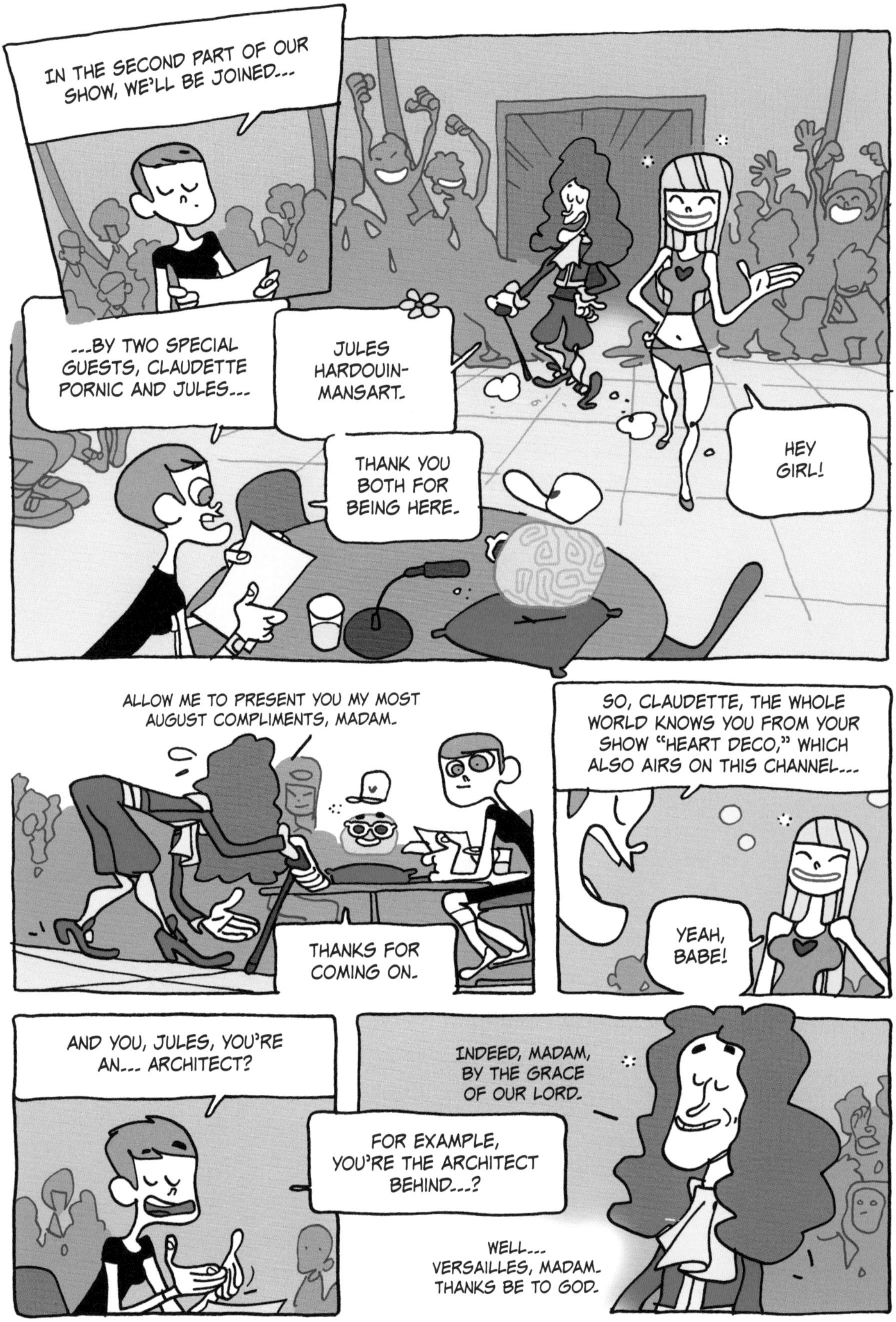
IN THE SECOND PART OF OUR SHOW, WE'LL BE JOINED...
...BY TWO SPECIAL GUESTS, CLAUDETTE PORNIC AND JULES...
JULES HARDOUIN-MANSART.
THANK YOU BOTH FOR BEING HERE.
HEY GIRL!
ALLOW ME TO PRESENT YOU MY MOST AUGUST COMPLIMENTS, MADAM.
THANKS FOR COMING ON.
SO, CLAUDETTE, THE WHOLE WORLD KNOWS YOU FROM YOUR SHOW "HEART DECO," WHICH ALSO AIRS ON THIS CHANNEL...
YEAH, BABE!
AND YOU, JULES, YOU'RE AN... ARCHITECT?
INDEED, MADAM, BY THE GRACE OF OUR LORD.
FOR EXAMPLE, YOU'RE THE ARCHITECT BEHIND...?
WELL... VERSAILLES, MADAM. THANKS BE TO GOD.

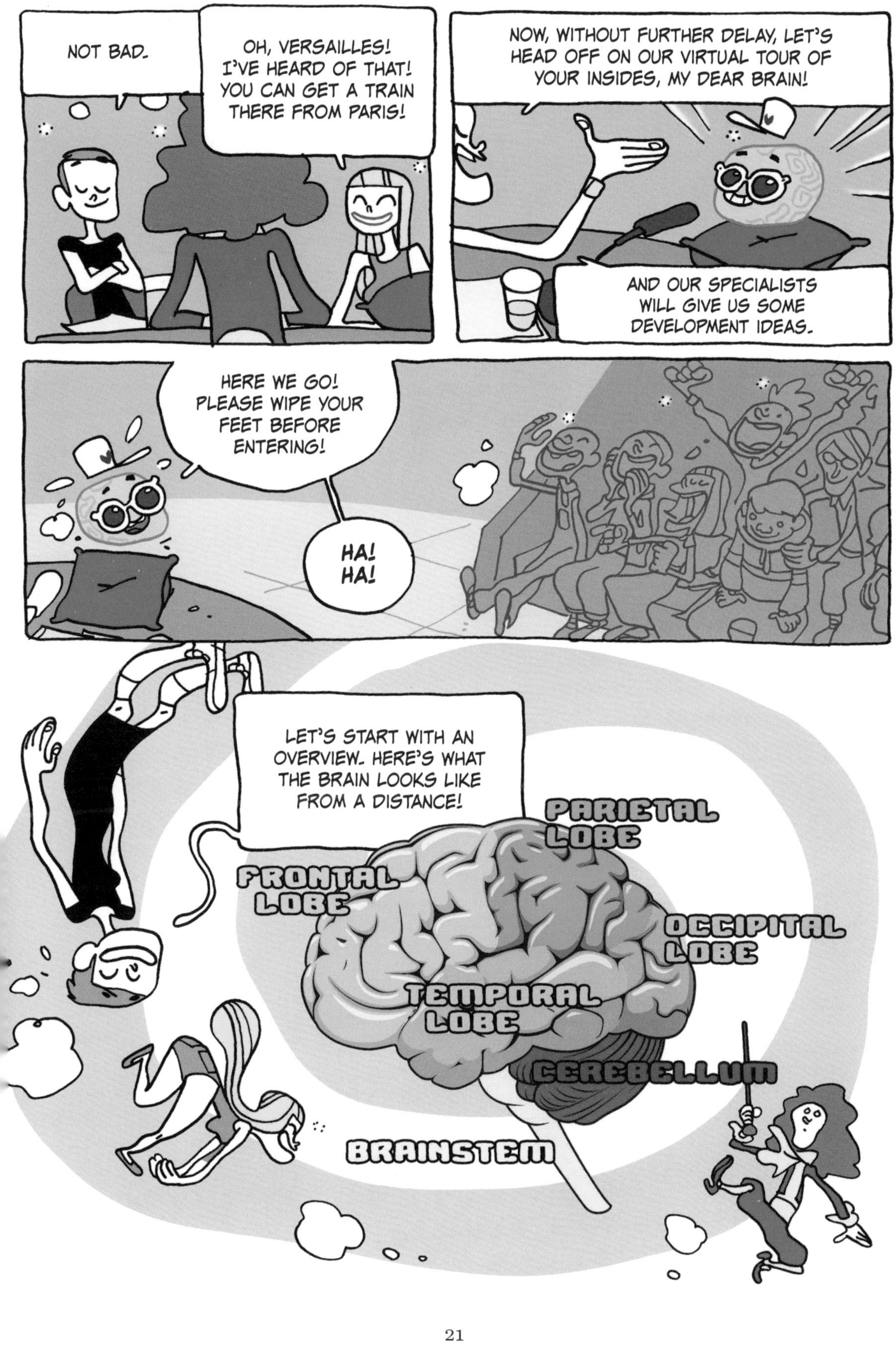
NOT BAD.
OH, VERSAILLES! I'VE HEARD OF THAT! YOU CAN GET A TRAIN THERE FROM PARIS!
NOW, WITHOUT FURTHER DELAY, LET'S HEAD OFF ON OUR VIRTUAL TOUR OF YOUR INSIDES, MY DEAR BRAIN!
AND OUR SPECIALISTS WILL GIVE US SOME DEVELOPMENT IDEAS.
HERE WE GO! PLEASE WIPE YOUR FEET BEFORE ENTERING!
HA! HA!
LET'S START WITH AN OVERVIEW. HERE'S WHAT THE BRAIN LOOKS LIKE FROM A DISTANCE!
PARIETAL LOBE
FRONTAL LOBE
OCCIPITAL LOBE
TEMPORAL LOBE
CEREBELLUM
BRAINSTEM

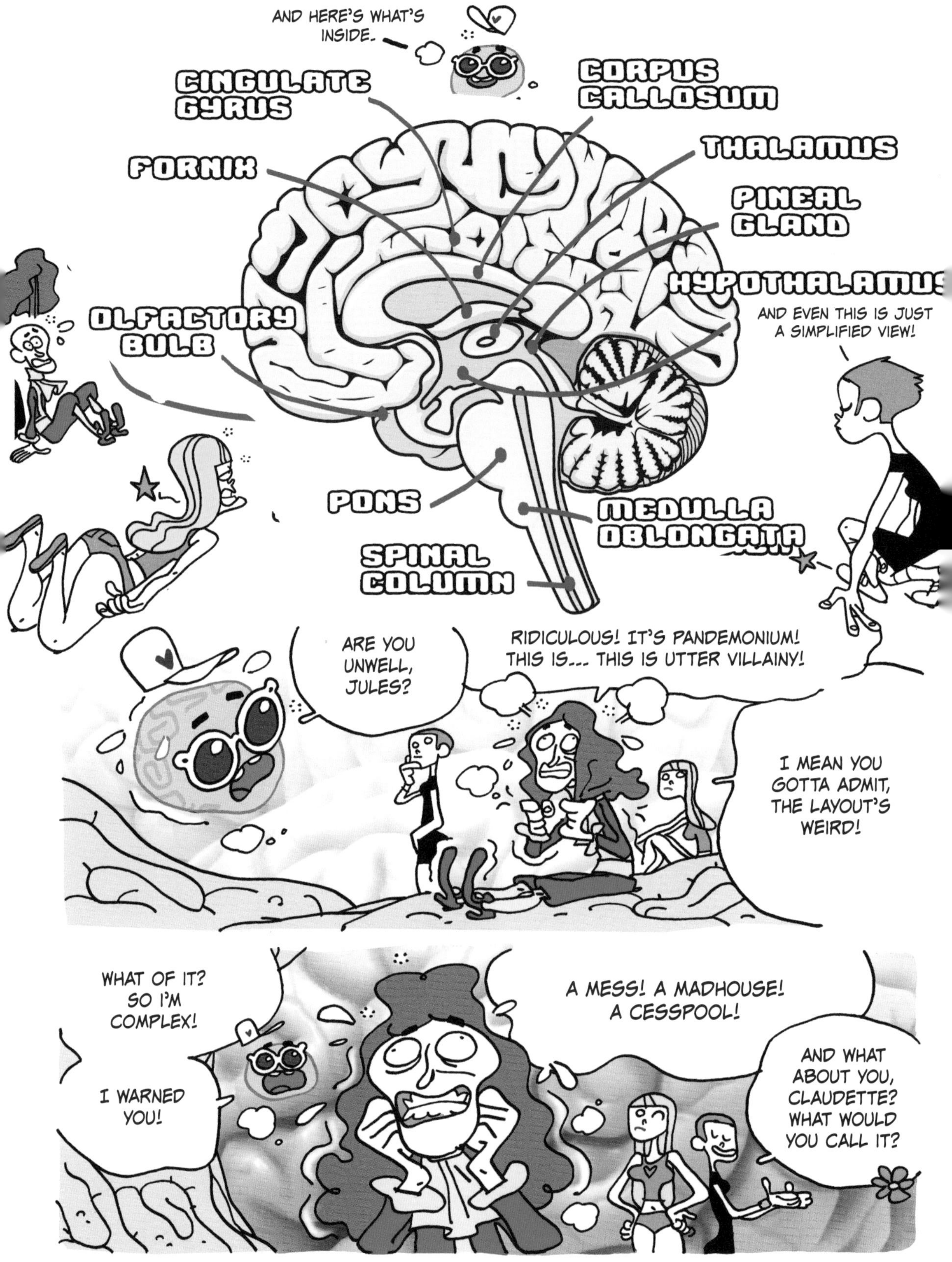
AND HERE'S WHAT'S INSIDE.
CINGULATE GYRUS
CORPUS CALLOSUM
THALAMUS
FORNIX
PINEAL GLAND
HYPOTHALAMUS
OLFACTORY BULB
AND EVEN THIS IS JUST A SIMPLIFIED VIEW!
PONS
MEDULLA OBLONGATA
SPINAL COLUMN
ARE YOU UNWELL, JULES?
RIDICULOUS! IT'S PANDEMONIUM! THIS IS... THIS IS UTTER VILLAINY!
I MEAN YOU GOTTA ADMIT, THE LAYOUT'S WEIRD!
WHAT OF IT? SO I'M COMPLEX!
I WARNED YOU!
A MESS! A MADHOUSE! A CESSPOOL!
AND WHAT ABOUT YOU, CLAUDETTE? WHAT WOULD YOU CALL IT?

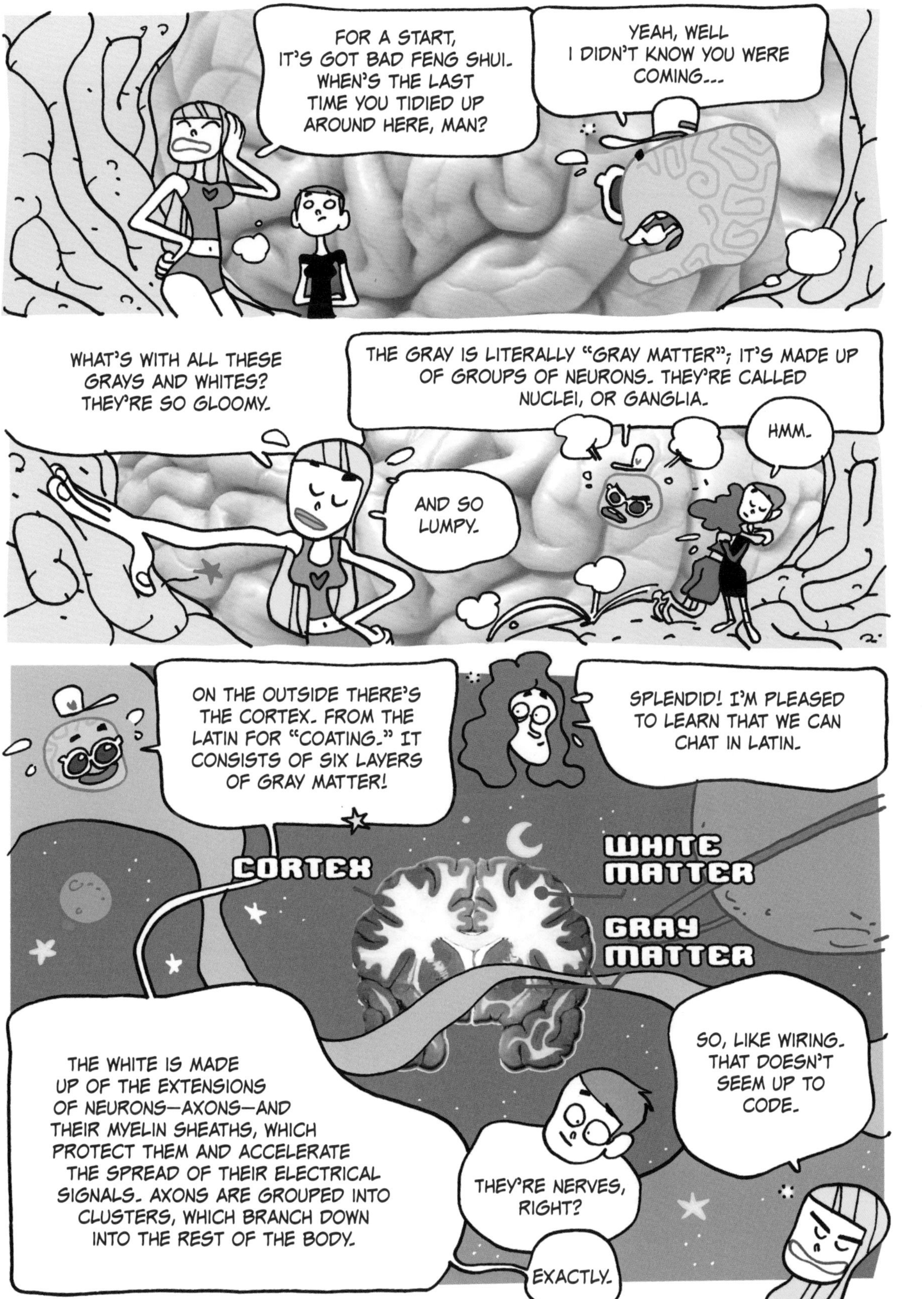
FOR A START, IT'S GOT BAD FENG SHUI. WHEN'S THE LAST TIME YOU TIDIED UP AROUND HERE, MAN?
YEAH, WELL I DIDN'T KNOW YOU WERE COMING...
WHAT'S WITH ALL THESE GRAYS AND WHITES? THEY'RE SO GLOOMY.
THE GRAY IS LITERALLY "GRAY MATTER"; IT'S MADE UP OF GROUPS OF NEURONS. THEY'RE CALLED NUCLEI, OR GANGLIA.
HMM.
AND SO LUMPY.
ON THE OUTSIDE THERE'S THE CORTEX. FROM THE LATIN FOR "COATING." IT CONSISTS OF SIX LAYERS OF GRAY MATTER!
SPLENDID! I'M PLEASED TO LEARN THAT WE CAN CHAT IN LATIN.
CORTEX
WHITE MATTER
GRAY MATTER
THE WHITE IS MADE UP OF THE EXTENSIONS OF NEURONS—AXONS—AND THEIR MYELIN SHEATHS, WHICH PROTECT THEM AND ACCELERATE THE SPREAD OF THEIR ELECTRICAL SIGNALS. AXONS ARE GROUPED INTO CLUSTERS, WHICH BRANCH DOWN INTO THE REST OF THE BODY.
SO, LIKE WIRING. THAT DOESN'T SEEM UP TO CODE.
THEY'RE NERVES, RIGHT?
EXACTLY.

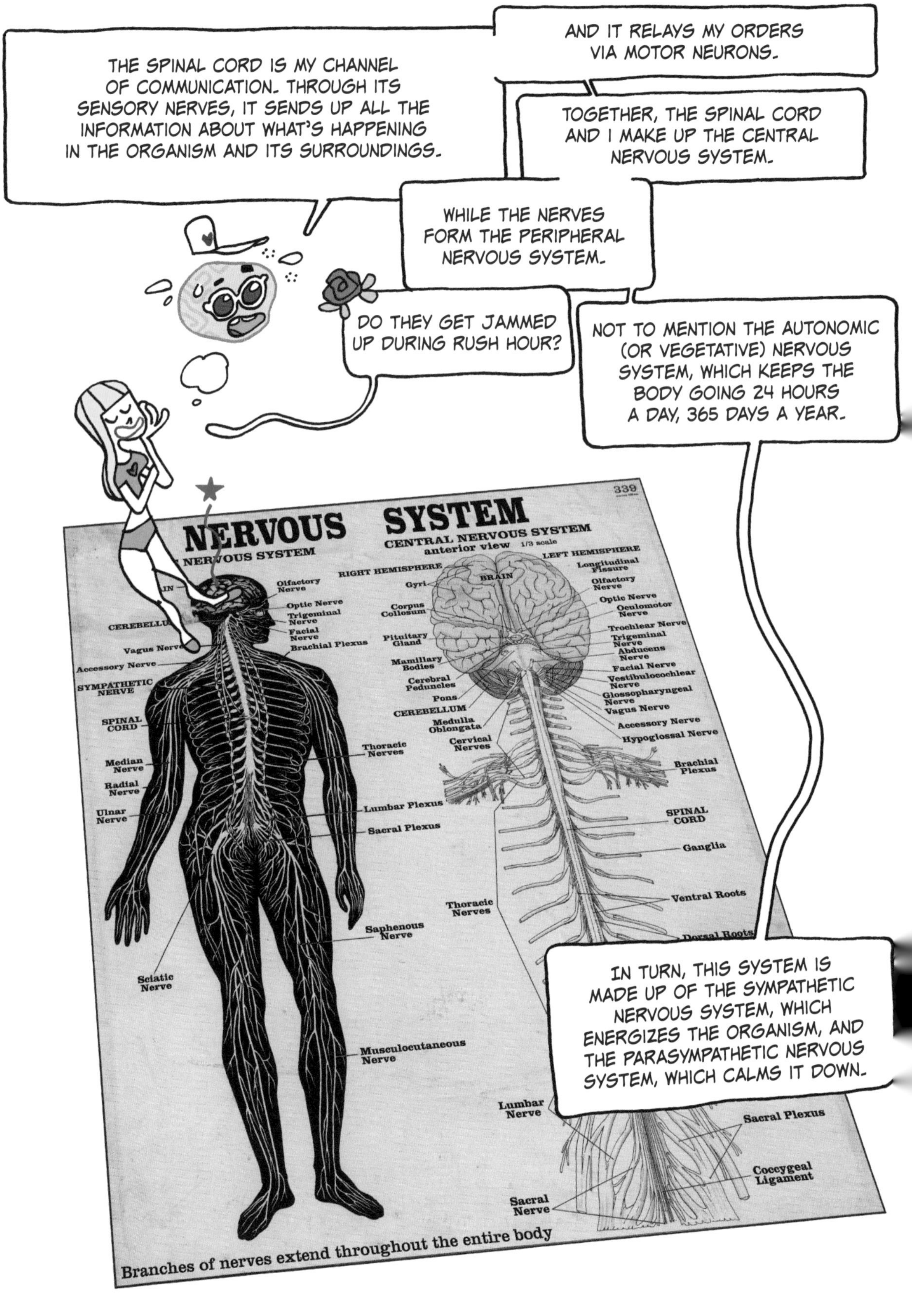
THE SPINAL CORD IS MY CHANNEL OF COMMUNICATION. THROUGH ITS SENSORY NERVES, IT SENDS UP ALL THE INFORMATION ABOUT WHAT'S HAPPENING IN THE ORGANISM AND ITS SURROUNDINGS.
AND IT RELAYS MY ORDERS VIA MOTOR NEURONS.
TOGETHER, THE SPINAL CORD AND I MAKE UP THE CENTRAL NERVOUS SYSTEM.
WHILE THE NERVES FORM THE PERIPHERAL NERVOUS SYSTEM.
DO THEY GET JAMMED UP DURING RUSH HOUR?
NOT TO MENTION THE AUTONOMIC (OR VEGETATIVE) NERVOUS SYSTEM, WHICH KEEPS THE BODY GOING 24 HOURS A DAY, 365 DAYS A YEAR.
IN TURN, THIS SYSTEM IS MADE UP OF THE SYMPATHETIC NERVOUS SYSTEM, WHICH ENERGIZES THE ORGANISM, AND THE PARASYMPATHETIC NERVOUS SYSTEM, WHICH CALMS IT DOWN.
339
NERVOUS SYSTEM
NERVOUS SYSTEM
CENTRAL NERVOUS SYSTEM
anterior view 1/3 scale
RIGHT HEMISPHERE
LEFT HEMISPHERE
BRAIN
Olfactory Nerve
Optic Nerve
Trigeminal Nerve
Facial Nerve
Brachial Plexus
CEREBELLUM
Vagus Nerve
Accessory Nerve
SYMPATHETIC NERVE
SPINAL CORD
Median Nerve
Radial Nerve
Ulnar Nerve
Thoracic Nerves
Lumbar Plexus
Sacral Plexus
Saphenous Nerve
Sciatic Nerve
Musculocutaneous Nerve
Gyri
Corpus Collosum
Pituitary Gland
Mamillary Bodies
Cerebral Peduncles
Pons
CEREBELLUM
Medulla Oblongata
Cervical Nerves
Longitudinal Fissure
Olfactory Nerve
Optic Nerve
Oculomotor Nerve
Trochlear Nerve
Trigeminal Nerve
Abducens Nerve
Facial Nerve
Vestibulocochlear Nerve
Glossopharyngeal Nerve
Vagus Nerve
Accessory Nerve
Hypoglossal Nerve
Brachial Plexus
SPINAL CORD
Ganglia
Ventral Roots
Dorsal Roots
Thoracic Nerves
Lumbar Nerve
Sacral Plexus
Coccygeal Ligament
Sacral Nerve
Branches of nerves extend throughout the entire body

BUT LET'S KEEP THINGS SIMPLE. SO WE'VE GOT TWO HEMISPHERES? AND THE LEFT IS VERBAL WHILE THE RIGHT IS VISUAL, RIGHT?

WHOA, HANG ON! FIRST OFF, LATERALIZATION ISN'T SPECIFIC TO HUMANS: IT'S ALSO PRESENT IN LOTS OF VERTEBRATES, INCLUDING FISH, BIRDS...

YES, BUT THE LEFT IS VERBAL WHILE THE RIGHT IS VISUAL, RIGHT?

TAKE A LOOK AT THIS HUGE MASS OF FIBERS THAT ENSURE THE TWO ARE IN CONSTANT COMMUNICATION: THE CORPUS CALLOSUM.

ALSO FROM THE LATIN.

YES, BUT THE LEFT IS V—

NO, IT'S NOT SO SIMPLE AS THAT. YOU HAVE TO BE WARY OF BLACK-AND-WHITE DICHOTOMIES WHEN TALKING ABOUT NEUROSCIENCE.

DICHOTO-WHATS? COME ON, SPEAK ENGLISH, KIDS!

FROM THE GREEK "DIKHOTOMIA," MADAM.

WHY DIDN'T YA SAY SO?

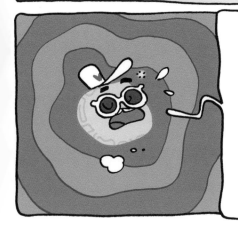

IN BROAD TERMS, THE LEFT HEMISPHERE IS MORE ANALYTICAL. IT PERCEIVES DETAILS, REASONS STEP BY STEP, THINKS AHEAD... THE RIGHT HEMISPHERE IS MORE HOLISTIC, MEANING IT PERCEIVES THINGS IN THEIR ENTIRETY, AND IS MORE INTUITIVE.

FROM THE GREEK "HOLOS."

HOLO? YOLO!

25

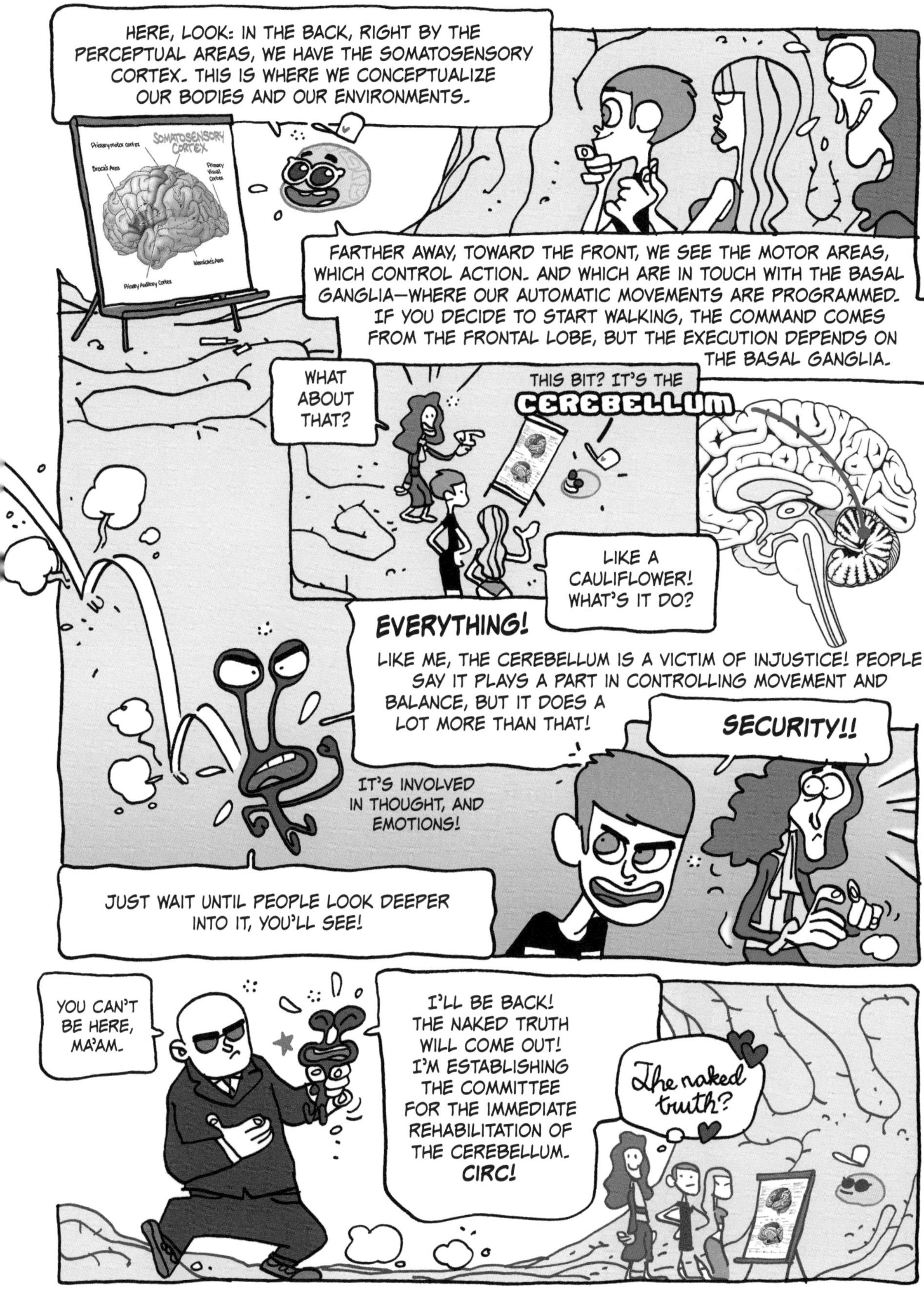
HERE, LOOK: IN THE BACK, RIGHT BY THE PERCEPTUAL AREAS, WE HAVE THE SOMATOSENSORY CORTEX. THIS IS WHERE WE CONCEPTUALIZE OUR BODIES AND OUR ENVIRONMENTS.
SOMATOSENSORY CORTEX
Primary motor cortex
Broca's Area
Primary Visual Cortex
Wernicke's Area
Primary Auditory Cortex
FARTHER AWAY, TOWARD THE FRONT, WE SEE THE MOTOR AREAS, WHICH CONTROL ACTION. AND WHICH ARE IN TOUCH WITH THE BASAL GANGLIA—WHERE OUR AUTOMATIC MOVEMENTS ARE PROGRAMMED. IF YOU DECIDE TO START WALKING, THE COMMAND COMES FROM THE FRONTAL LOBE, BUT THE EXECUTION DEPENDS ON THE BASAL GANGLIA.
WHAT ABOUT THAT?
THIS BIT? IT'S THE CEREBELLUM
LIKE A CAULIFLOWER! WHAT'S IT DO?
EVERYTHING!
LIKE ME, THE CEREBELLUM IS A VICTIM OF INJUSTICE! PEOPLE SAY IT PLAYS A PART IN CONTROLLING MOVEMENT AND BALANCE, BUT IT DOES A LOT MORE THAN THAT!
IT'S INVOLVED IN THOUGHT, AND EMOTIONS!
SECURITY!!
JUST WAIT UNTIL PEOPLE LOOK DEEPER INTO IT, YOU'LL SEE!
YOU CAN'T BE HERE, MA'AM.
I'LL BE BACK! THE NAKED TRUTH WILL COME OUT! I'M ESTABLISHING THE COMMITTEE FOR THE IMMEDIATE REHABILITATION OF THE CEREBELLUM. CIRC!
The naked truth?

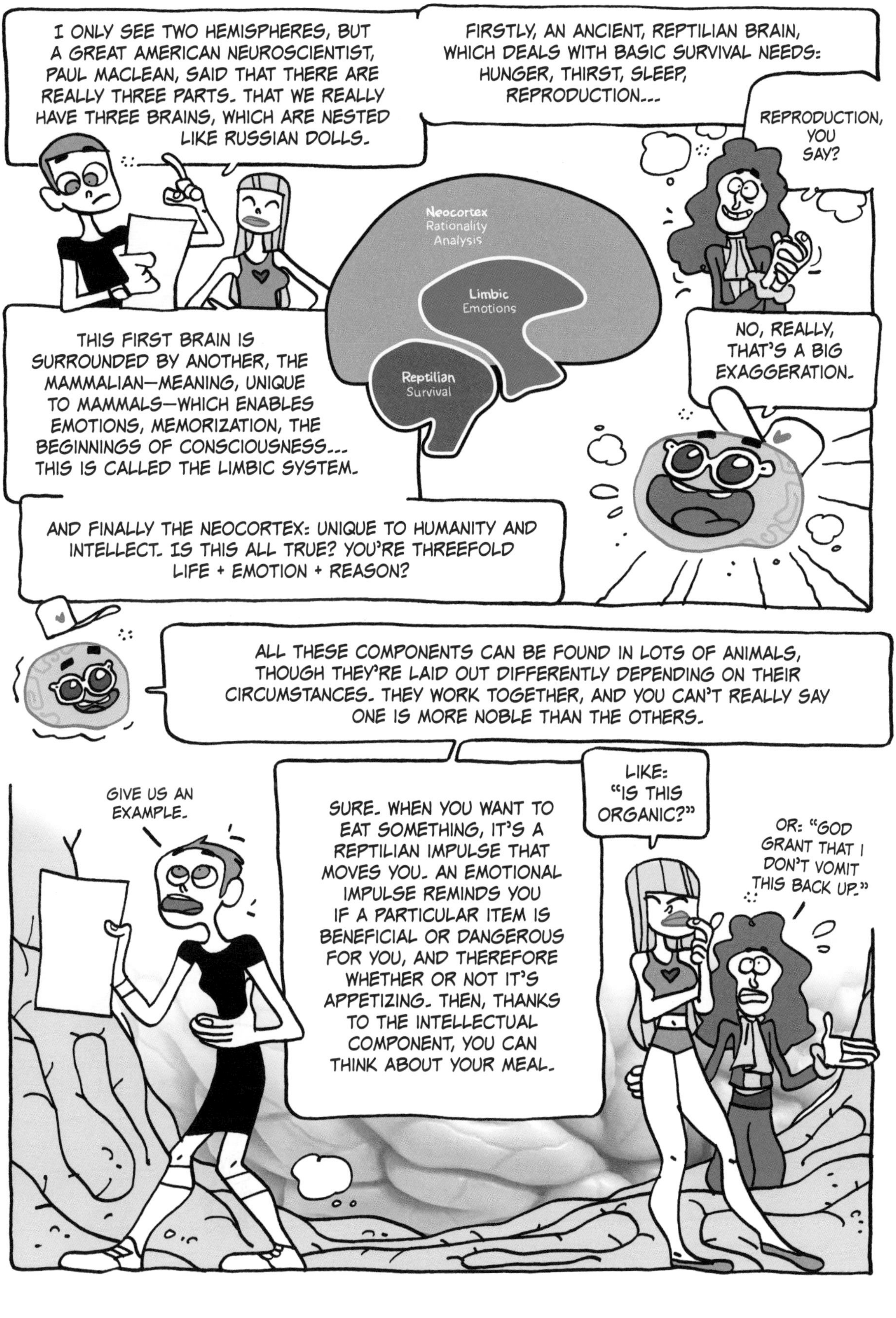
I ONLY SEE TWO HEMISPHERES, BUT A GREAT AMERICAN NEUROSCIENTIST, PAUL MACLEAN, SAID THAT THERE ARE REALLY THREE PARTS. THAT WE REALLY HAVE THREE BRAINS, WHICH ARE NESTED LIKE RUSSIAN DOLLS.
FIRSTLY, AN ANCIENT, REPTILIAN BRAIN, WHICH DEALS WITH BASIC SURVIVAL NEEDS: HUNGER, THIRST, SLEEP, REPRODUCTION...
REPRODUCTION, YOU SAY?
Neocortex
Rationality
Analysis
Limbic
Emotions
Reptilian
Survival
THIS FIRST BRAIN IS SURROUNDED BY ANOTHER, THE MAMMALIAN—MEANING, UNIQUE TO MAMMALS—WHICH ENABLES EMOTIONS, MEMORIZATION, THE BEGINNINGS OF CONSCIOUSNESS... THIS IS CALLED THE LIMBIC SYSTEM.
NO, REALLY, THAT'S A BIG EXAGGERATION.
AND FINALLY THE NEOCORTEX: UNIQUE TO HUMANITY AND INTELLECT. IS THIS ALL TRUE? YOU'RE THREEFOLD LIFE + EMOTION + REASON?
ALL THESE COMPONENTS CAN BE FOUND IN LOTS OF ANIMALS, THOUGH THEY'RE LAID OUT DIFFERENTLY DEPENDING ON THEIR CIRCUMSTANCES. THEY WORK TOGETHER, AND YOU CAN'T REALLY SAY ONE IS MORE NOBLE THAN THE OTHERS.
GIVE US AN EXAMPLE.
SURE. WHEN YOU WANT TO EAT SOMETHING, IT'S A REPTILIAN IMPULSE THAT MOVES YOU. AN EMOTIONAL IMPULSE REMINDS YOU IF A PARTICULAR ITEM IS BENEFICIAL OR DANGEROUS FOR YOU, AND THEREFORE WHETHER OR NOT IT'S APPETIZING. THEN, THANKS TO THE INTELLECTUAL COMPONENT, YOU CAN THINK ABOUT YOUR MEAL.
LIKE: "IS THIS ORGANIC?"
OR: "GOD GRANT THAT I DON'T VOMIT THIS BACK UP."

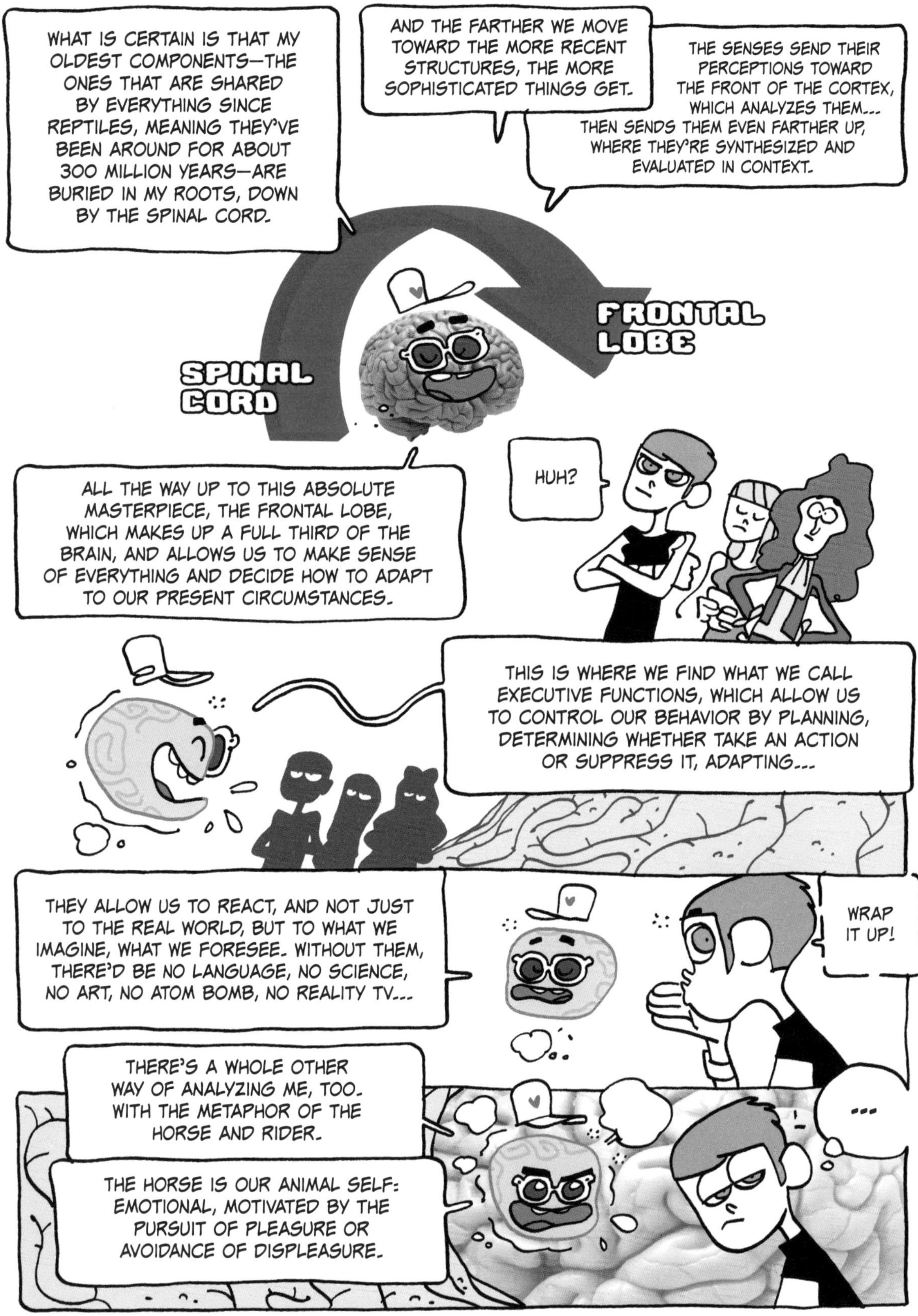
WHAT IS CERTAIN IS THAT MY OLDEST COMPONENTS—THE ONES THAT ARE SHARED BY EVERYTHING SINCE REPTILES, MEANING THEY'VE BEEN AROUND FOR ABOUT 300 MILLION YEARS—ARE BURIED IN MY ROOTS, DOWN BY THE SPINAL CORD.
AND THE FARTHER WE MOVE TOWARD THE MORE RECENT STRUCTURES, THE MORE SOPHISTICATED THINGS GET.
THE SENSES SEND THEIR PERCEPTIONS TOWARD THE FRONT OF THE CORTEX, WHICH ANALYZES THEM... THEN SENDS THEM EVEN FARTHER UP, WHERE THEY'RE SYNTHESIZED AND EVALUATED IN CONTEXT.
SPINAL CORD
FRONTAL LOBE
ALL THE WAY UP TO THIS ABSOLUTE MASTERPIECE, THE FRONTAL LOBE, WHICH MAKES UP A FULL THIRD OF THE BRAIN, AND ALLOWS US TO MAKE SENSE OF EVERYTHING AND DECIDE HOW TO ADAPT TO OUR PRESENT CIRCUMSTANCES.
HUH?
THIS IS WHERE WE FIND WHAT WE CALL EXECUTIVE FUNCTIONS, WHICH ALLOW US TO CONTROL OUR BEHAVIOR BY PLANNING, DETERMINING WHETHER TAKE AN ACTION OR SUPPRESS IT, ADAPTING...
THEY ALLOW US TO REACT, AND NOT JUST TO THE REAL WORLD, BUT TO WHAT WE IMAGINE, WHAT WE FORESEE. WITHOUT THEM, THERE'D BE NO LANGUAGE, NO SCIENCE, NO ART, NO ATOM BOMB, NO REALITY TV...
WRAP IT UP!
THERE'S A WHOLE OTHER WAY OF ANALYZING ME, TOO. WITH THE METAPHOR OF THE HORSE AND RIDER.
THE HORSE IS OUR ANIMAL SELF: EMOTIONAL, MOTIVATED BY THE PURSUIT OF PLEASURE OR AVOIDANCE OF DISPLEASURE.
...

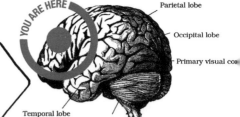

THE RIDER, WHO SEES THE PATH AHEAD AND DIRECTS THEIR MOVEMENT ACCORDINGLY, IS THE FRONTAL LOBE.

Parietal lobe

Occipital lobe

Primary visual cor[...]

Temporal lobe

RIDER:
CONSCIOUSNESS, LANGUAGE, THOUGHT.

HORSE:
THE AUTOMATIC, EMOTIONAL, VISCERAL BRAIN.

NOT TO BE PRESUMPTUOUS, MR. BRAIN, BUT... IS NOTHING WITHIN YOU RATIONALLY LAID OUT, THEN?

NOPE! HE'S LIKE AN OLD HOARDER!

AND IT'S ALL SO TIGHTLY PACKED TOGETHER! YOU'RE CLAUSTROPHOBIC, MR. BRAIN!

JULES, YOU'RE AN ARCHITECT. IMAGINE, IF YOU WILL, THAT I'M LIKE A CATHEDRAL THAT IS CALLED BEAUTIFUL AND STRIKING EVEN THOUGH IT IS MADE UP OF A MEDIEVAL CRYPT, A ROMAN NAVE, GOTHIC CHAPELS, AND STAINED GLASS, AND A BIT OF CONTEMPORARY ART TO GENERATE BUZZ.

THE VARIANCE AMONG MY COMPONENTS, WHICH ORIGINATED IN DIFFERENT EPOCHS, DOES NOT PREVENT THEM FROM WORKING TOGETHER IN HARMONY, LIKE A MARINE LANDSCAPE IN WHICH THE AGE-OLD SEA APPEARS ALONGSIDE A MODERN FISHERMAN'S HUT...

OR A CHIP SHACK.

GENERATE WHAT?

"BUZZ." YOU KNOW, "HYPE."

AH.

FISH & CHIPS

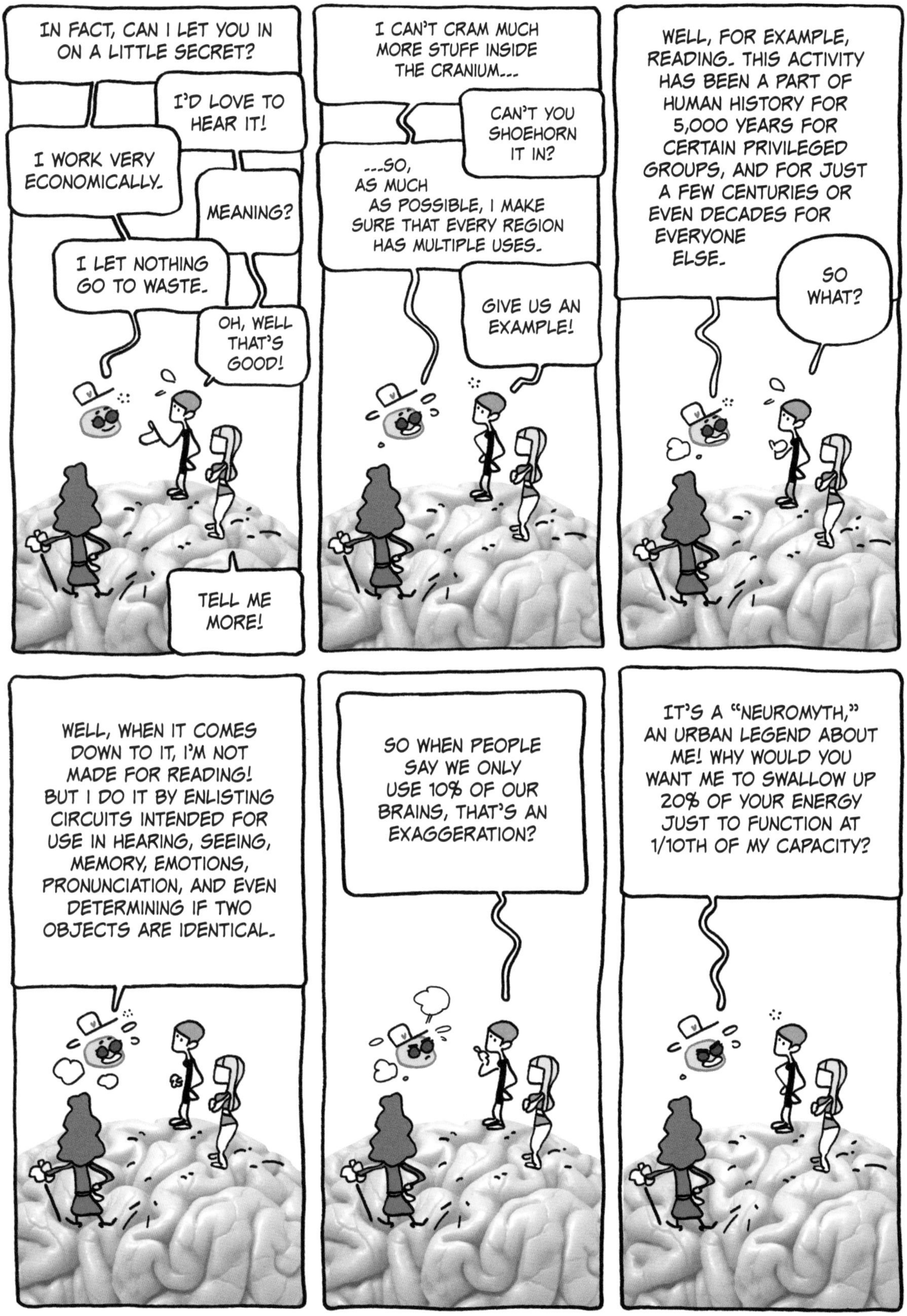
IN FACT, CAN I LET YOU IN ON A LITTLE SECRET?
I'D LOVE TO HEAR IT!
I WORK VERY ECONOMICALLY.
MEANING?
I LET NOTHING GO TO WASTE.
OH, WELL THAT'S GOOD!
TELL ME MORE!
I CAN'T CRAM MUCH MORE STUFF INSIDE THE CRANIUM...
CAN'T YOU SHOEHORN IT IN?
...SO, AS MUCH AS POSSIBLE, I MAKE SURE THAT EVERY REGION HAS MULTIPLE USES.
GIVE US AN EXAMPLE!
WELL, FOR EXAMPLE, READING. THIS ACTIVITY HAS BEEN A PART OF HUMAN HISTORY FOR 5,000 YEARS FOR CERTAIN PRIVILEGED GROUPS, AND FOR JUST A FEW CENTURIES OR EVEN DECADES FOR EVERYONE ELSE.
SO WHAT?
WELL, WHEN IT COMES DOWN TO IT, I'M NOT MADE FOR READING! BUT I DO IT BY ENLISTING CIRCUITS INTENDED FOR USE IN HEARING, SEEING, MEMORY, EMOTIONS, PRONUNCIATION, AND EVEN DETERMINING IF TWO OBJECTS ARE IDENTICAL.
SO WHEN PEOPLE SAY WE ONLY USE 10% OF OUR BRAINS, THAT'S AN EXAGGERATION?
IT'S A "NEUROMYTH," AN URBAN LEGEND ABOUT ME! WHY WOULD YOU WANT ME TO SWALLOW UP 20% OF YOUR ENERGY JUST TO FUNCTION AT 1/10TH OF MY CAPACITY?

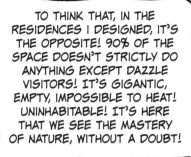

TO THINK THAT, IN THE RESIDENCES I DESIGNED, IT'S THE OPPOSITE! 90% OF THE SPACE DOESN'T STRICTLY DO ANYTHING EXCEPT DAZZLE VISITORS! IT'S GIGANTIC, EMPTY, IMPOSSIBLE TO HEAT! UNINHABITABLE! IT'S HERE THAT WE SEE THE MASTERY OF NATURE, WITHOUT A DOUBT!

GET AHOLD OF YOURSELF, YOU CRYBABY, HONESTLY!

AND YOU! YOU CAN'T KEEP LIVING IN THIS DUMP! TAKE MY CARD.

BUT I DON'T HAVE ARMS....

NO PROBLEM, I'LL SHOOT YOU AN EMAIL.

AND THEN, WHY WOULD NATURAL SELECTION HAVE MADE ME THIS WAY, IF THE VAST MAJORITY OF ME DID NOTHING? NOTHING USEFUL FOR SURVIVAL OR DAY-TO-DAY LIFE?

AH, NATURAL SELECTION, LET'S TALK ABOUT THAT! I'VE HEARD IT SAID THAT YOU'RE OPTIMIZED TO SURVIVE IN AN ENVIRONMENT THAT NO LONGER EXISTS! WHAT'S THAT SUPPOSED TO MEAN?

THAT'S RIGHT. IT'S LIKE BILLY JOEL SANG. THE NEOLITHIC PERIOD TURNED HUMANITY UPSIDE DOWN ABOUT 10,000 YEARS AGO, BUT AS FOR ME, REALLY....

I "DON'T GO CHANGING...."

SOMETIMES I GET NOSTALGIC FOR THE OLD DAYS, BUT I'VE HAD TO ADAPT.

AND I'M GREAT AT DIGGING UP GLUCOSE, SINCE 200,000 YEARS AGO, ON THE SAVANNA, IT WAS IN SHORT SUPPLY!

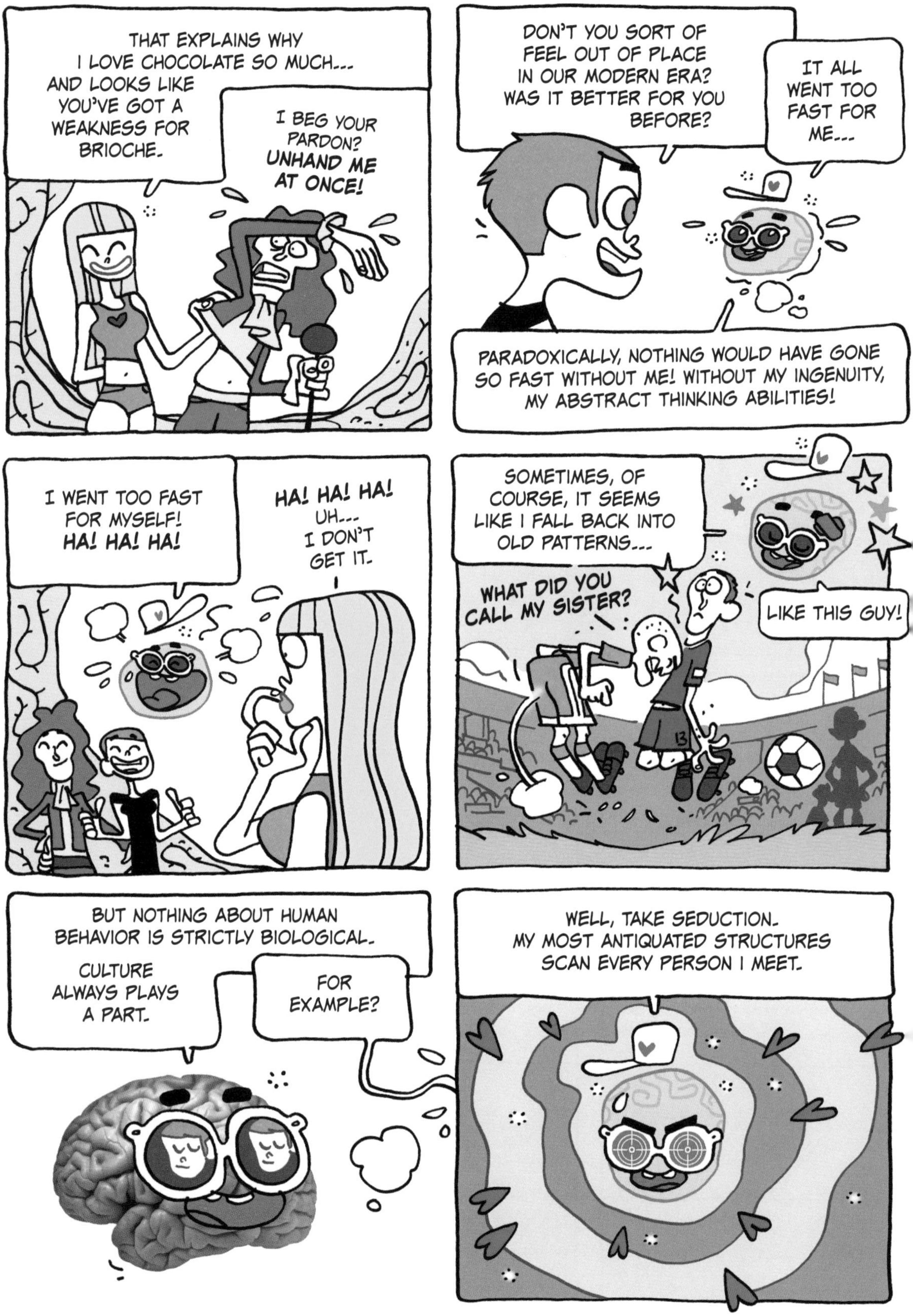
THAT EXPLAINS WHY I LOVE CHOCOLATE SO MUCH...
AND LOOKS LIKE YOU'VE GOT A WEAKNESS FOR BRIOCHE.
I BEG YOUR PARDON? UNHAND ME AT ONCE!
DON'T YOU SORT OF FEEL OUT OF PLACE IN OUR MODERN ERA? WAS IT BETTER FOR YOU BEFORE?
IT ALL WENT TOO FAST FOR ME...
PARADOXICALLY, NOTHING WOULD HAVE GONE SO FAST WITHOUT ME! WITHOUT MY INGENUITY, MY ABSTRACT THINKING ABILITIES!
I WENT TOO FAST FOR MYSELF! HA! HA! HA!
HA! HA! HA! UH... I DON'T GET IT.
SOMETIMES, OF COURSE, IT SEEMS LIKE I FALL BACK INTO OLD PATTERNS...
WHAT DID YOU CALL MY SISTER?
LIKE THIS GUY!
BUT NOTHING ABOUT HUMAN BEHAVIOR IS STRICTLY BIOLOGICAL.
CULTURE ALWAYS PLAYS A PART.
FOR EXAMPLE?
WELL, TAKE SEDUCTION. MY MOST ANTIQUATED STRUCTURES SCAN EVERY PERSON I MEET.

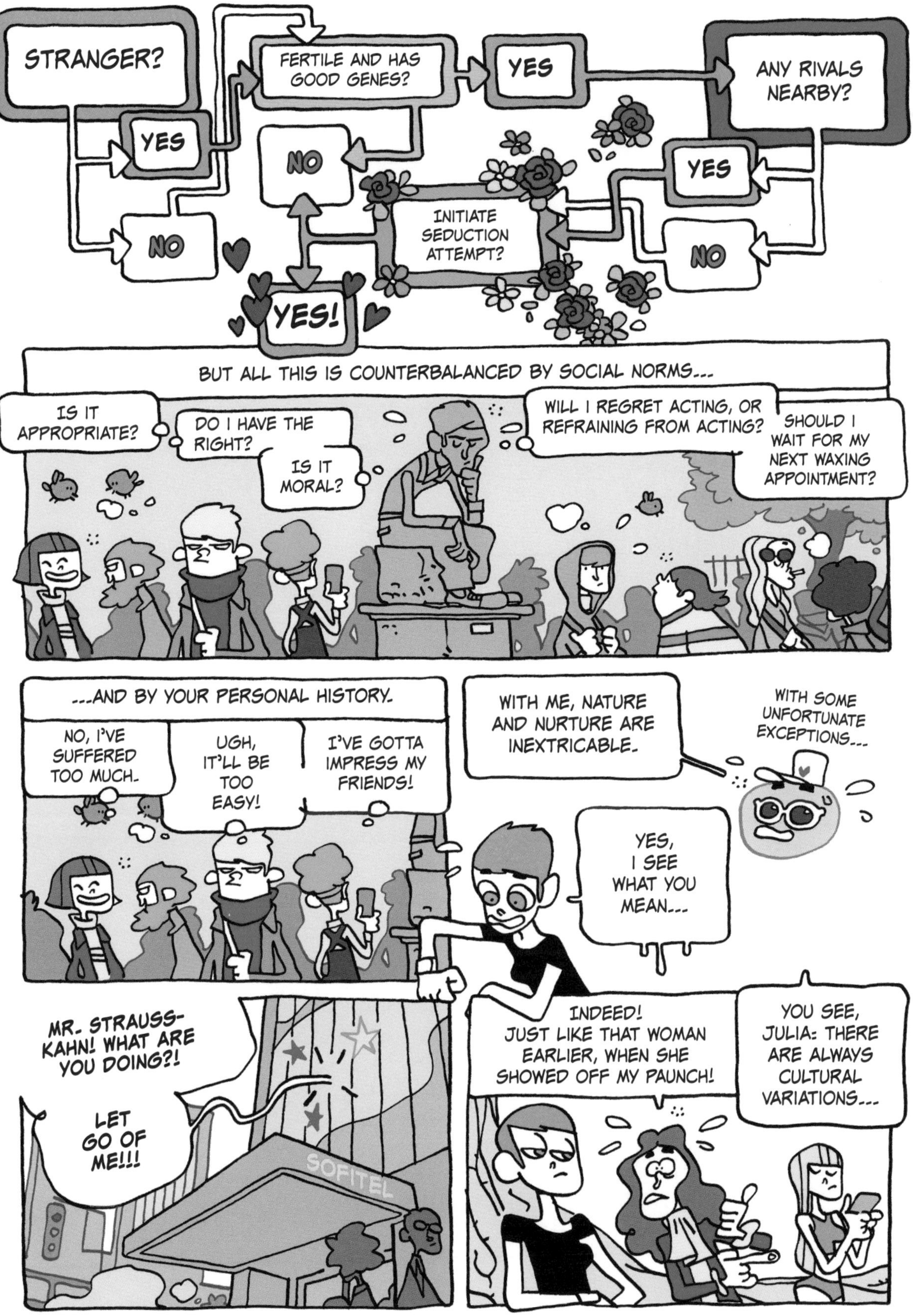
STRANGER?
YES
NO
FERTILE AND HAS GOOD GENES?
NO
YES
ANY RIVALS NEARBY?
YES
NO
INITIATE SEDUCTION ATTEMPT?
YES!
BUT ALL THIS IS COUNTERBALANCED BY SOCIAL NORMS...
IS IT APPROPRIATE?
DO I HAVE THE RIGHT?
IS IT MORAL?
WILL I REGRET ACTING, OR REFRAINING FROM ACTING?
SHOULD I WAIT FOR MY NEXT WAXING APPOINTMENT?
...AND BY YOUR PERSONAL HISTORY.
NO, I'VE SUFFERED TOO MUCH.
UGH, IT'LL BE TOO EASY!
I'VE GOTTA IMPRESS MY FRIENDS!
WITH ME, NATURE AND NURTURE ARE INEXTRICABLE.
WITH SOME UNFORTUNATE EXCEPTIONS...
YES, I SEE WHAT YOU MEAN...
MR. STRAUSS-KAHN! WHAT ARE YOU DOING?!
LET GO OF ME!!!
SOFITEL
INDEED! JUST LIKE THAT WOMAN EARLIER, WHEN SHE SHOWED OFF MY PAUNCH!
YOU SEE, JULIA: THERE ARE ALWAYS CULTURAL VARIATIONS...

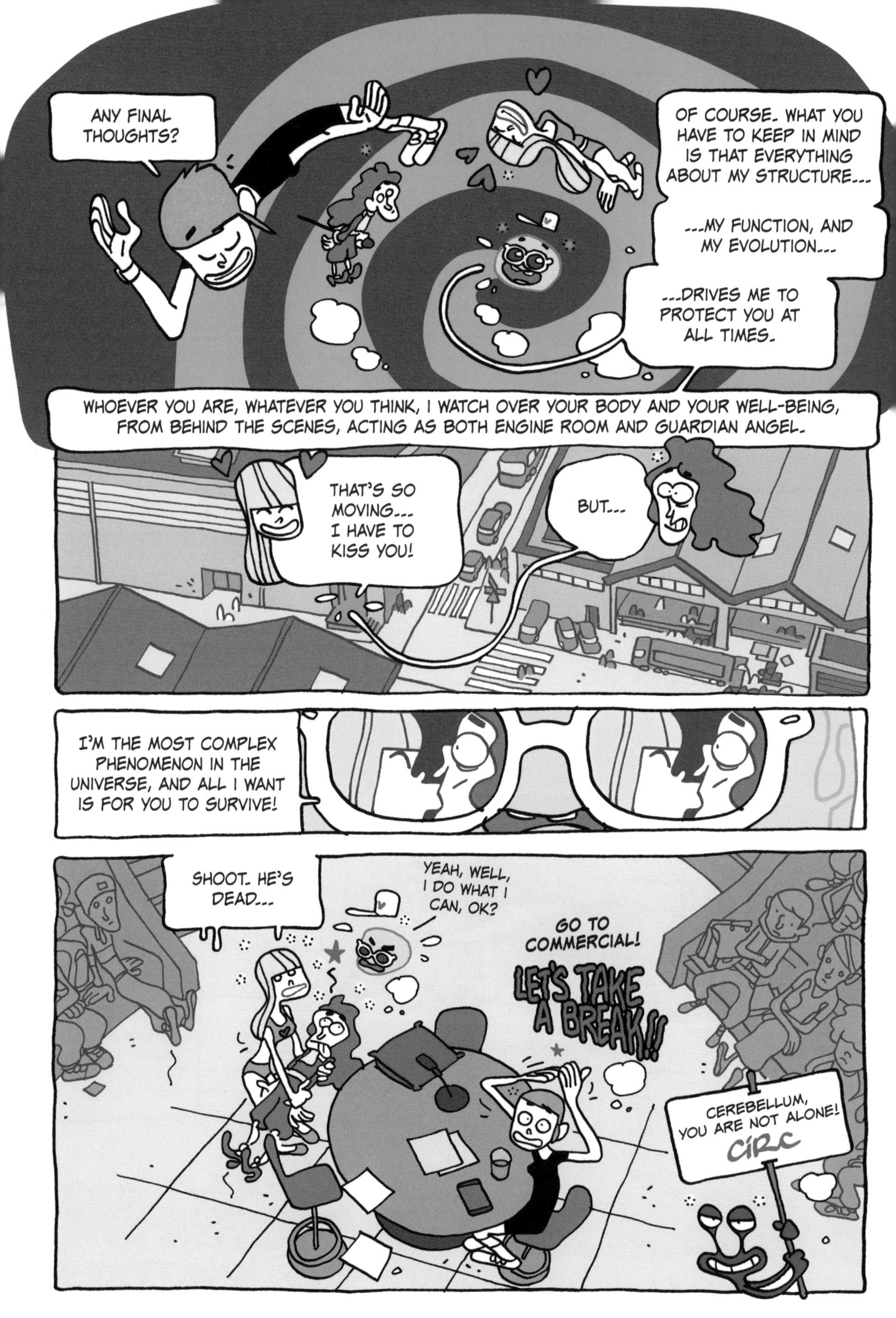
ANY FINAL THOUGHTS?
OF COURSE. WHAT YOU HAVE TO KEEP IN MIND IS THAT EVERYTHING ABOUT MY STRUCTURE...
...MY FUNCTION, AND MY EVOLUTION...
...DRIVES ME TO PROTECT YOU AT ALL TIMES.
WHOEVER YOU ARE, WHATEVER YOU THINK, I WATCH OVER YOUR BODY AND YOUR WELL-BEING, FROM BEHIND THE SCENES, ACTING AS BOTH ENGINE ROOM AND GUARDIAN ANGEL.
THAT'S SO MOVING... I HAVE TO KISS YOU!
BUT...
I'M THE MOST COMPLEX PHENOMENON IN THE UNIVERSE, AND ALL I WANT IS FOR YOU TO SURVIVE!
SHOOT. HE'S DEAD...
YEAH, WELL, I DO WHAT I CAN, OK?
GO TO COMMERCIAL!
LET'S TAKE A BREAK!!!
CEREBELLUM, YOU ARE NOT ALONE!
CIRC

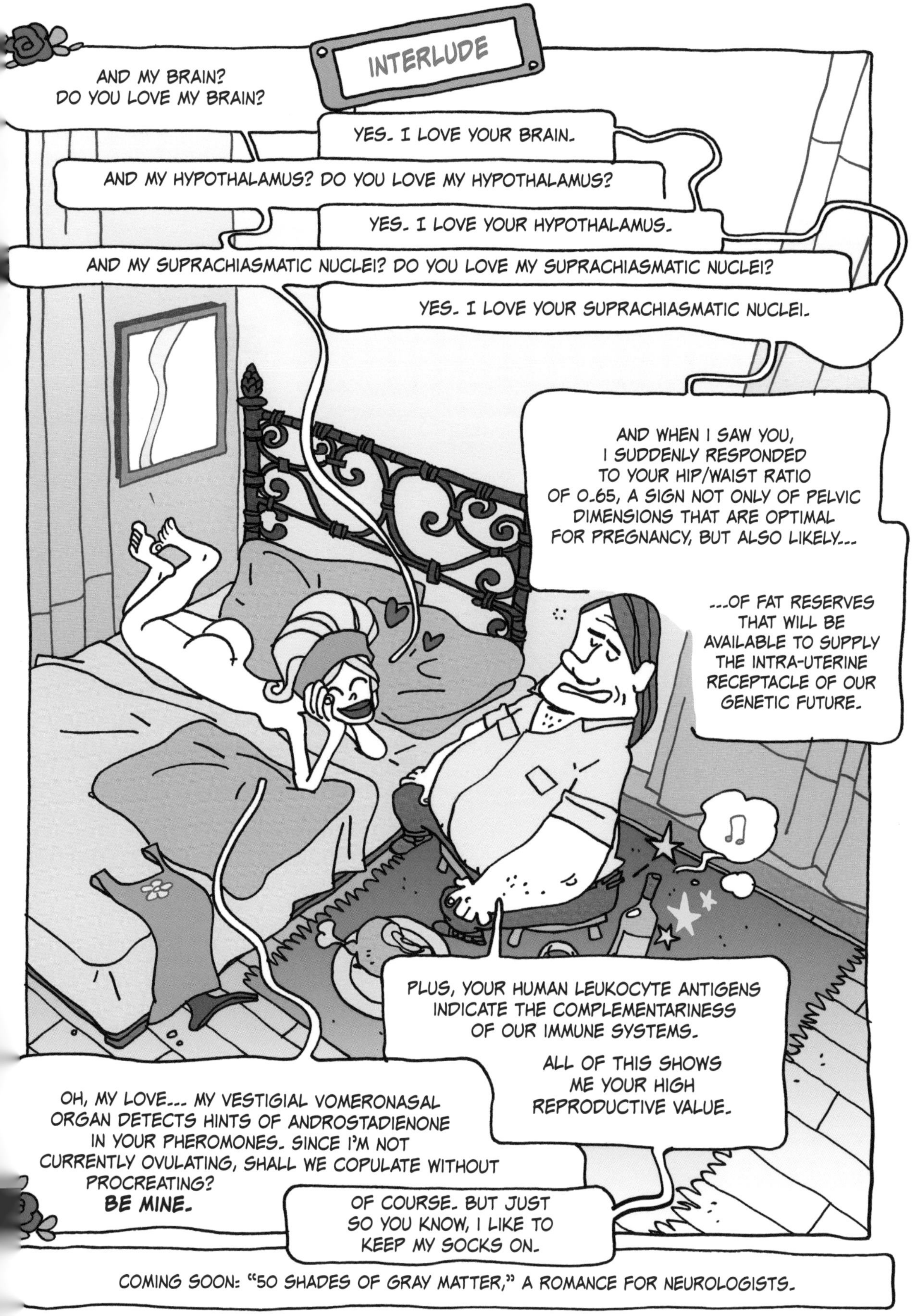
INTERLUDE
AND MY BRAIN? DO YOU LOVE MY BRAIN?
YES. I LOVE YOUR BRAIN.
AND MY HYPOTHALAMUS? DO YOU LOVE MY HYPOTHALAMUS?
YES. I LOVE YOUR HYPOTHALAMUS.
AND MY SUPRACHIASMATIC NUCLEI? DO YOU LOVE MY SUPRACHIASMATIC NUCLEI?
YES. I LOVE YOUR SUPRACHIASMATIC NUCLEI.
AND WHEN I SAW YOU, I SUDDENLY RESPONDED TO YOUR HIP/WAIST RATIO OF 0.65, A SIGN NOT ONLY OF PELVIC DIMENSIONS THAT ARE OPTIMAL FOR PREGNANCY, BUT ALSO LIKELY...
...OF FAT RESERVES THAT WILL BE AVAILABLE TO SUPPLY THE INTRA-UTERINE RECEPTACLE OF OUR GENETIC FUTURE.
PLUS, YOUR HUMAN LEUKOCYTE ANTIGENS INDICATE THE COMPLEMENTARINESS OF OUR IMMUNE SYSTEMS.
ALL OF THIS SHOWS ME YOUR HIGH REPRODUCTIVE VALUE.
OH, MY LOVE... MY VESTIGIAL VOMERONASAL ORGAN DETECTS HINTS OF ANDROSTADIENONE IN YOUR PHEROMONES. SINCE I'M NOT CURRENTLY OVULATING, SHALL WE COPULATE WITHOUT PROCREATING? **BE MINE.**
OF COURSE. BUT JUST SO YOU KNOW, I LIKE TO KEEP MY SOCKS ON.
COMING SOON: "50 SHADES OF GRAY MATTER," A ROMANCE FOR NEUROLOGISTS.

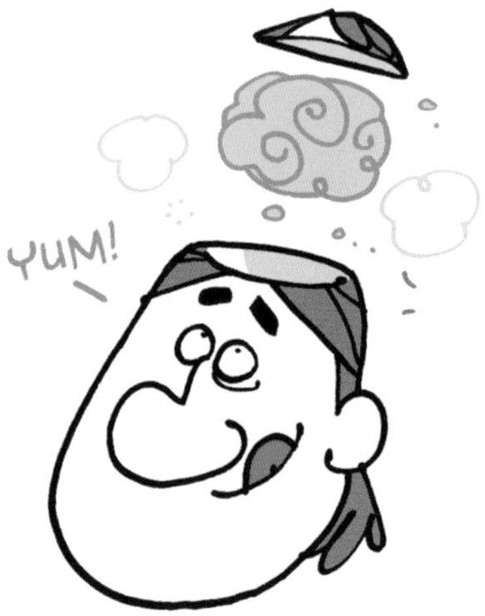
YUM!

Chapter 3

PERCEPTION IS A CONSTRUCTION

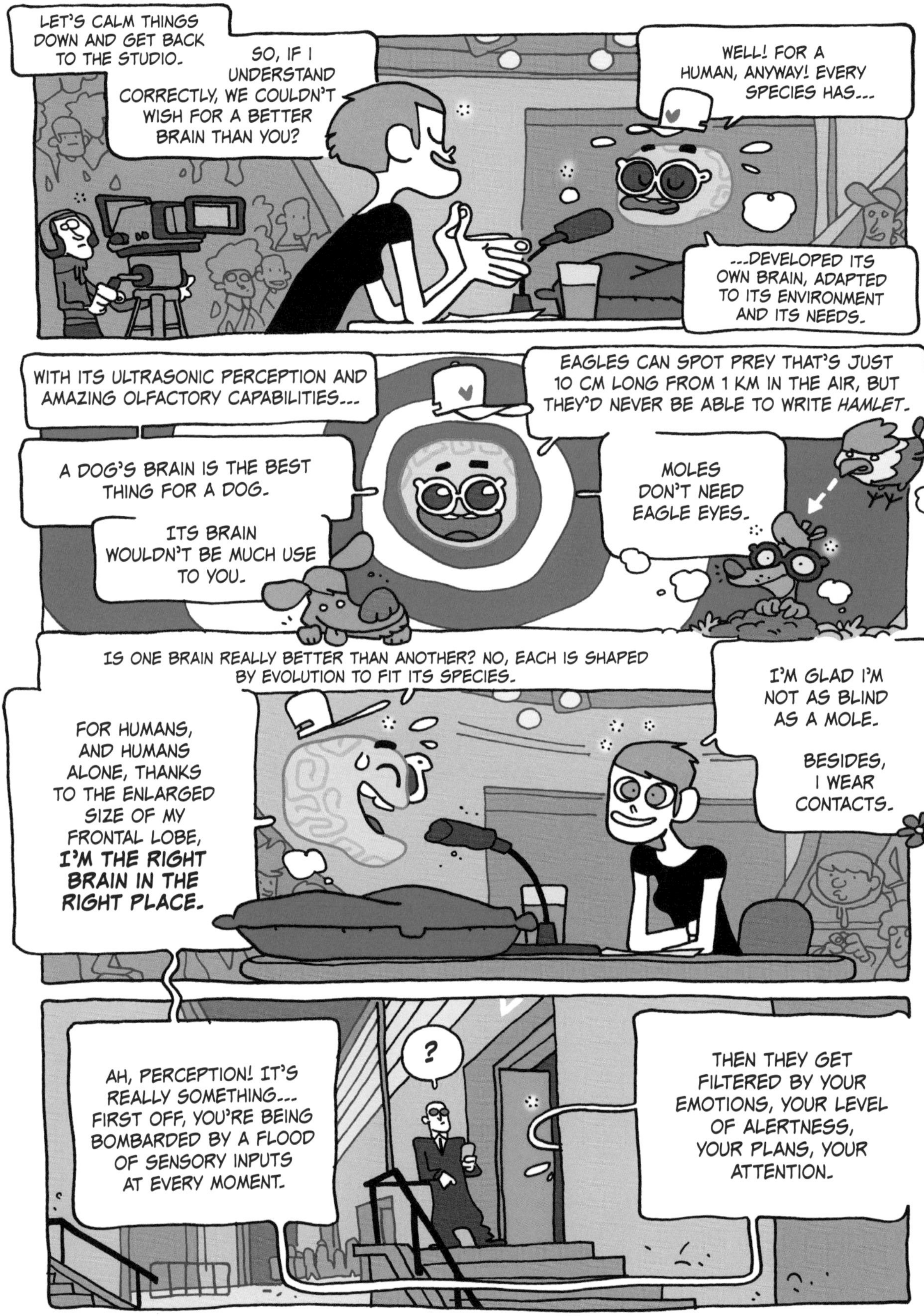
LET'S CALM THINGS DOWN AND GET BACK TO THE STUDIO.
SO, IF I UNDERSTAND CORRECTLY, WE COULDN'T WISH FOR A BETTER BRAIN THAN YOU?
WELL! FOR A HUMAN, ANYWAY! EVERY SPECIES HAS...
...DEVELOPED ITS OWN BRAIN, ADAPTED TO ITS ENVIRONMENT AND ITS NEEDS.
WITH ITS ULTRASONIC PERCEPTION AND AMAZING OLFACTORY CAPABILITIES...
A DOG'S BRAIN IS THE BEST THING FOR A DOG.
ITS BRAIN WOULDN'T BE MUCH USE TO YOU.
EAGLES CAN SPOT PREY THAT'S JUST 10 CM LONG FROM 1 KM IN THE AIR, BUT THEY'D NEVER BE ABLE TO WRITE *HAMLET*.
MOLES DON'T NEED EAGLE EYES.
IS ONE BRAIN REALLY BETTER THAN ANOTHER? NO, EACH IS SHAPED BY EVOLUTION TO FIT ITS SPECIES.
FOR HUMANS, AND HUMANS ALONE, THANKS TO THE ENLARGED SIZE OF MY FRONTAL LOBE, **I'M THE RIGHT BRAIN IN THE RIGHT PLACE.**
I'M GLAD I'M NOT AS BLIND AS A MOLE.
BESIDES, I WEAR CONTACTS.
AH, PERCEPTION! IT'S REALLY SOMETHING... FIRST OFF, YOU'RE BEING BOMBARDED BY A FLOOD OF SENSORY INPUTS AT EVERY MOMENT.
?
THEN THEY GET FILTERED BY YOUR EMOTIONS, YOUR LEVEL OF ALERTNESS, YOUR PLANS, YOUR ATTENTION.

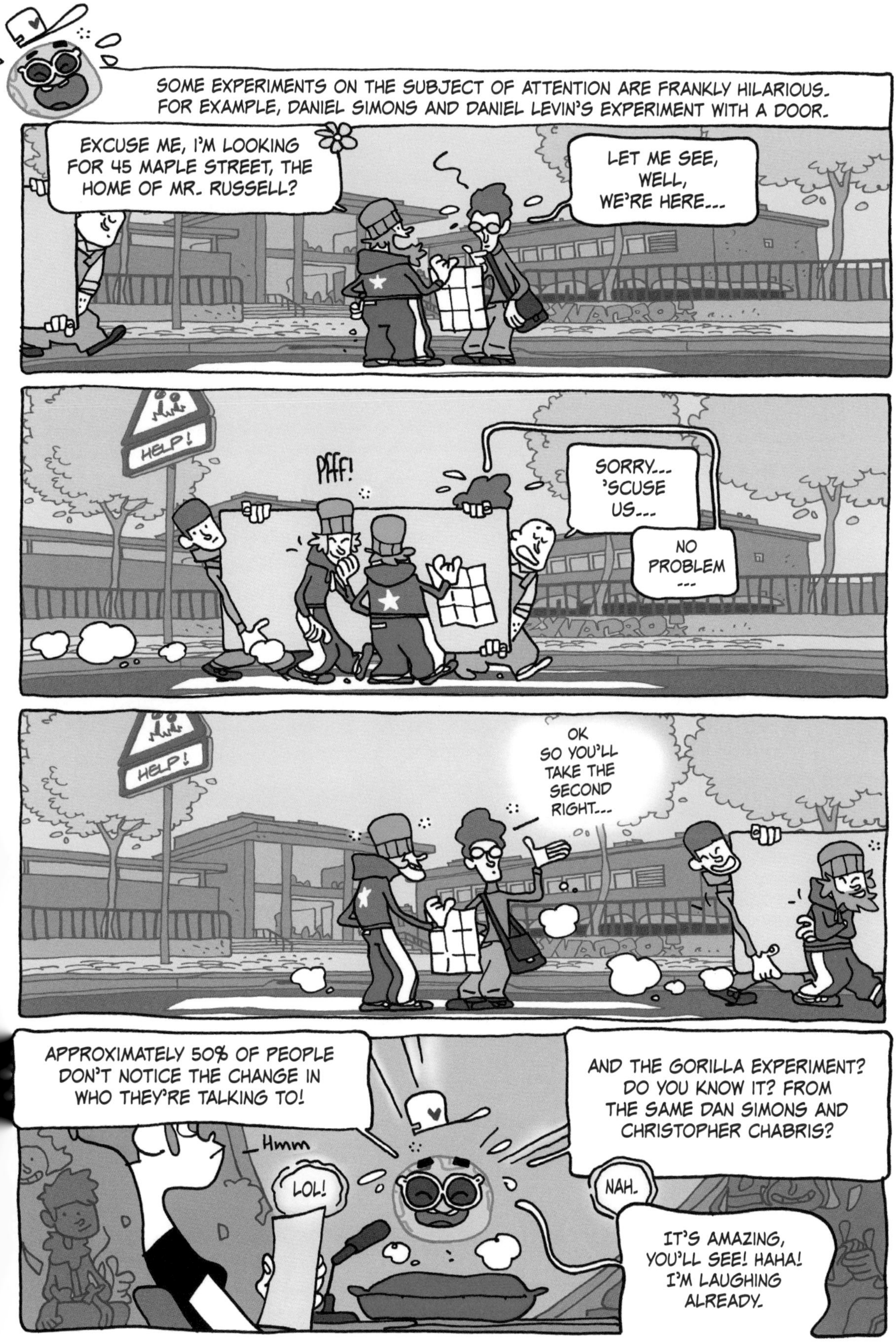

SOME EXPERIMENTS ON THE SUBJECT OF ATTENTION ARE FRANKLY HILARIOUS. FOR EXAMPLE, DANIEL SIMONS AND DANIEL LEVIN'S EXPERIMENT WITH A DOOR.
EXCUSE ME, I'M LOOKING FOR 45 MAPLE STREET, THE HOME OF MR. RUSSELL?
LET ME SEE, WELL, WE'RE HERE...
HELP!
PFFF!
SORRY... 'SCUSE US...
NO PROBLEM...
HELP!
OK SO YOU'LL TAKE THE SECOND RIGHT...
APPROXIMATELY 50% OF PEOPLE DON'T NOTICE THE CHANGE IN WHO THEY'RE TALKING TO!
Hmm
LOL!
AND THE GORILLA EXPERIMENT? DO YOU KNOW IT? FROM THE SAME DAN SIMONS AND CHRISTOPHER CHABRIS?
NAH.
IT'S AMAZING, YOU'LL SEE! HAHA! I'M LAUGHING ALREADY.

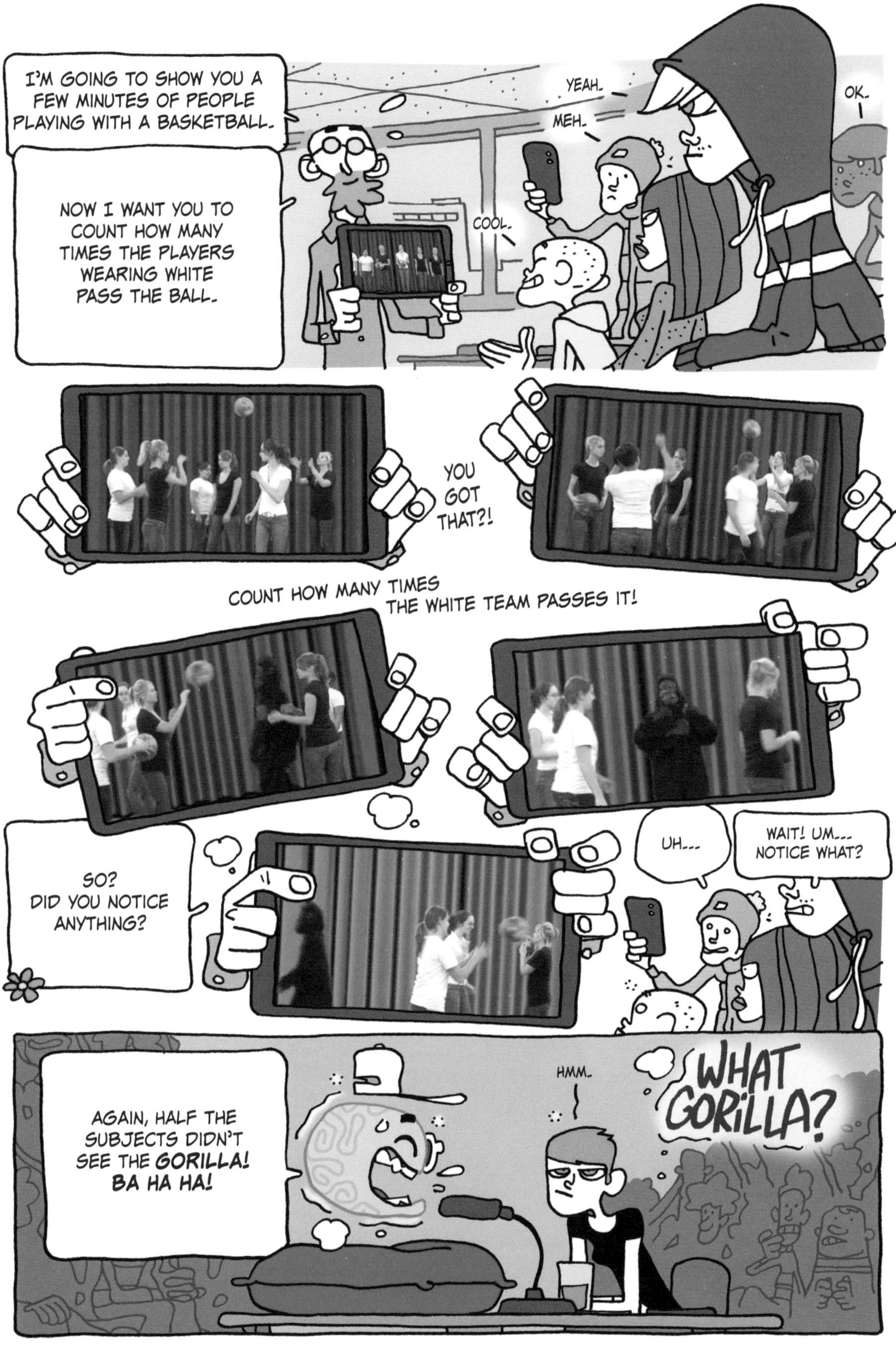
I'M GOING TO SHOW YOU A FEW MINUTES OF PEOPLE PLAYING WITH A BASKETBALL.
NOW I WANT YOU TO COUNT HOW MANY TIMES THE PLAYERS WEARING WHITE PASS THE BALL.
YEAH.
MEH.
OK.
COOL.
YOU GOT THAT?!
COUNT HOW MANY TIMES THE WHITE TEAM PASSES IT!
SO? DID YOU NOTICE ANYTHING?
UH...
WAIT! UM... NOTICE WHAT?
AGAIN, HALF THE SUBJECTS DIDN'T SEE THE **GORILLA! BA HA HA!**
HMM.
WHAT GORILLA?

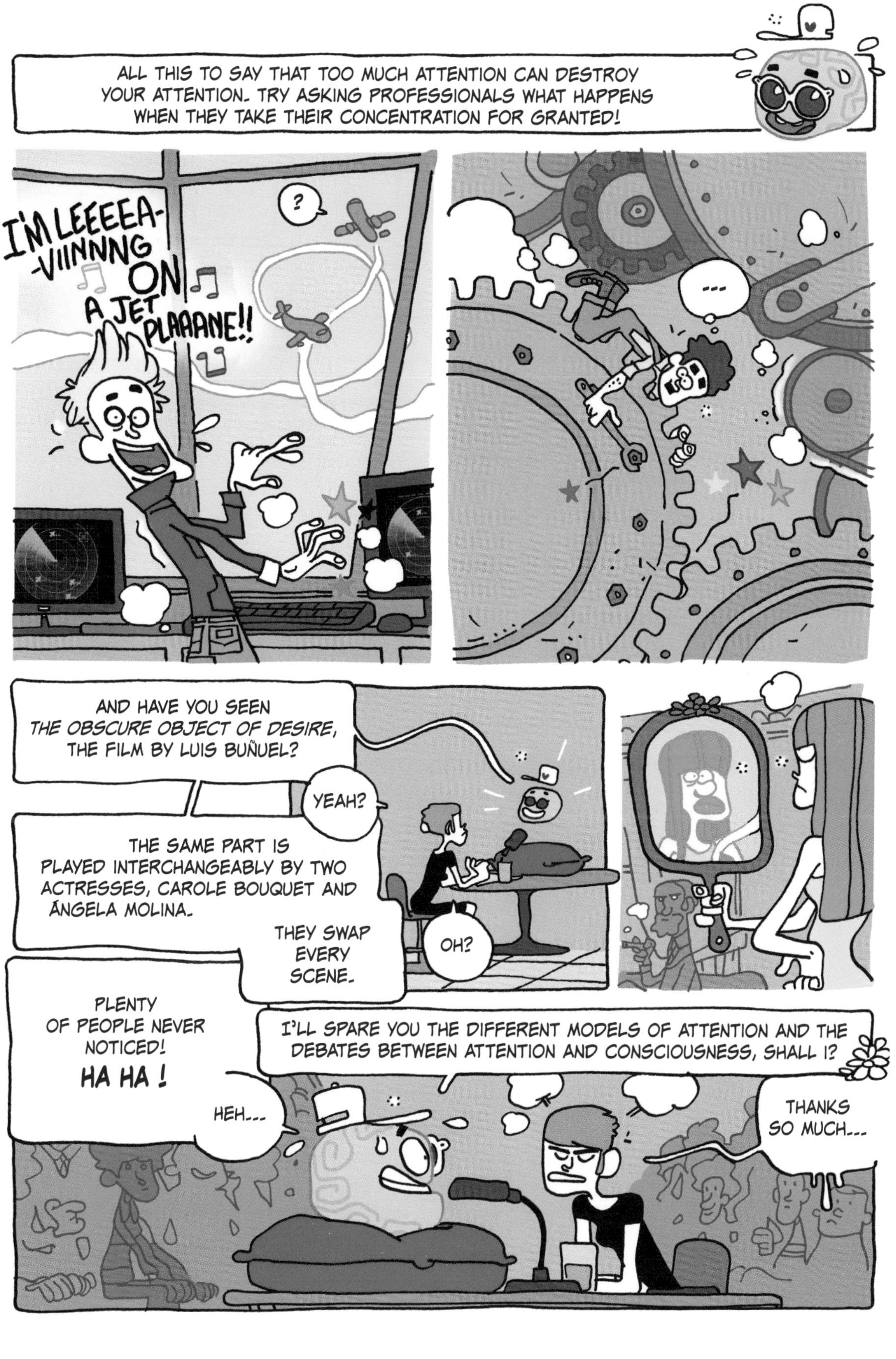

ALL THIS TO SAY THAT TOO MUCH ATTENTION CAN DESTROY YOUR ATTENTION. TRY ASKING PROFESSIONALS WHAT HAPPENS WHEN THEY TAKE THEIR CONCENTRATION FOR GRANTED!
I'M LEEEEA-VIINNNG ON A JET PLAAANE!!
?
...
AND HAVE YOU SEEN *THE OBSCURE OBJECT OF DESIRE*, THE FILM BY LUIS BUÑUEL?
YEAH?
THE SAME PART IS PLAYED INTERCHANGEABLY BY TWO ACTRESSES, CAROLE BOUQUET AND ÁNGELA MOLINA.
THEY SWAP EVERY SCENE.
OH?
PLENTY OF PEOPLE NEVER NOTICED! **HA HA !**
I'LL SPARE YOU THE DIFFERENT MODELS OF ATTENTION AND THE DEBATES BETWEEN ATTENTION AND CONSCIOUSNESS, SHALL I?
HEH...
THANKS SO MUCH...

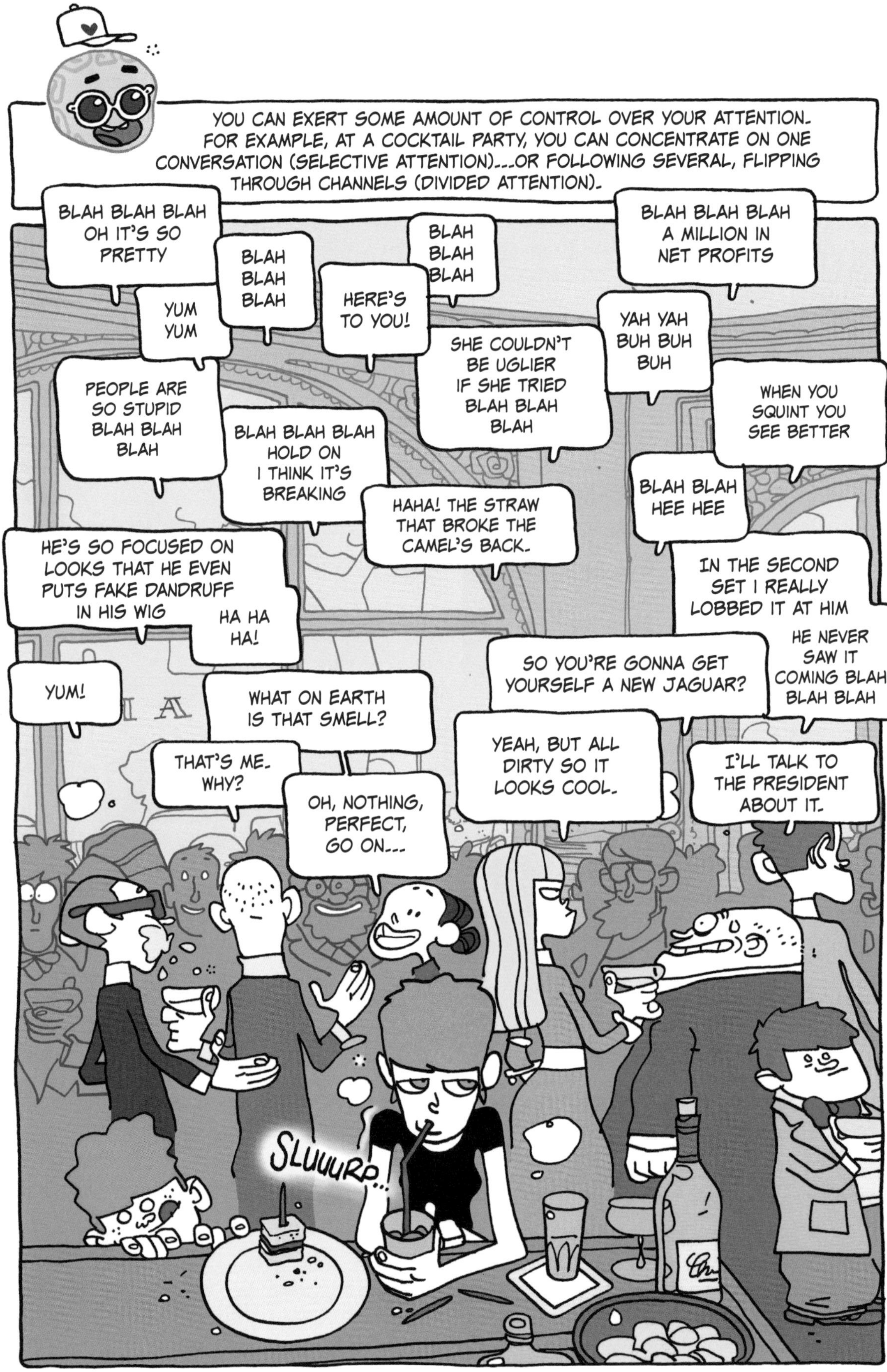
YOU CAN EXERT SOME AMOUNT OF CONTROL OVER YOUR ATTENTION. FOR EXAMPLE, AT A COCKTAIL PARTY, YOU CAN CONCENTRATE ON ONE CONVERSATION (SELECTIVE ATTENTION)...OR FOLLOWING SEVERAL, FLIPPING THROUGH CHANNELS (DIVIDED ATTENTION).
BLAH BLAH BLAH OH IT'S SO PRETTY
YUM YUM
BLAH BLAH BLAH
HERE'S TO YOU!
BLAH BLAH BLAH
BLAH BLAH BLAH A MILLION IN NET PROFITS
YAH YAH BUH BUH BUH
SHE COULDN'T BE UGLIER IF SHE TRIED BLAH BLAH BLAH
PEOPLE ARE SO STUPID BLAH BLAH BLAH
BLAH BLAH BLAH HOLD ON I THINK IT'S BREAKING
WHEN YOU SQUINT YOU SEE BETTER
BLAH BLAH HEE HEE
HAHA! THE STRAW THAT BROKE THE CAMEL'S BACK.
HE'S SO FOCUSED ON LOOKS THAT HE EVEN PUTS FAKE DANDRUFF IN HIS WIG
HA HA HA!
IN THE SECOND SET I REALLY LOBBED IT AT HIM
HE NEVER SAW IT COMING BLAH BLAH BLAH
SO YOU'RE GONNA GET YOURSELF A NEW JAGUAR?
YUM!
WHAT ON EARTH IS THAT SMELL?
THAT'S ME. WHY?
OH, NOTHING, PERFECT, GO ON...
YEAH, BUT ALL DIRTY SO IT LOOKS COOL.
I'LL TALK TO THE PRESIDENT ABOUT IT.
SLUUURP...

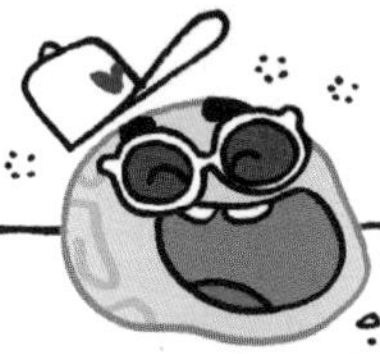

BUT ATTENTION IS LARGELY AUTOMATED, FUNCTIONING MOST OFTEN WITHOUT YOU CONSCIOUSLY NOTICING. FOR EXAMPLE, AT THAT SAME COCKTAIL PARTY, YOU'RE VERY LIKELY, WITHOUT EVEN TRYING, TO CATCH YOUR OWN NAME OVER THE HUBBUB, EVEN FROM THE OTHER SIDE OF THE ROOM.

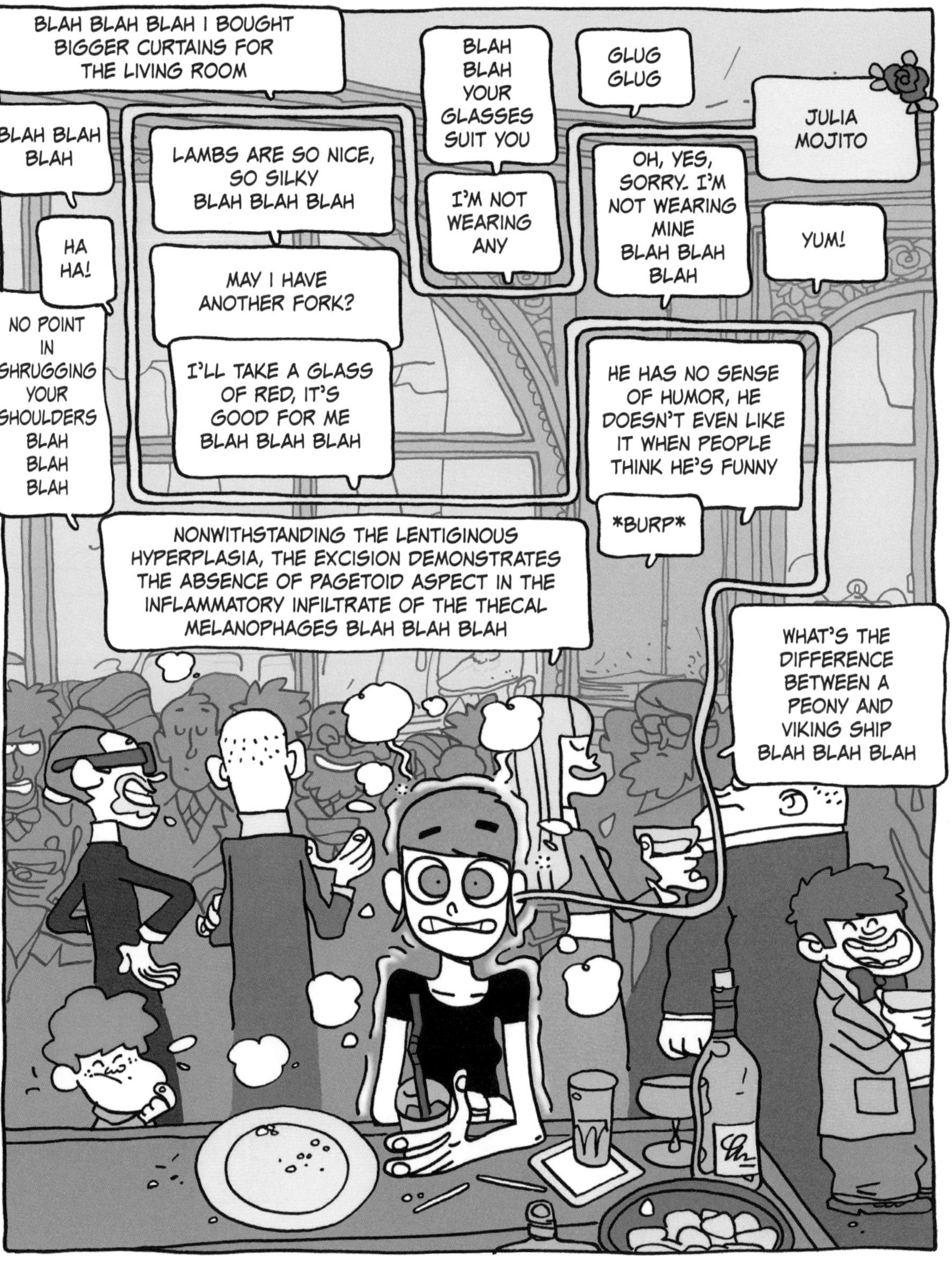
BLAH BLAH BLAH I BOUGHT BIGGER CURTAINS FOR THE LIVING ROOM
BLAH BLAH YOUR GLASSES SUIT YOU
GLUG GLUG
JULIA MOJITO
BLAH BLAH BLAH
LAMBS ARE SO NICE, SO SILKY BLAH BLAH BLAH
I'M NOT WEARING ANY
OH, YES, SORRY. I'M NOT WEARING MINE BLAH BLAH BLAH
YUM!
HA HA!
MAY I HAVE ANOTHER FORK?
NO POINT IN SHRUGGING YOUR SHOULDERS BLAH BLAH BLAH
I'LL TAKE A GLASS OF RED, IT'S GOOD FOR ME BLAH BLAH BLAH
HE HAS NO SENSE OF HUMOR, HE DOESN'T EVEN LIKE IT WHEN PEOPLE THINK HE'S FUNNY
BURP
NONWITHSTANDING THE LENTIGINOUS HYPERPLASIA, THE EXCISION DEMONSTRATES THE ABSENCE OF PAGETOID ASPECT IN THE INFLAMMATORY INFILTRATE OF THE THECAL MELANOPHAGES BLAH BLAH BLAH
WHAT'S THE DIFFERENCE BETWEEN A PEONY AND VIKING SHIP BLAH BLAH BLAH

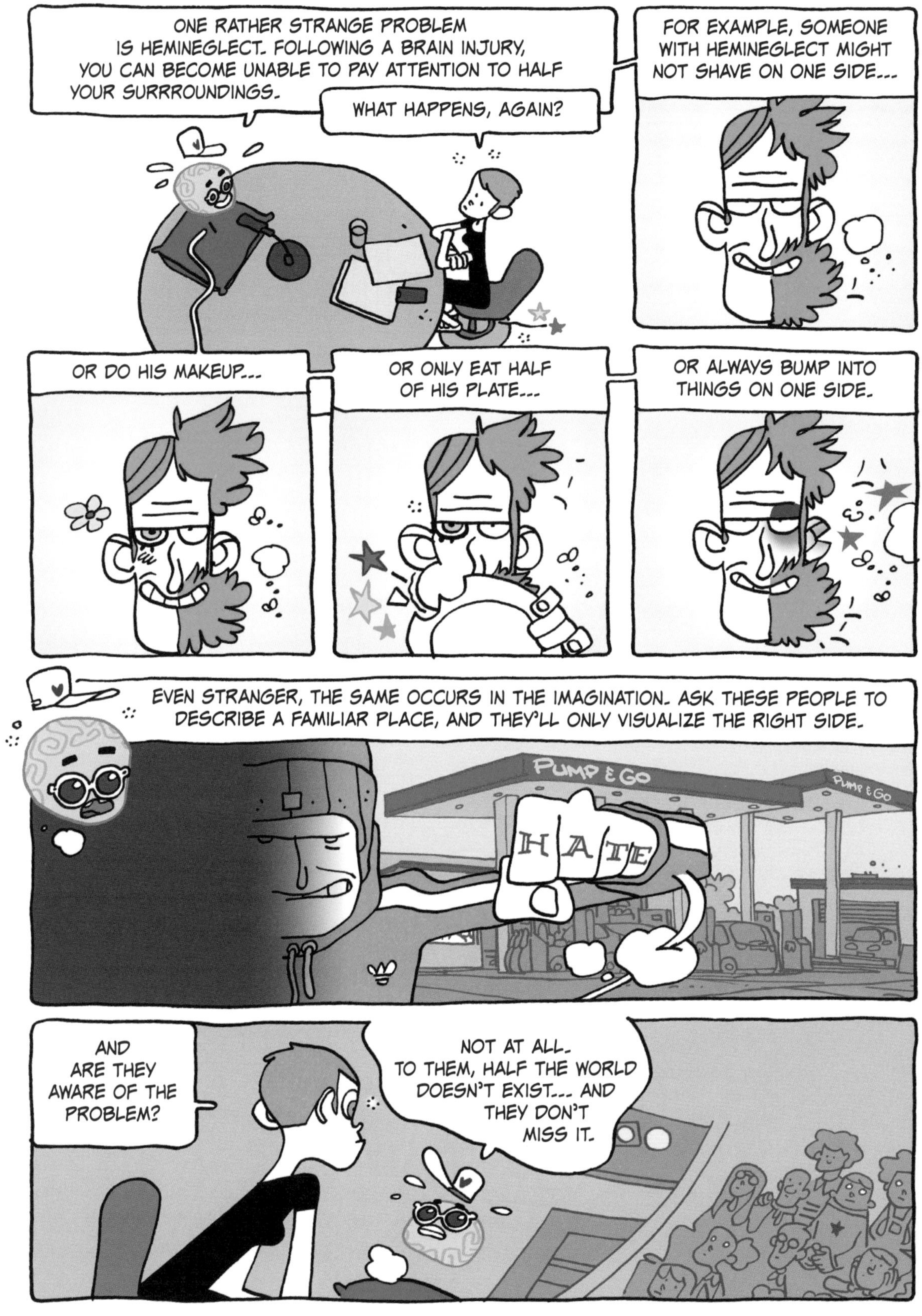

ONE RATHER STRANGE PROBLEM IS HEMINEGLECT. FOLLOWING A BRAIN INJURY, YOU CAN BECOME UNABLE TO PAY ATTENTION TO HALF YOUR SURRROUNDINGS.
WHAT HAPPENS, AGAIN?
FOR EXAMPLE, SOMEONE WITH HEMINEGLECT MIGHT NOT SHAVE ON ONE SIDE...
OR DO HIS MAKEUP...
OR ONLY EAT HALF OF HIS PLATE...
OR ALWAYS BUMP INTO THINGS ON ONE SIDE.
EVEN STRANGER, THE SAME OCCURS IN THE IMAGINATION. ASK THESE PEOPLE TO DESCRIBE A FAMILIAR PLACE, AND THEY'LL ONLY VISUALIZE THE RIGHT SIDE.
PUMP & GO
PUMP & GO
HATE
AND ARE THEY AWARE OF THE PROBLEM?
NOT AT ALL. TO THEM, HALF THE WORLD DOESN'T EXIST... AND THEY DON'T MISS IT.

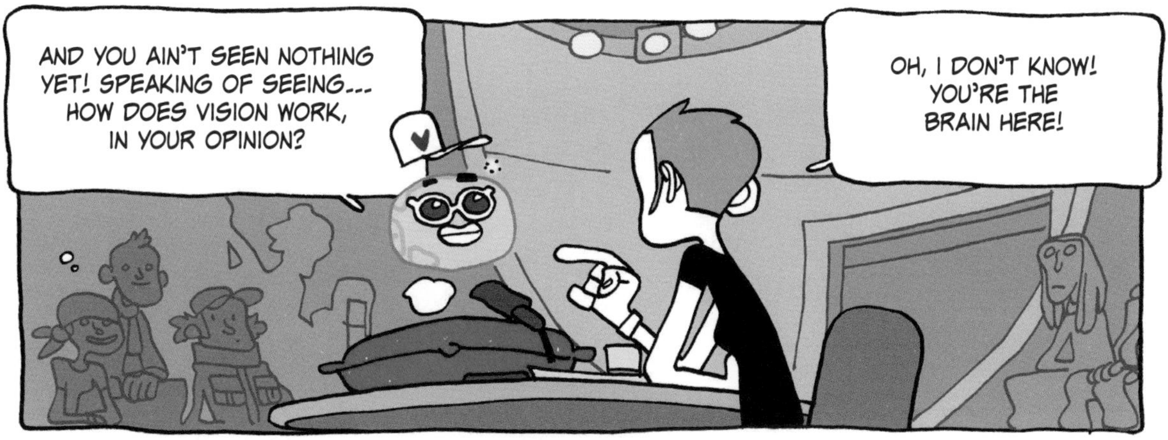
AND YOU AIN'T SEEN NOTHING YET! SPEAKING OF SEEING... HOW DOES VISION WORK, IN YOUR OPINION?
OH, I DON'T KNOW! YOU'RE THE BRAIN HERE!

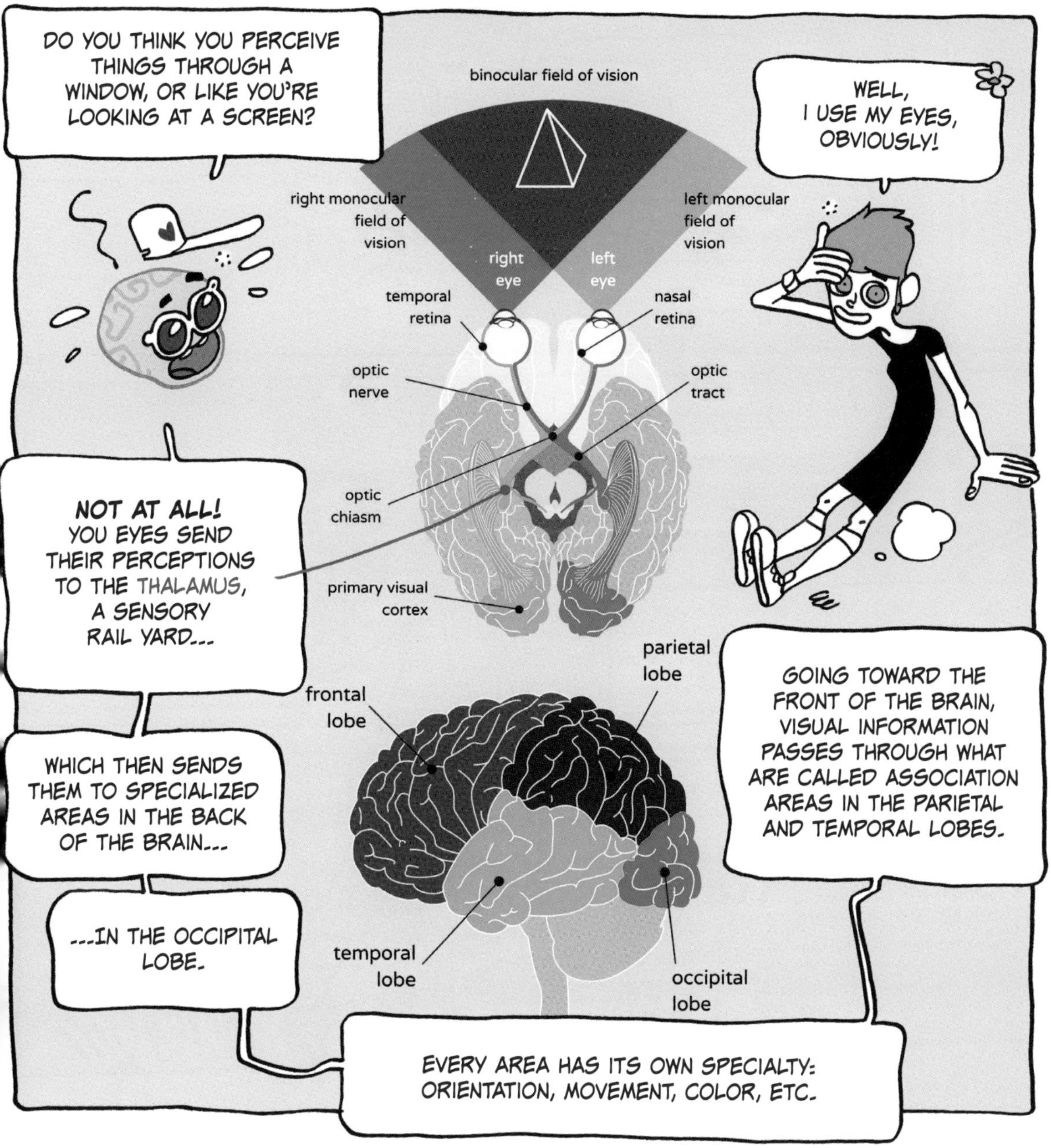
DO YOU THINK YOU PERCEIVE THINGS THROUGH A WINDOW, OR LIKE YOU'RE LOOKING AT A SCREEN?
binocular field of vision
WELL, I USE MY EYES, OBVIOUSLY!
right monocular field of vision
left monocular field of vision
right eye
left eye
temporal retina
nasal retina
optic nerve
optic tract
optic chiasm
primary visual cortex
NOT AT ALL! YOU EYES SEND THEIR PERCEPTIONS TO THE THALAMUS, A SENSORY RAIL YARD...
parietal lobe
frontal lobe
GOING TOWARD THE FRONT OF THE BRAIN, VISUAL INFORMATION PASSES THROUGH WHAT ARE CALLED ASSOCIATION AREAS IN THE PARIETAL AND TEMPORAL LOBES.
WHICH THEN SENDS THEM TO SPECIALIZED AREAS IN THE BACK OF THE BRAIN...
...IN THE OCCIPITAL LOBE.
temporal lobe
occipital lobe
EVERY AREA HAS ITS OWN SPECIALTY: ORIENTATION, MOVEMENT, COLOR, ETC.

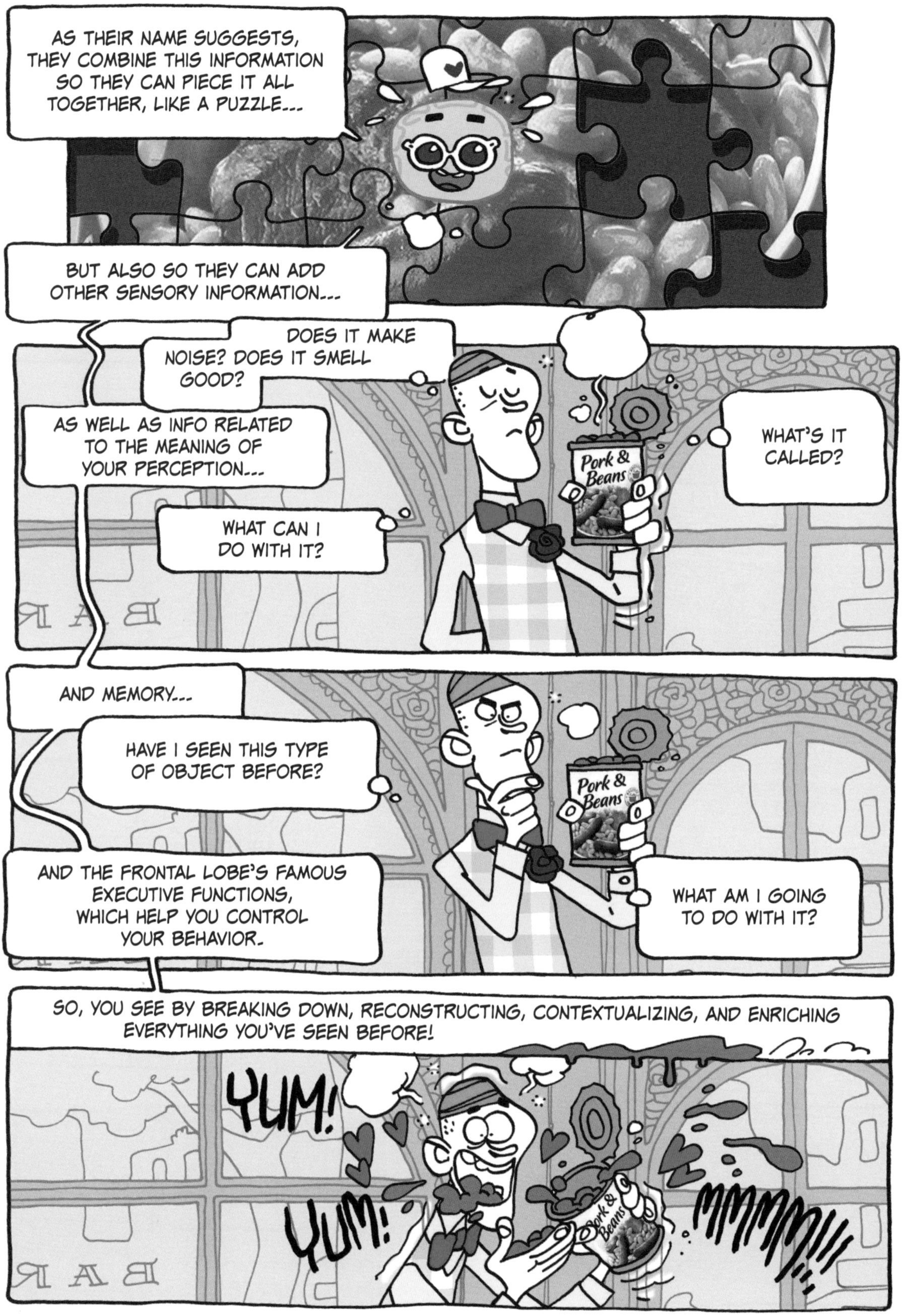
AS THEIR NAME SUGGESTS, THEY COMBINE THIS INFORMATION SO THEY CAN PIECE IT ALL TOGETHER, LIKE A PUZZLE...
BUT ALSO SO THEY CAN ADD OTHER SENSORY INFORMATION...
DOES IT MAKE NOISE? DOES IT SMELL GOOD?
AS WELL AS INFO RELATED TO THE MEANING OF YOUR PERCEPTION...
WHAT'S IT CALLED?
WHAT CAN I DO WITH IT?
Pork & Beans
AND MEMORY...
HAVE I SEEN THIS TYPE OF OBJECT BEFORE?
AND THE FRONTAL LOBE'S FAMOUS EXECUTIVE FUNCTIONS, WHICH HELP YOU CONTROL YOUR BEHAVIOR.
WHAT AM I GOING TO DO WITH IT?
SO, YOU SEE BY BREAKING DOWN, RECONSTRUCTING, CONTEXTUALIZING, AND ENRICHING EVERYTHING YOU'VE SEEN BEFORE!
YUM!
YUM!
MMMM!!!

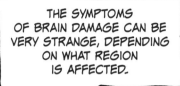

THE SYMPTOMS OF BRAIN DAMAGE CAN BE VERY STRANGE, DEPENDING ON WHAT REGION IS AFFECTED.

DO YOU WANT THE RED PILL OR THE BLUE PILL?

THERE HAVE BEEN CASES WHO COULD NO LONGER RECOGNIZE COLORS...

UH... WHICH ONE'S RED?

...OR MOTION...

LOOK OUT! THAT CAR'S COMING RIGHT AT YOU!

NO IT'S NOT, IT'S PARKED!

...SHAPES...

OTHERS HAVE A REDUCED FIELD OF VISION...

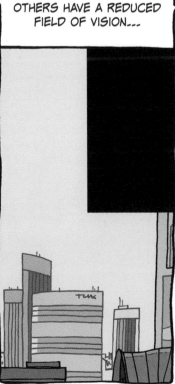

OR EVEN SEE HALLUCINATIONS IN CERTAIN SPOTS.

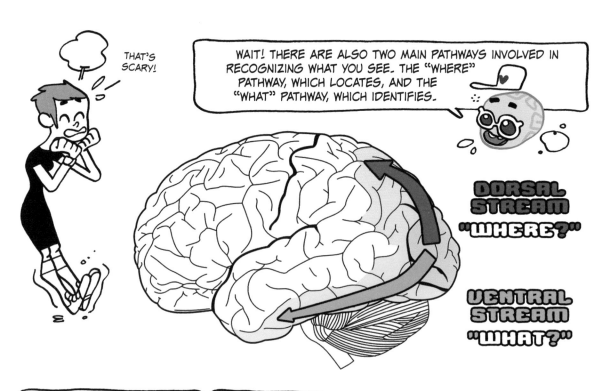

THAT'S SCARY!

WAIT! THERE ARE ALSO TWO MAIN PATHWAYS INVOLVED IN RECOGNIZING WHAT YOU SEE. THE "WHERE" PATHWAY, WHICH LOCATES, AND THE "WHAT" PATHWAY, WHICH IDENTIFIES.

DORSAL STREAM "WHERE?"

VENTRAL STREAM "WHAT?"

WHEN THE "WHERE" PATHWAY IS DAMAGED, WELL, YOU MAY FIND YOURSELF UNABLE TO LOCATE AN OBJECT...

WHAT IS THIS?

A TOOTHBRUSH.

NOW GRAB IT, PLEASE.

WHEN THE "WHAT" PATHWAY IS DAMAGED, YOU MAY NO LONGER RECOGNIZE OBJECTS RIGHT IN FRONT OF YOU.

WHAT IS THIS?

NO IDEA... A KEY?

CLOSE YOUR EYES AND FEEL IT...

OH! IT'S A TOOTHBRUSH!

BUT LOOK! HER TACTILE PERCEPTIONS ARE STILL INTACT! SO SHE CAN RECOGNIZE THE OBJECT BY TOUCHING IT!

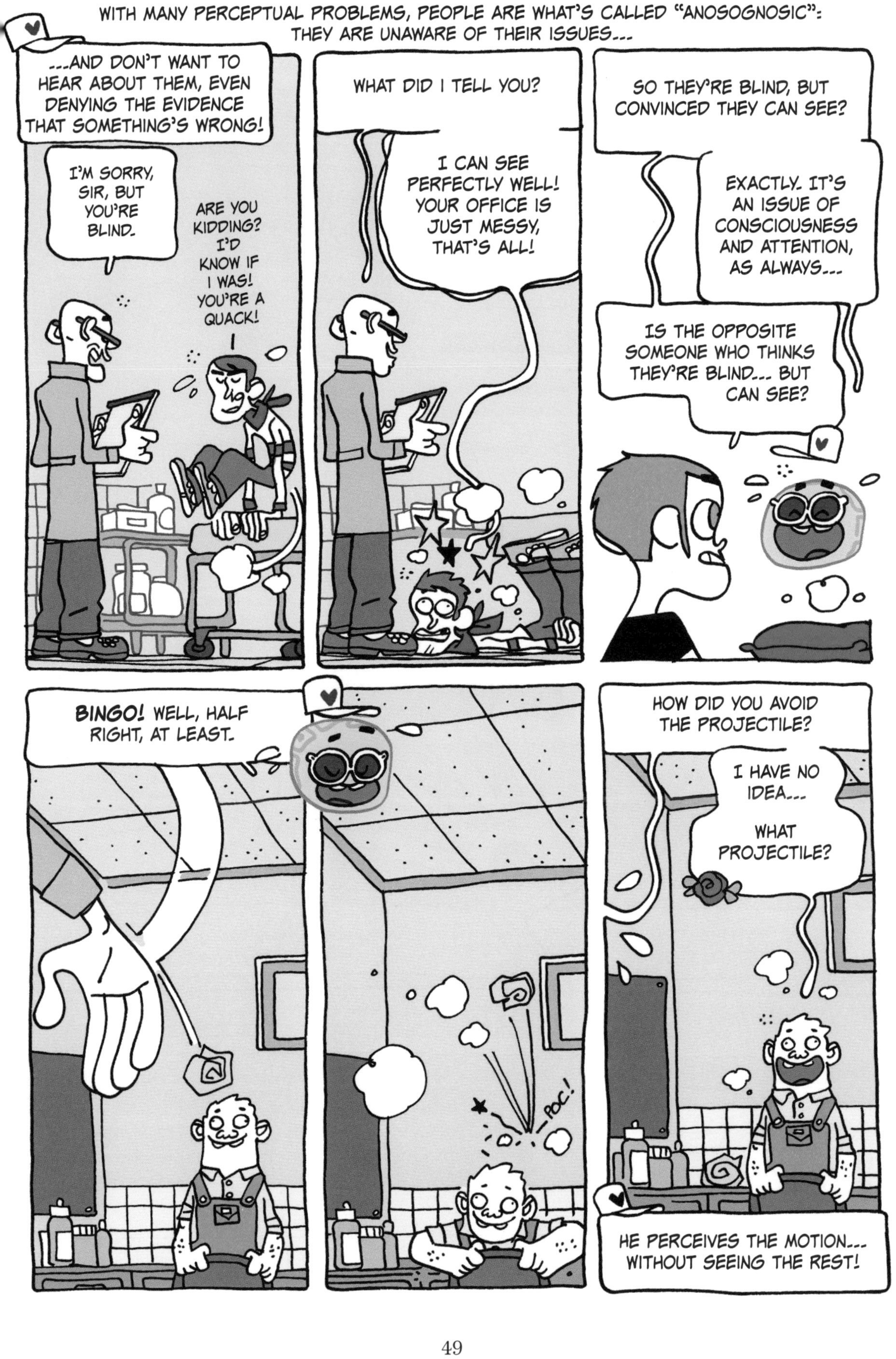
WITH MANY PERCEPTUAL PROBLEMS, PEOPLE ARE WHAT'S CALLED "ANOSOGNOSIC": THEY ARE UNAWARE OF THEIR ISSUES...
...AND DON'T WANT TO HEAR ABOUT THEM, EVEN DENYING THE EVIDENCE THAT SOMETHING'S WRONG!
I'M SORRY, SIR, BUT YOU'RE BLIND.
ARE YOU KIDDING? I'D KNOW IF I WAS! YOU'RE A QUACK!
WHAT DID I TELL YOU?
I CAN SEE PERFECTLY WELL! YOUR OFFICE IS JUST MESSY, THAT'S ALL!
SO THEY'RE BLIND, BUT CONVINCED THEY CAN SEE?
EXACTLY. IT'S AN ISSUE OF CONSCIOUSNESS AND ATTENTION, AS ALWAYS...
IS THE OPPOSITE SOMEONE WHO THINKS THEY'RE BLIND... BUT CAN SEE?
BINGO! WELL, HALF RIGHT, AT LEAST.
POC!
HOW DID YOU AVOID THE PROJECTILE?
I HAVE NO IDEA...
WHAT PROJECTILE?
HE PERCEIVES THE MOTION... WITHOUT SEEING THE REST!

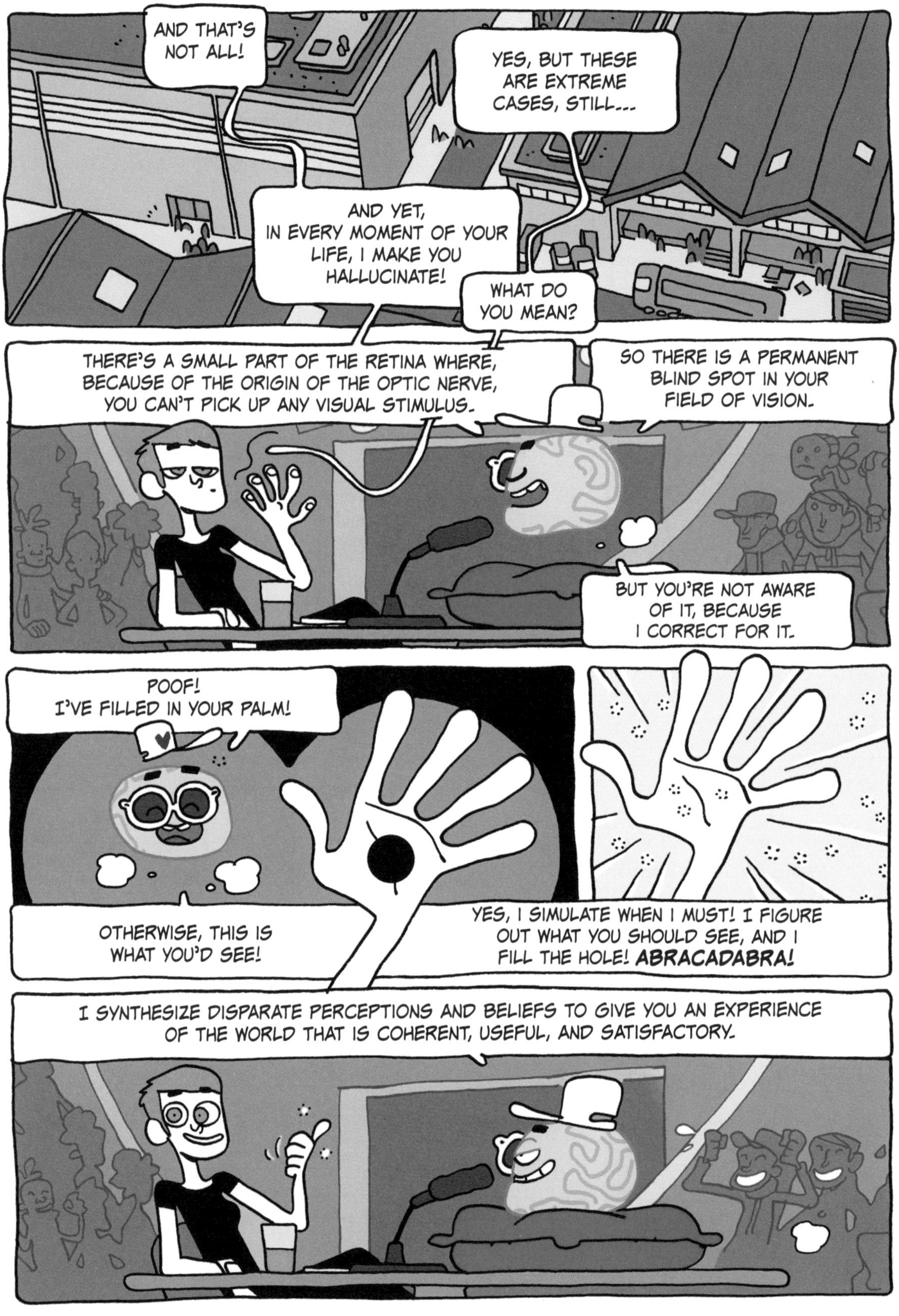
AND THAT'S NOT ALL!
YES, BUT THESE ARE EXTREME CASES, STILL...
AND YET, IN EVERY MOMENT OF YOUR LIFE, I MAKE YOU HALLUCINATE!
WHAT DO YOU MEAN?
THERE'S A SMALL PART OF THE RETINA WHERE, BECAUSE OF THE ORIGIN OF THE OPTIC NERVE, YOU CAN'T PICK UP ANY VISUAL STIMULUS.
SO THERE IS A PERMANENT BLIND SPOT IN YOUR FIELD OF VISION.
BUT YOU'RE NOT AWARE OF IT, BECAUSE I CORRECT FOR IT.
POOF! I'VE FILLED IN YOUR PALM!
OTHERWISE, THIS IS WHAT YOU'D SEE!
YES, I SIMULATE WHEN I MUST! I FIGURE OUT WHAT YOU SHOULD SEE, AND I FILL THE HOLE! ABRACADABRA!
I SYNTHESIZE DISPARATE PERCEPTIONS AND BELIEFS TO GIVE YOU AN EXPERIENCE OF THE WORLD THAT IS COHERENT, USEFUL, AND SATISFACTORY.

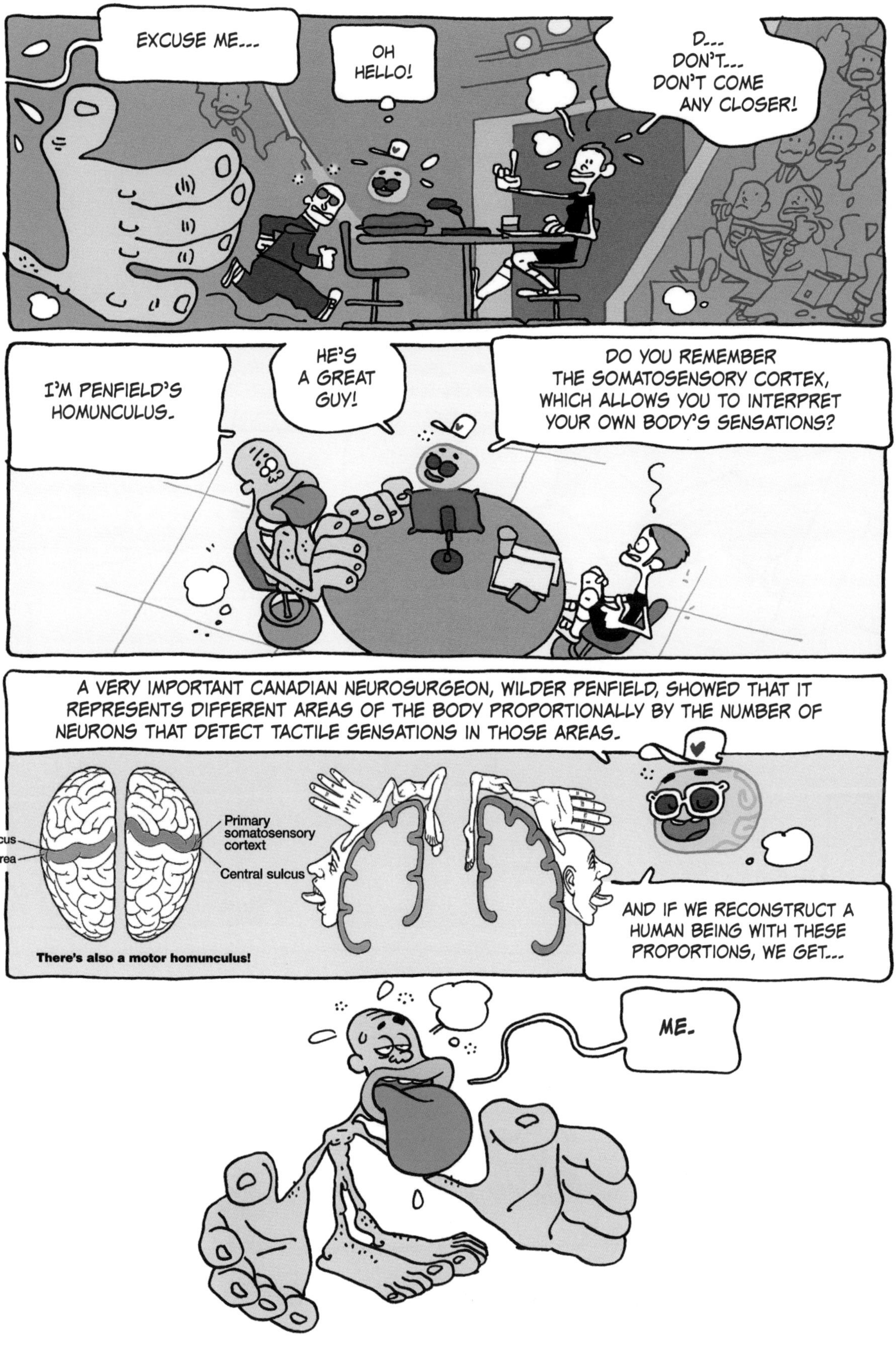

EXCUSE ME...
OH HELLO!
D... DON'T... DON'T COME ANY CLOSER!
I'M PENFIELD'S HOMUNCULUS.
HE'S A GREAT GUY!
DO YOU REMEMBER THE SOMATOSENSORY CORTEX, WHICH ALLOWS YOU TO INTERPRET YOUR OWN BODY'S SENSATIONS?
A VERY IMPORTANT CANADIAN NEUROSURGEON, WILDER PENFIELD, SHOWED THAT IT REPRESENTS DIFFERENT AREAS OF THE BODY PROPORTIONALLY BY THE NUMBER OF NEURONS THAT DETECT TACTILE SENSATIONS IN THOSE AREAS.
area
Primary somatosensory cortext
Central sulcus
There's also a motor homunculus!
AND IF WE RECONSTRUCT A HUMAN BEING WITH THESE PROPORTIONS, WE GET...
ME.

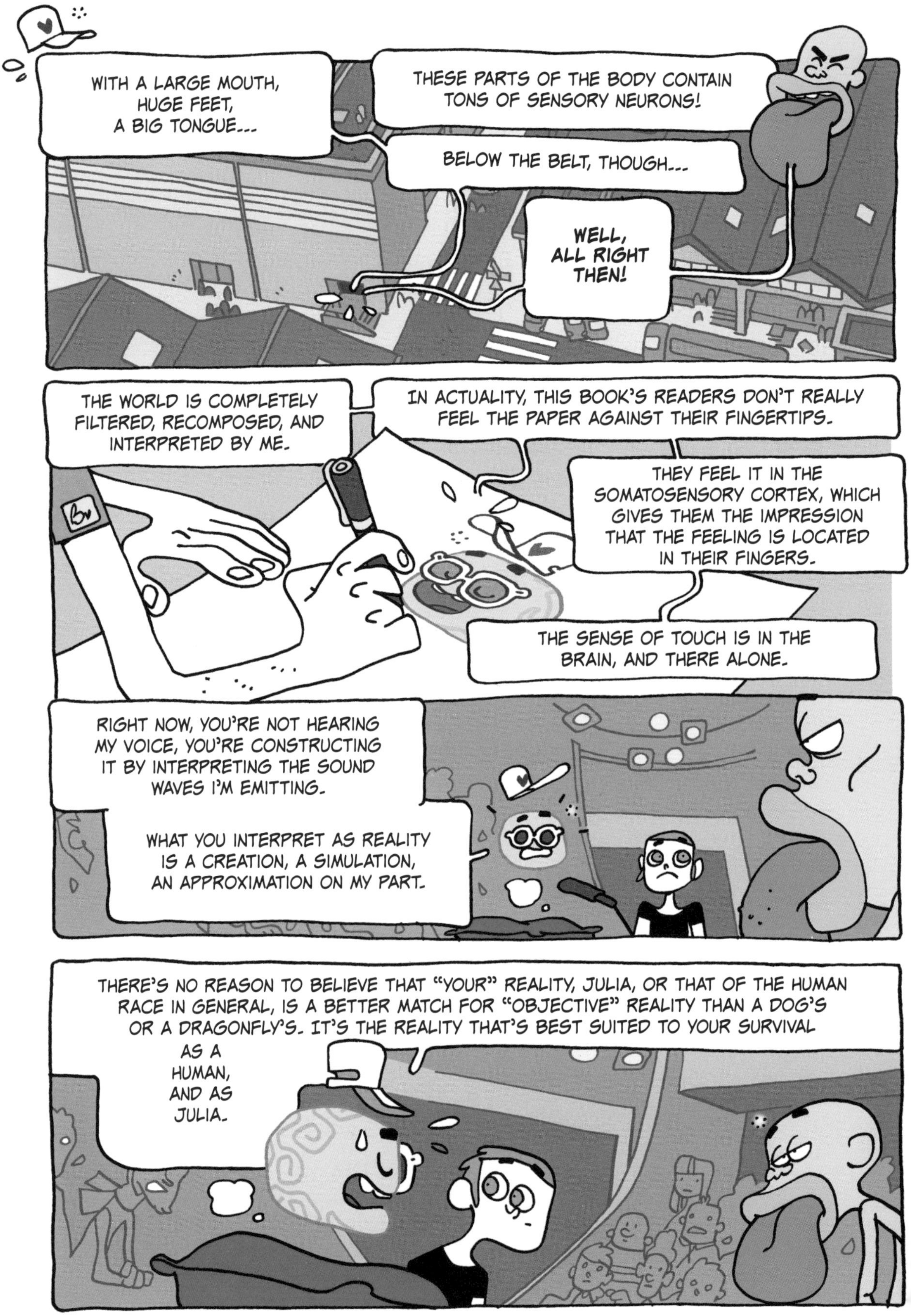
WITH A LARGE MOUTH, HUGE FEET, A BIG TONGUE...
THESE PARTS OF THE BODY CONTAIN TONS OF SENSORY NEURONS!
BELOW THE BELT, THOUGH...
WELL, ALL RIGHT THEN!
THE WORLD IS COMPLETELY FILTERED, RECOMPOSED, AND INTERPRETED BY ME.
IN ACTUALITY, THIS BOOK'S READERS DON'T REALLY FEEL THE PAPER AGAINST THEIR FINGERTIPS.
THEY FEEL IT IN THE SOMATOSENSORY CORTEX, WHICH GIVES THEM THE IMPRESSION THAT THE FEELING IS LOCATED IN THEIR FINGERS.
THE SENSE OF TOUCH IS IN THE BRAIN, AND THERE ALONE.
RIGHT NOW, YOU'RE NOT HEARING MY VOICE, YOU'RE CONSTRUCTING IT BY INTERPRETING THE SOUND WAVES I'M EMITTING.
WHAT YOU INTERPRET AS REALITY IS A CREATION, A SIMULATION, AN APPROXIMATION ON MY PART.
THERE'S NO REASON TO BELIEVE THAT "YOUR" REALITY, JULIA, OR THAT OF THE HUMAN RACE IN GENERAL, IS A BETTER MATCH FOR "OBJECTIVE" REALITY THAN A DOG'S OR A DRAGONFLY'S. IT'S THE REALITY THAT'S BEST SUITED TO YOUR SURVIVAL
AS A HUMAN, AND AS JULIA.

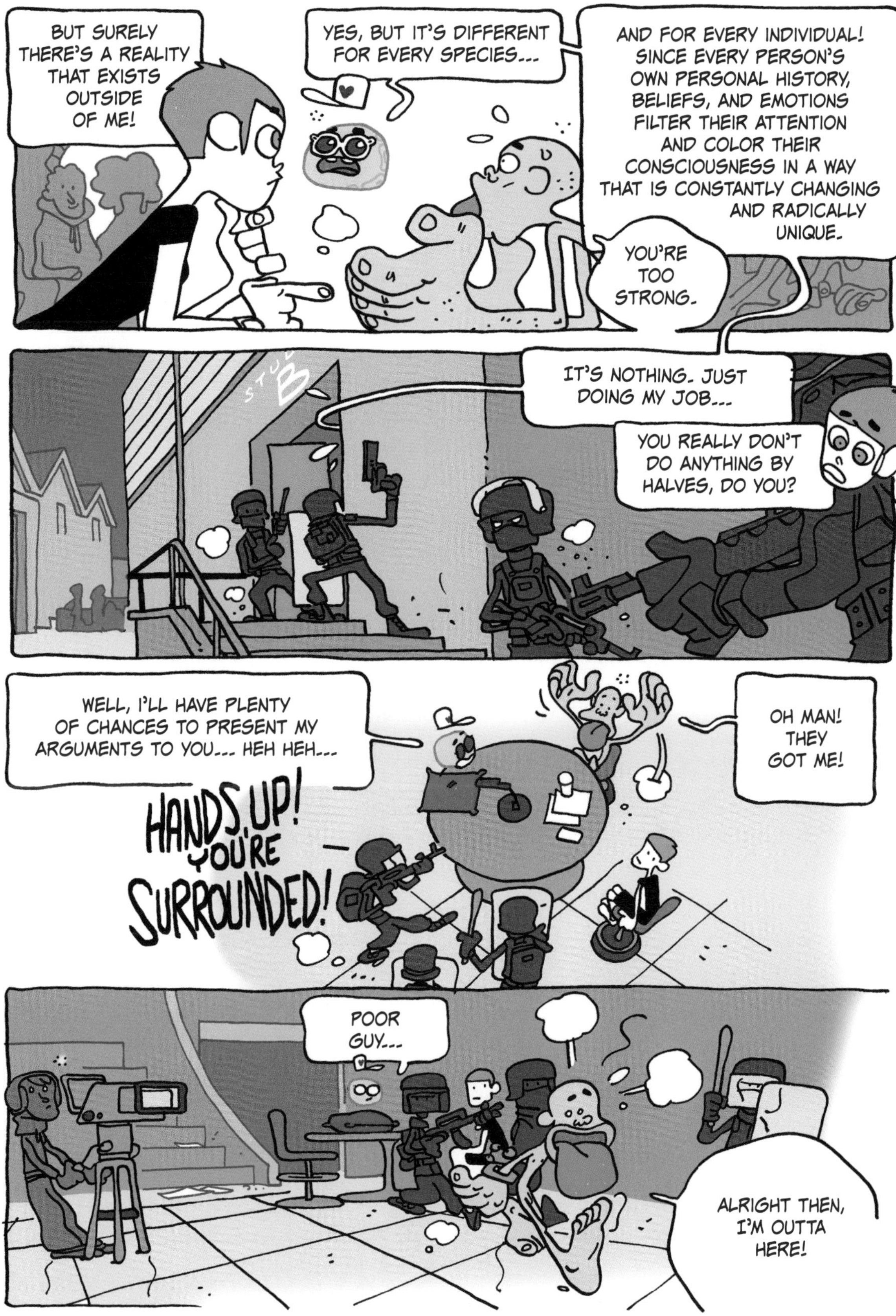

BUT SURELY THERE'S A REALITY THAT EXISTS OUTSIDE OF ME!
YES, BUT IT'S DIFFERENT FOR EVERY SPECIES...
AND FOR EVERY INDIVIDUAL! SINCE EVERY PERSON'S OWN PERSONAL HISTORY, BELIEFS, AND EMOTIONS FILTER THEIR ATTENTION AND COLOR THEIR CONSCIOUSNESS IN A WAY THAT IS CONSTANTLY CHANGING AND RADICALLY UNIQUE.
YOU'RE TOO STRONG.
STUD B
IT'S NOTHING. JUST DOING MY JOB...
YOU REALLY DON'T DO ANYTHING BY HALVES, DO YOU?
WELL, I'LL HAVE PLENTY OF CHANCES TO PRESENT MY ARGUMENTS TO YOU... HEH HEH...
OH MAN! THEY GOT ME!
HANDS UP! YOU'RE SURROUNDED!
POOR GUY...
ALRIGHT THEN, I'M OUTTA HERE!

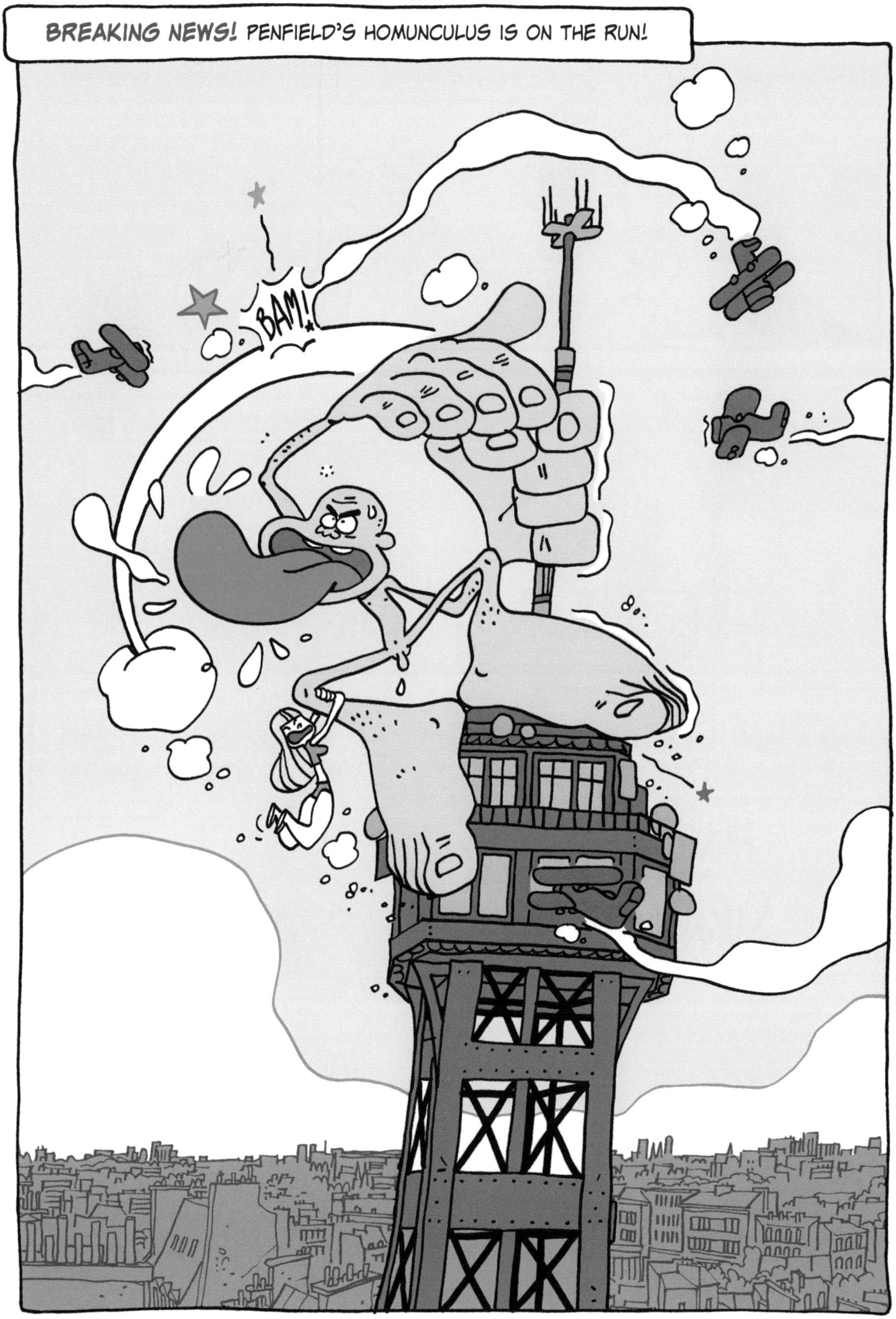
BREAKING NEWS! PENFIELD'S HOMUNCULUS IS ON THE RUN!
BAM!

Chapter 4

EMOTIONS,
THE COLORS OF LIFE

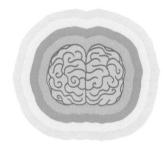

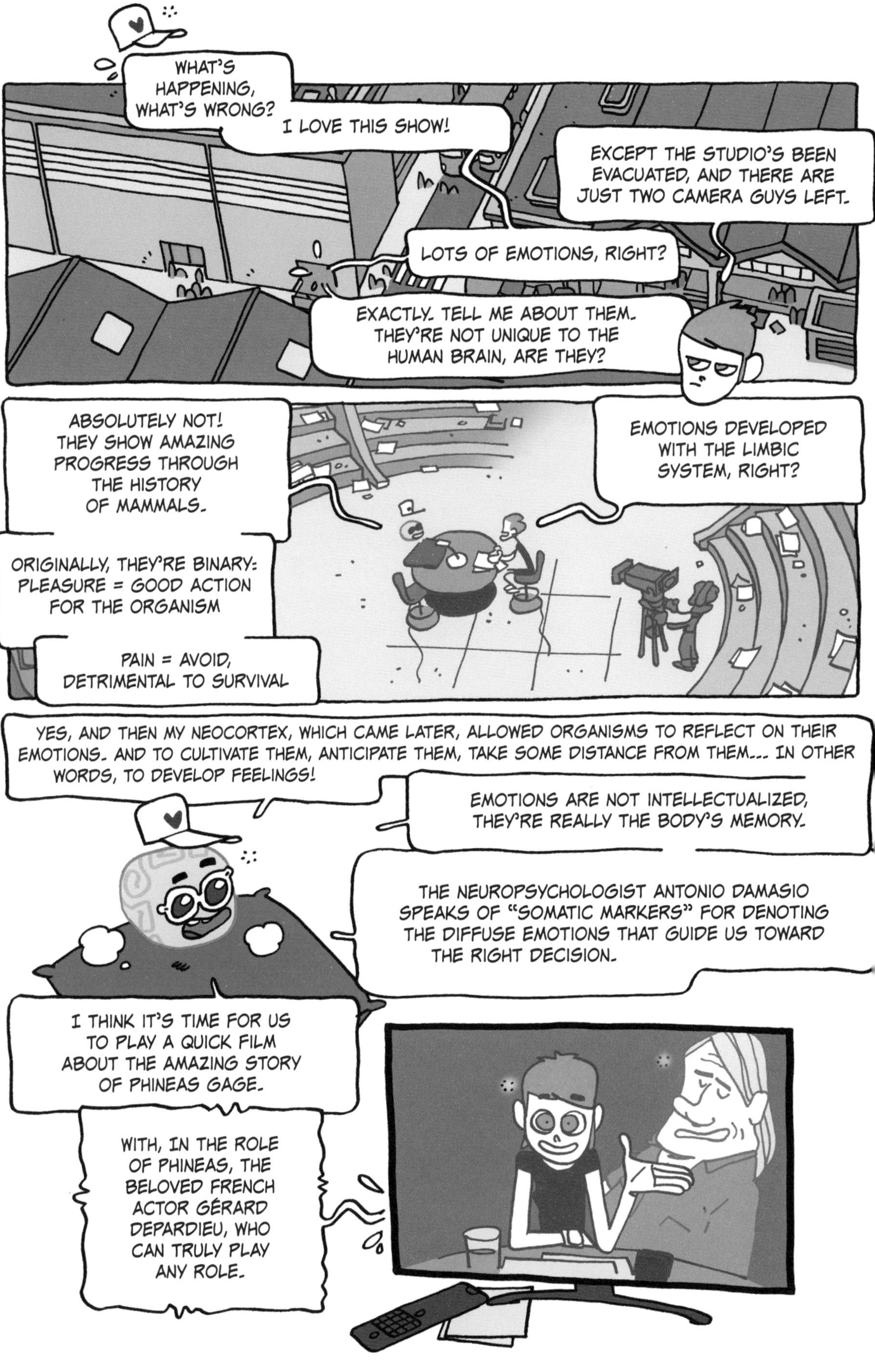
WHAT'S HAPPENING, WHAT'S WRONG?
I LOVE THIS SHOW!
EXCEPT THE STUDIO'S BEEN EVACUATED, AND THERE ARE JUST TWO CAMERA GUYS LEFT.
LOTS OF EMOTIONS, RIGHT?
EXACTLY. TELL ME ABOUT THEM. THEY'RE NOT UNIQUE TO THE HUMAN BRAIN, ARE THEY?
ABSOLUTELY NOT! THEY SHOW AMAZING PROGRESS THROUGH THE HISTORY OF MAMMALS.
EMOTIONS DEVELOPED WITH THE LIMBIC SYSTEM, RIGHT?
ORIGINALLY, THEY'RE BINARY: PLEASURE = GOOD ACTION FOR THE ORGANISM
PAIN = AVOID, DETRIMENTAL TO SURVIVAL
YES, AND THEN MY NEOCORTEX, WHICH CAME LATER, ALLOWED ORGANISMS TO REFLECT ON THEIR EMOTIONS. AND TO CULTIVATE THEM, ANTICIPATE THEM, TAKE SOME DISTANCE FROM THEM... IN OTHER WORDS, TO DEVELOP FEELINGS!
EMOTIONS ARE NOT INTELLECTUALIZED, THEY'RE REALLY THE BODY'S MEMORY.
THE NEUROPSYCHOLOGIST ANTONIO DAMASIO SPEAKS OF "SOMATIC MARKERS" FOR DENOTING THE DIFFUSE EMOTIONS THAT GUIDE US TOWARD THE RIGHT DECISION.
I THINK IT'S TIME FOR US TO PLAY A QUICK FILM ABOUT THE AMAZING STORY OF PHINEAS GAGE.
WITH, IN THE ROLE OF PHINEAS, THE BELOVED FRENCH ACTOR GÉRARD DEPARDIEU, WHO CAN TRULY PLAY ANY ROLE.

ONCE UPON A TIME....

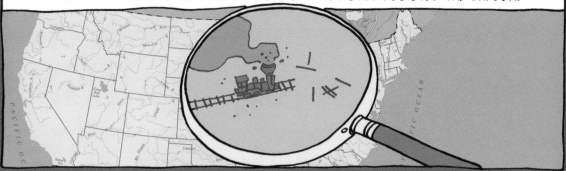

1848. THE WHOLE OF AMERICA WAS CROSSED BY THE UNION PACIFIC RAILROAD. WELL, EXCEPT FOR A BIT OF TERRITORY IN VERMONT THAT STILL RESISTED THE INVADER.

PHINEAS GAGE WORKED IN RAILWAY CONSTRUCTION. HE WAS A FOREMAN. A GOOD GUY. RESPECTED.

HELLO, MY FRIENDS! MAY GOD BLESS YOU!

HELLO, MR. GAGE! ANOTHER BEAUTIFUL DAY!

SO VERY GOOD LOOKING....

MY, HOW HANDSOME YOU ARE....

HIS JOB WAS TO PREPARE EXPLOSIVES TO BREAK UP LARGE ROCKS THAT WERE IN THE WAY.

WHAT'S THAT YOU'RE DOING, PHINEAS?

JOHN PETER PERNOW

THIS PART'S DELICATE.... I PACK THE POWDER IN WITH MY TAMPING IRON.... THEN I LIGHT THE MATCH....

KHH!!

HURRY UP, AFTER THIS I HAVE TO DO A STORY ON A SAUSAGE EXHIBITION....

BOM

AND THERE IT GOES! WE'RE CLEAR! LAY THE TRACK!

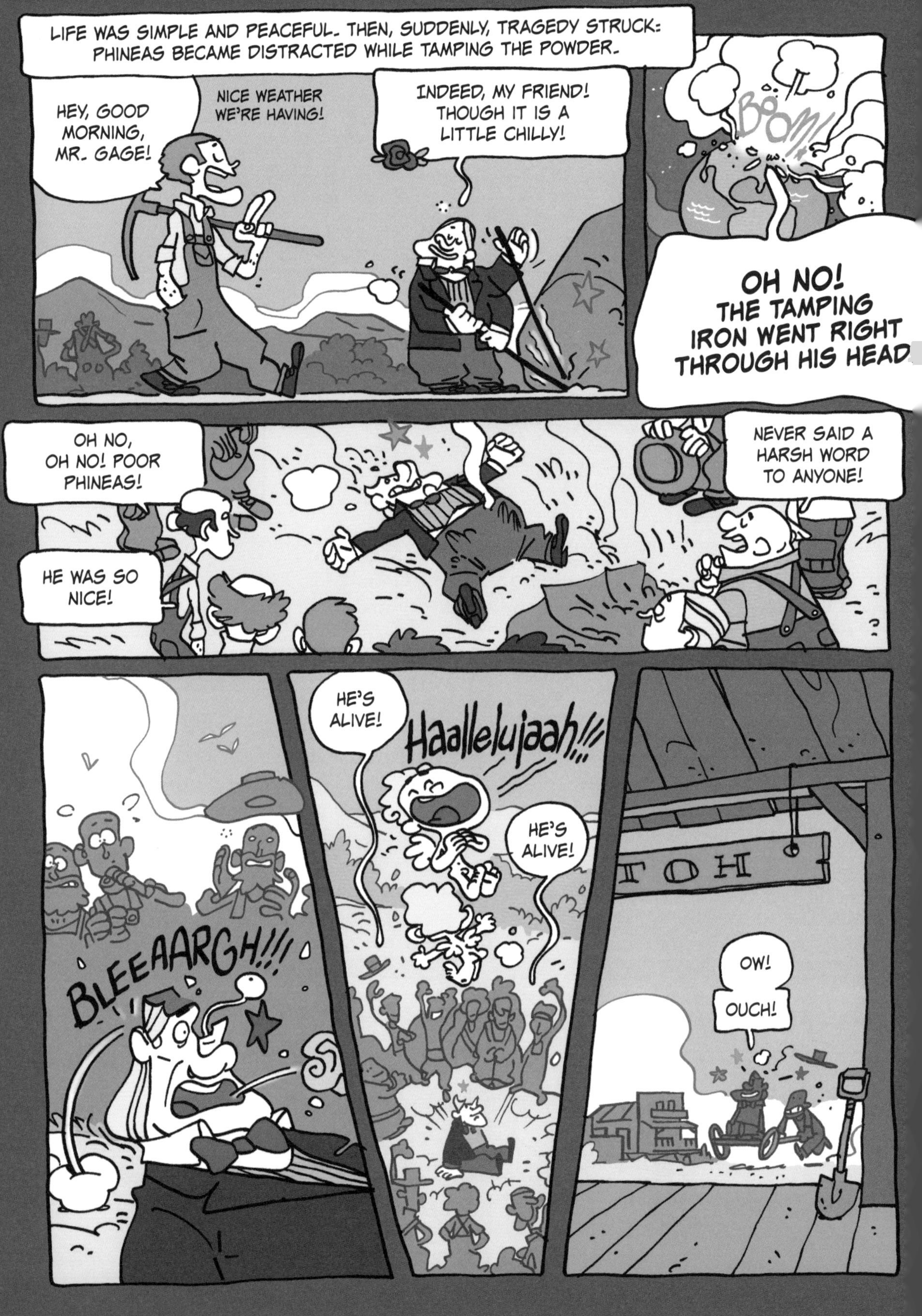
LIFE WAS SIMPLE AND PEACEFUL. THEN, SUDDENLY, TRAGEDY STRUCK: PHINEAS BECAME DISTRACTED WHILE TAMPING THE POWDER.
HEY, GOOD MORNING, MR. GAGE!
NICE WEATHER WE'RE HAVING!
INDEED, MY FRIEND! THOUGH IT IS A LITTLE CHILLY!
BOOM!
OH NO! THE TAMPING IRON WENT RIGHT THROUGH HIS HEAD
OH NO, OH NO! POOR PHINEAS!
HE WAS SO NICE!
NEVER SAID A HARSH WORD TO ANYONE!
BLEEAARGH!!!
HE'S ALIVE!
Haallelujaah!!!
HE'S ALIVE!
HOT
OW!
OUCH!

THREE DAYS LATER...

HE'S GOING TO LIVE!

HOORAY!

YES, HE'D LIVE. BUT WHAT KIND OF LIFE?

GOOD MORNING, MR. GAGE!

Grumble

IT'S GOOD TO SEE YOU!

LEAVE ME ALONE, ALL OF YOU!!!

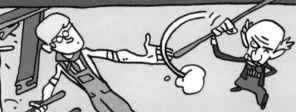

H- HERE'S YOUR TAMPING IRON, SIR.

TAKE THAT! AND THAT!

AND THAT! GO ON!

UH.... IS THAT A GOOD IDEA?

AAAH! AAAH! I'M CUCKOO!!!

ACCORDING TO THE DOCTOR...

GAGE IS NO LONGER GAGE.

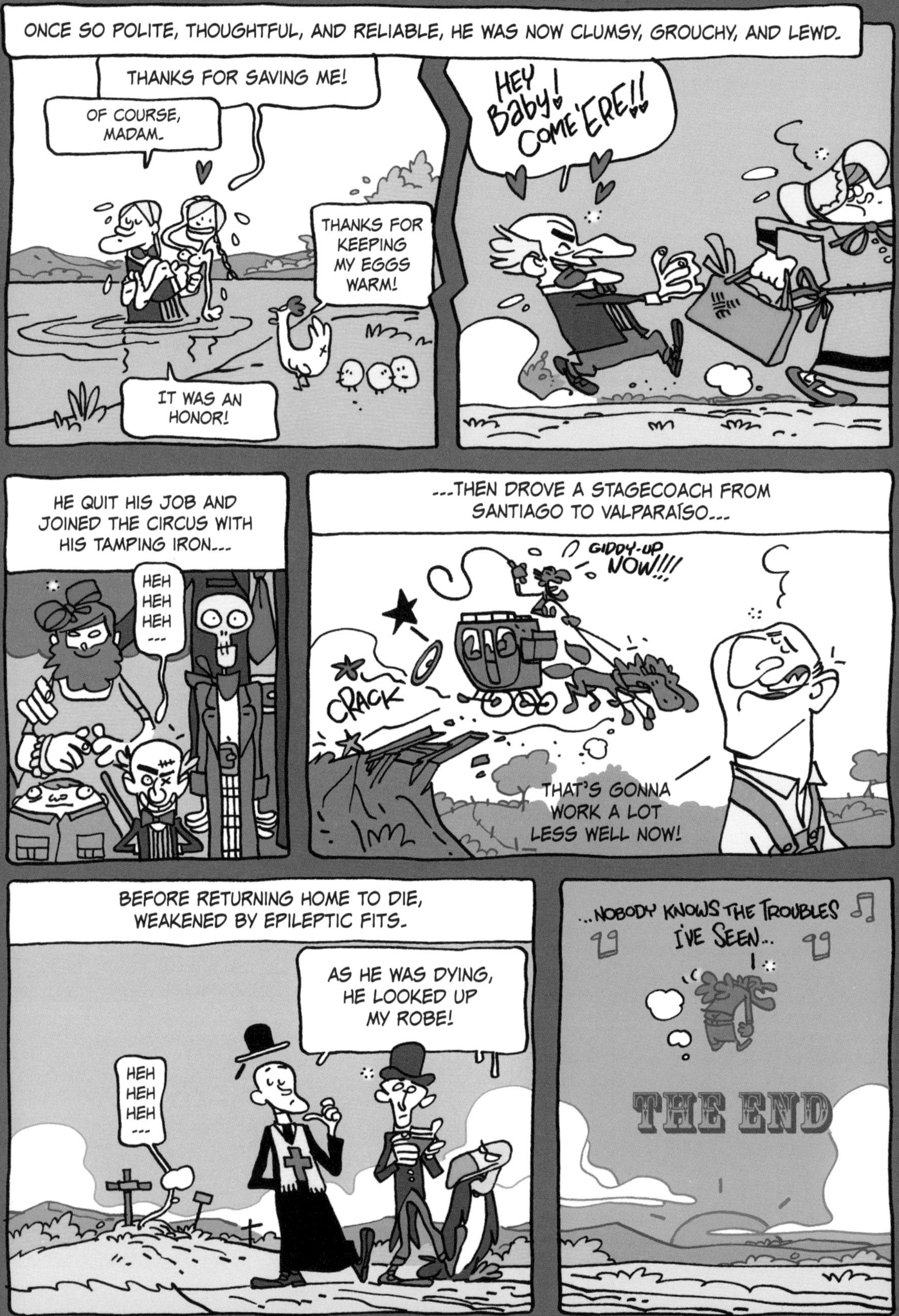

ONCE SO POLITE, THOUGHTFUL, AND RELIABLE, HE WAS NOW CLUMSY, GROUCHY, AND LEWD.
THANKS FOR SAVING ME!
OF COURSE, MADAM.
THANKS FOR KEEPING MY EGGS WARM!
IT WAS AN HONOR!
HEY Baby! COME 'ERE!!
HE QUIT HIS JOB AND JOINED THE CIRCUS WITH HIS TAMPING IRON...
HEH HEH HEH ...
...THEN DROVE A STAGECOACH FROM SANTIAGO TO VALPARAÍSO...
GIDDY-UP NOW!!!!
CRACK
THAT'S GONNA WORK A LOT LESS WELL NOW!
BEFORE RETURNING HOME TO DIE, WEAKENED BY EPILEPTIC FITS.
AS HE WAS DYING, HE LOOKED UP MY ROBE!
HEH HEH HEH ...
...NOBODY KNOWS THE TROUBLES I'VE SEEN...
THE END

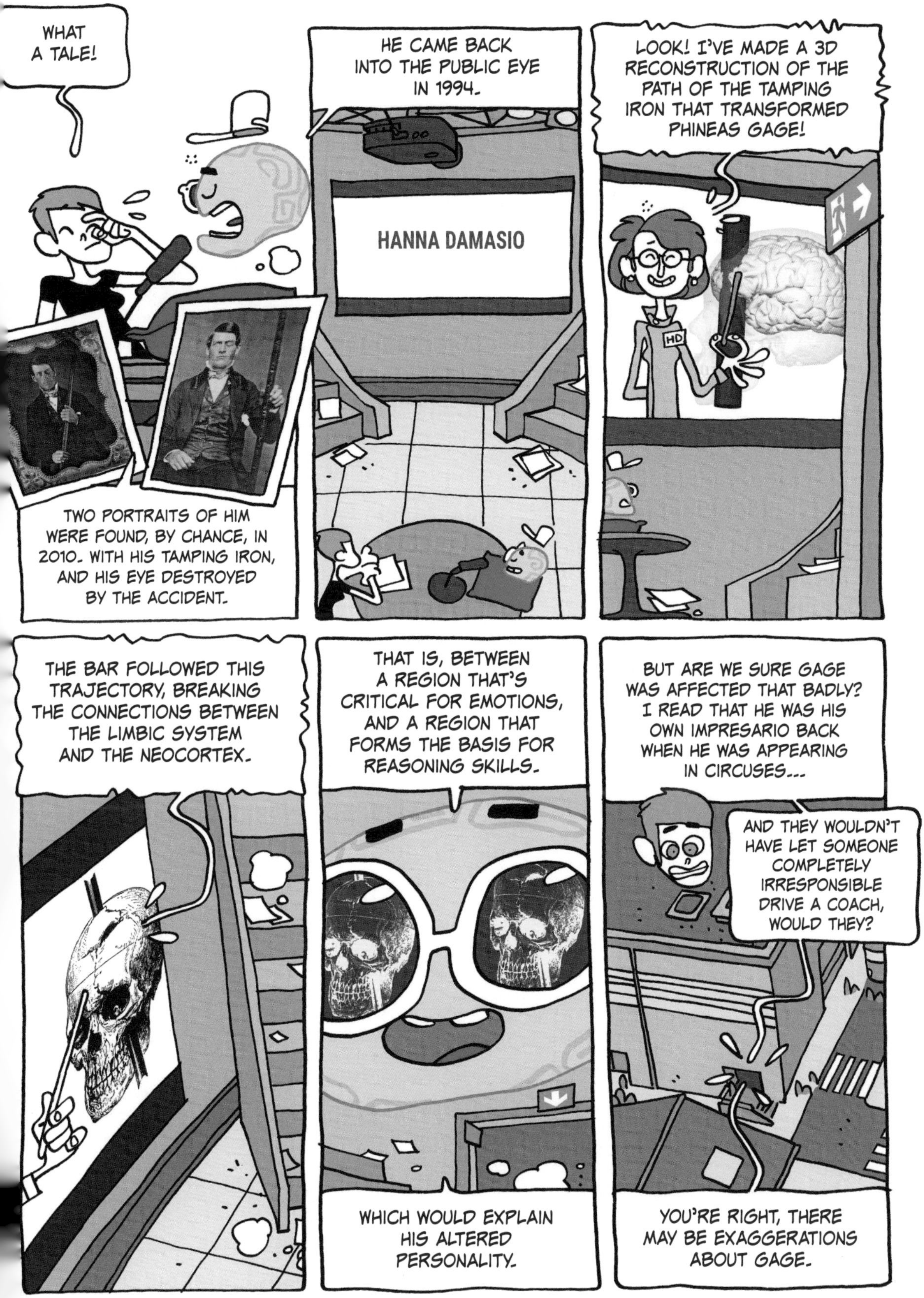

WHAT A TALE!
TWO PORTRAITS OF HIM WERE FOUND, BY CHANCE, IN 2010. WITH HIS TAMPING IRON, AND HIS EYE DESTROYED BY THE ACCIDENT.
HE CAME BACK INTO THE PUBLIC EYE IN 1994.
HANNA DAMASIO
LOOK! I'VE MADE A 3D RECONSTRUCTION OF THE PATH OF THE TAMPING IRON THAT TRANSFORMED PHINEAS GAGE!
HD
THE BAR FOLLOWED THIS TRAJECTORY, BREAKING THE CONNECTIONS BETWEEN THE LIMBIC SYSTEM AND THE NEOCORTEX.
THAT IS, BETWEEN A REGION THAT'S CRITICAL FOR EMOTIONS, AND A REGION THAT FORMS THE BASIS FOR REASONING SKILLS.
WHICH WOULD EXPLAIN HIS ALTERED PERSONALITY.
BUT ARE WE SURE GAGE WAS AFFECTED THAT BADLY? I READ THAT HE WAS HIS OWN IMPRESARIO BACK WHEN HE WAS APPEARING IN CIRCUSES...
AND THEY WOULDN'T HAVE LET SOMEONE COMPLETELY IRRESPONSIBLE DRIVE A COACH, WOULD THEY?
YOU'RE RIGHT, THERE MAY BE EXAGGERATIONS ABOUT GAGE.

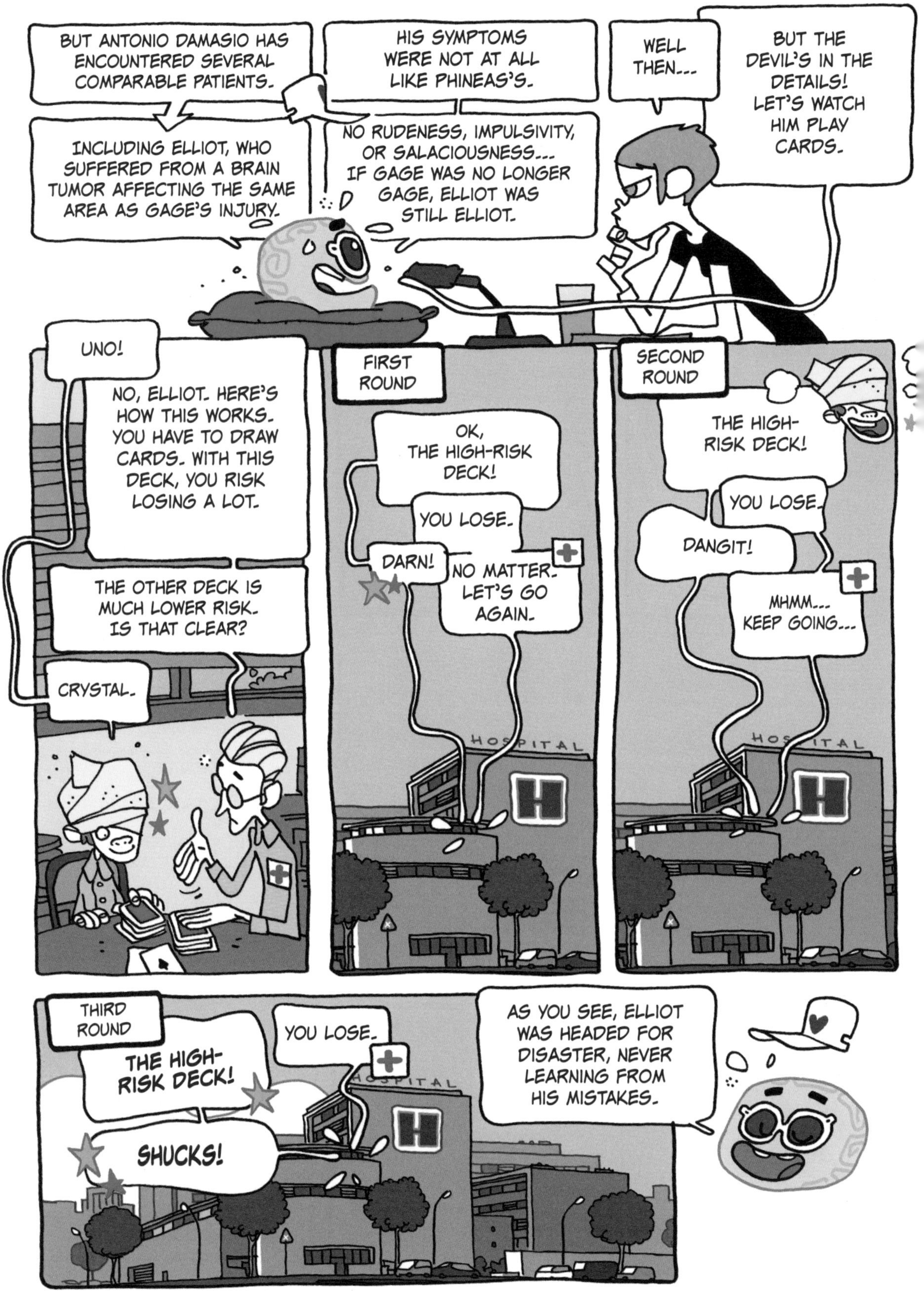

BUT ANTONIO DAMASIO HAS ENCOUNTERED SEVERAL COMPARABLE PATIENTS.
INCLUDING ELLIOT, WHO SUFFERED FROM A BRAIN TUMOR AFFECTING THE SAME AREA AS GAGE'S INJURY.
HIS SYMPTOMS WERE NOT AT ALL LIKE PHINEAS'S.
NO RUDENESS, IMPULSIVITY, OR SALACIOUSNESS... IF GAGE WAS NO LONGER GAGE, ELLIOT WAS STILL ELLIOT.
WELL THEN...
BUT THE DEVIL'S IN THE DETAILS! LET'S WATCH HIM PLAY CARDS.
UNO!
NO, ELLIOT. HERE'S HOW THIS WORKS. YOU HAVE TO DRAW CARDS. WITH THIS DECK, YOU RISK LOSING A LOT.
THE OTHER DECK IS MUCH LOWER RISK. IS THAT CLEAR?
CRYSTAL.
FIRST ROUND
OK, THE HIGH-RISK DECK!
YOU LOSE.
DARN!
NO MATTER. LET'S GO AGAIN.
HOSPITAL
SECOND ROUND
THE HIGH-RISK DECK!
YOU LOSE.
DANGIT!
MHMM... KEEP GOING...
HOSPITAL
THIRD ROUND
THE HIGH-RISK DECK!
YOU LOSE.
SHUCKS!
HOSPITAL
AS YOU SEE, ELLIOT WAS HEADED FOR DISASTER, NEVER LEARNING FROM HIS MISTAKES.

ANOTHER EXAMPLE?
WHEN DO YOU WANT TO HAVE OUR NEXT APPOINTMENT?
...
UH...
...
MONDAY? MONDAY WOULD WORK. I HAVE TIME.
VERY WELL. MONDAY.
ACTUALLY... TUESDAY. TUESDAY'S BETTER.
THERE'S LESS TRAFFIC.
TUESDAY. VERY WELL.
WELL, FRIDAY IS PRACTICALLY THE WEEKEND, AND I'LL BE ABLE TO REST AFTERWARD...
GRR... ALRIGHT.
BUT THEN...
GRAAAHHH!!

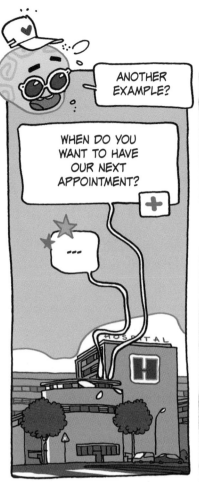

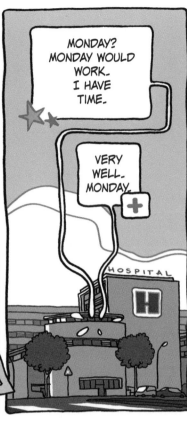

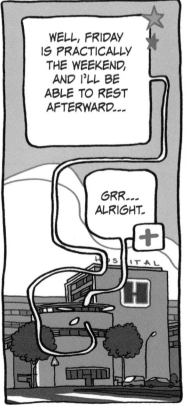

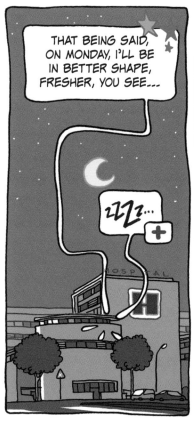

THAT BEING SAID, ON MONDAY, I'LL BE IN BETTER SHAPE, FRESHER, YOU SEE....

ZZZZ...

OR MAYBE THURSDAY. I LIKE THURSDAYS. I MET MY WIFE ON A THURSDAY.

YOU OK, DR. DAMASIO?

I'LL GO GET YOU A COFFEE.

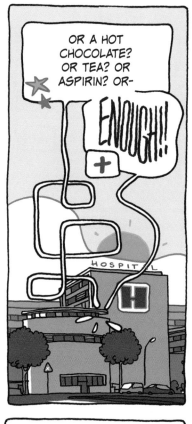

OR A HOT CHOCOLATE? OR TEA? OR ASPIRIN? OR—

ENOUGH!!

OK, BUT YOU'RE EXAGGERATING, RIGHT?

I KNOW I JOKE AROUND.... BUT SERIOUSLY, BY KNOWING WITHOUT FEELING, ELLIOTT HAD BECOME UNABLE TO MAKE VERY SIMPLE CHOICES.

THE MORAL: WITHOUT "SOMATIC MARKERS," OR THE ABILITY TO CALL UPON THE EMOTIONS THAT THE BODY LEARNS THROUGH EXPERIENCE....

YOU'RE POOR AT LEARNING, REASONING, AND CHOOSING?

RIGHT. EMOTION AND REASON ARE NOT IN OPPOSITION, THEY'RE COMPLEMENTARY.

SO WE DON'T HAVE NOBLE REASON ON ONE SIDE, AND PARASITIC, PRIMITIVE EMOTIONS ON THE OTHER?

THAT'S WHAT DAMASIO TERMED "DESCARTES' ERROR." THE MIND NEEDS THE BODY. IT ARISES FROM THE BODY, AND CAN'T DISAVOW IT.

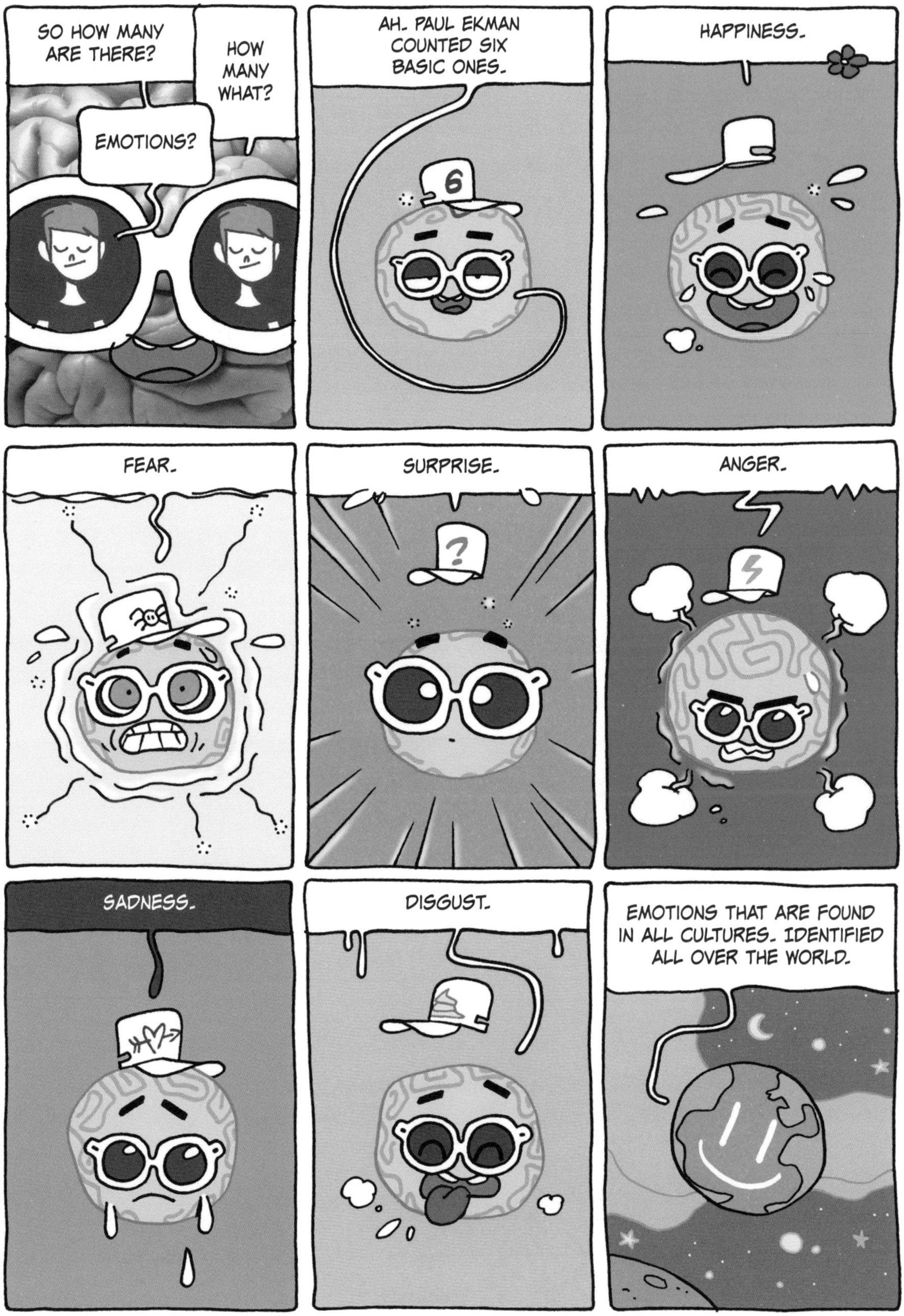
SO HOW MANY ARE THERE?
HOW MANY WHAT?
EMOTIONS?
AH. PAUL EKMAN COUNTED SIX BASIC ONES.
6
HAPPINESS.
FEAR.
SURPRISE.
?
ANGER.
SADNESS.
DISGUST.
EMOTIONS THAT ARE FOUND IN ALL CULTURES. IDENTIFIED ALL OVER THE WORLD.

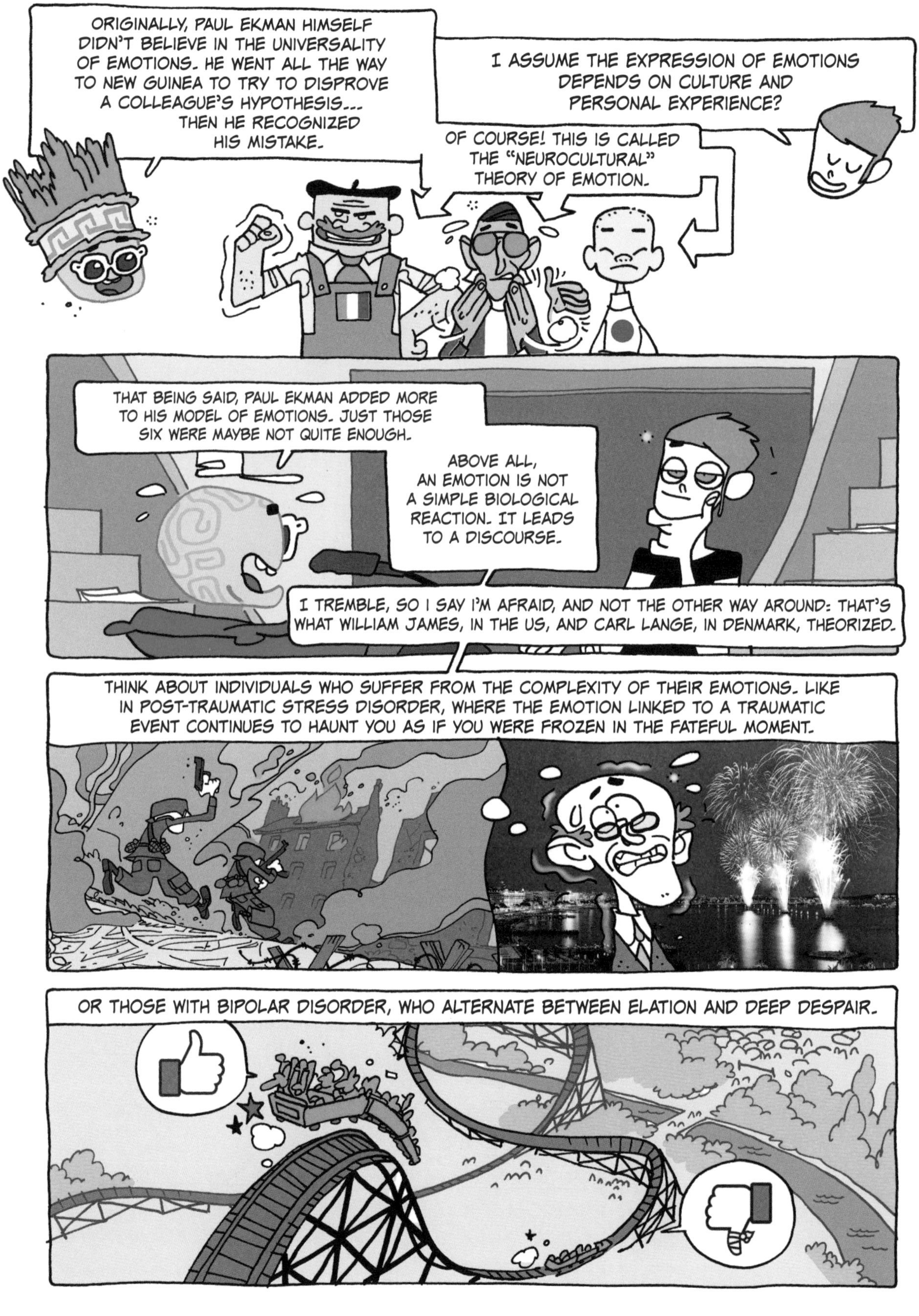
ORIGINALLY, PAUL EKMAN HIMSELF DIDN'T BELIEVE IN THE UNIVERSALITY OF EMOTIONS. HE WENT ALL THE WAY TO NEW GUINEA TO TRY TO DISPROVE A COLLEAGUE'S HYPOTHESIS... THEN HE RECOGNIZED HIS MISTAKE.
I ASSUME THE EXPRESSION OF EMOTIONS DEPENDS ON CULTURE AND PERSONAL EXPERIENCE?
OF COURSE! THIS IS CALLED THE "NEUROCULTURAL" THEORY OF EMOTION.
THAT BEING SAID, PAUL EKMAN ADDED MORE TO HIS MODEL OF EMOTIONS. JUST THOSE SIX WERE MAYBE NOT QUITE ENOUGH.
ABOVE ALL, AN EMOTION IS NOT A SIMPLE BIOLOGICAL REACTION. IT LEADS TO A DISCOURSE.
I TREMBLE, SO I SAY I'M AFRAID, AND NOT THE OTHER WAY AROUND: THAT'S WHAT WILLIAM JAMES, IN THE US, AND CARL LANGE, IN DENMARK, THEORIZED.
THINK ABOUT INDIVIDUALS WHO SUFFER FROM THE COMPLEXITY OF THEIR EMOTIONS. LIKE IN POST-TRAUMATIC STRESS DISORDER, WHERE THE EMOTION LINKED TO A TRAUMATIC EVENT CONTINUES TO HAUNT YOU AS IF YOU WERE FROZEN IN THE FATEFUL MOMENT.
OR THOSE WITH BIPOLAR DISORDER, WHO ALTERNATE BETWEEN ELATION AND DEEP DESPAIR.

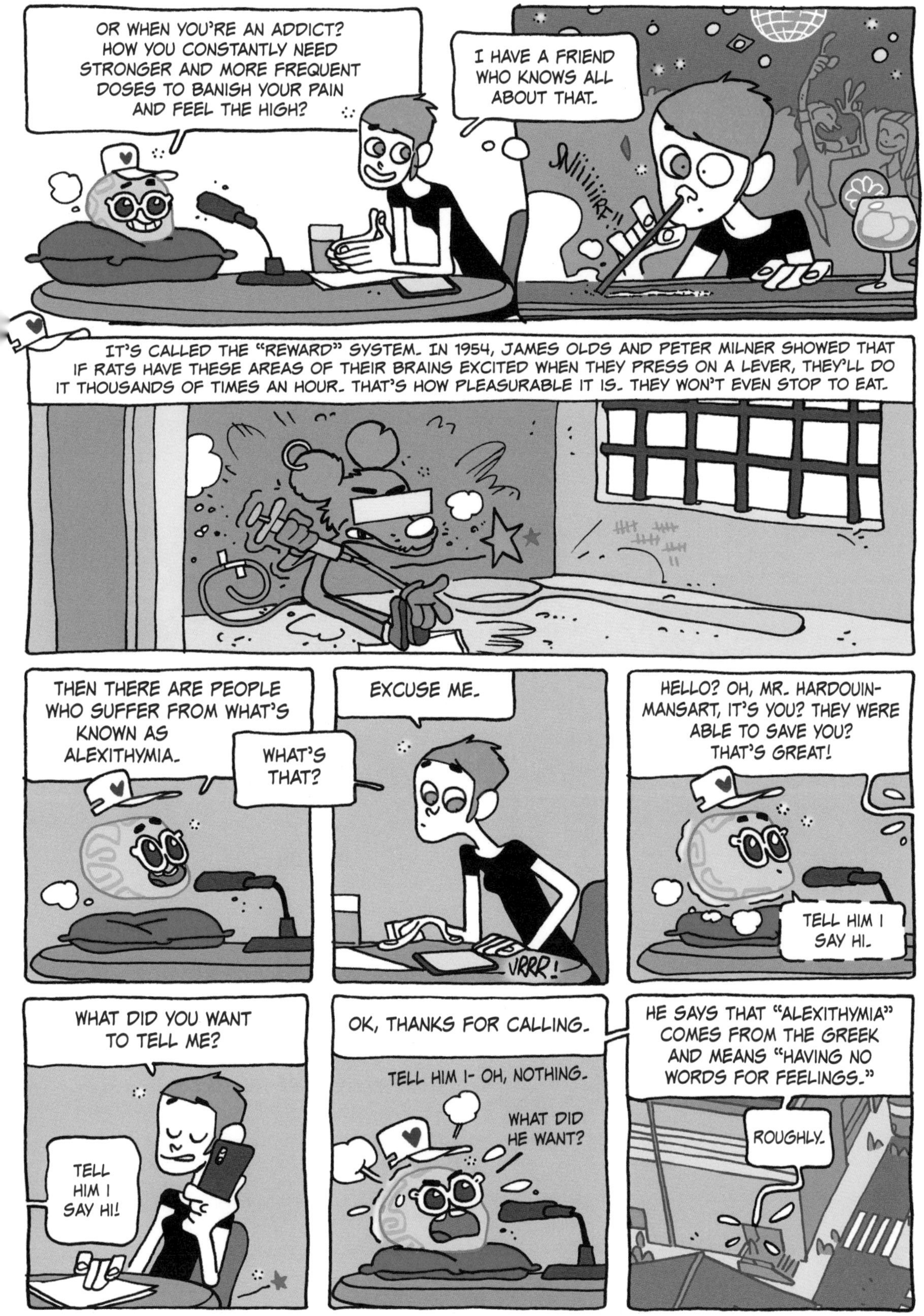

OR WHEN YOU'RE AN ADDICT? HOW YOU CONSTANTLY NEED STRONGER AND MORE FREQUENT DOSES TO BANISH YOUR PAIN AND FEEL THE HIGH?
I HAVE A FRIEND WHO KNOWS ALL ABOUT THAT.
SNIIIIIIRF!!
IT'S CALLED THE "REWARD" SYSTEM. IN 1954, JAMES OLDS AND PETER MILNER SHOWED THAT IF RATS HAVE THESE AREAS OF THEIR BRAINS EXCITED WHEN THEY PRESS ON A LEVER, THEY'LL DO IT THOUSANDS OF TIMES AN HOUR. THAT'S HOW PLEASURABLE IT IS. THEY WON'T EVEN STOP TO EAT.
THEN THERE ARE PEOPLE WHO SUFFER FROM WHAT'S KNOWN AS ALEXITHYMIA.
WHAT'S THAT?
EXCUSE ME.
VRRR!
HELLO? OH, MR. HARDOUIN-MANSART, IT'S YOU? THEY WERE ABLE TO SAVE YOU? THAT'S GREAT!
TELL HIM I SAY HI.
WHAT DID YOU WANT TO TELL ME?
TELL HIM I SAY HI!
OK, THANKS FOR CALLING.
TELL HIM I- OH, NOTHING.
WHAT DID HE WANT?
HE SAYS THAT "ALEXITHYMIA" COMES FROM THE GREEK AND MEANS "HAVING NO WORDS FOR FEELINGS."
ROUGHLY.

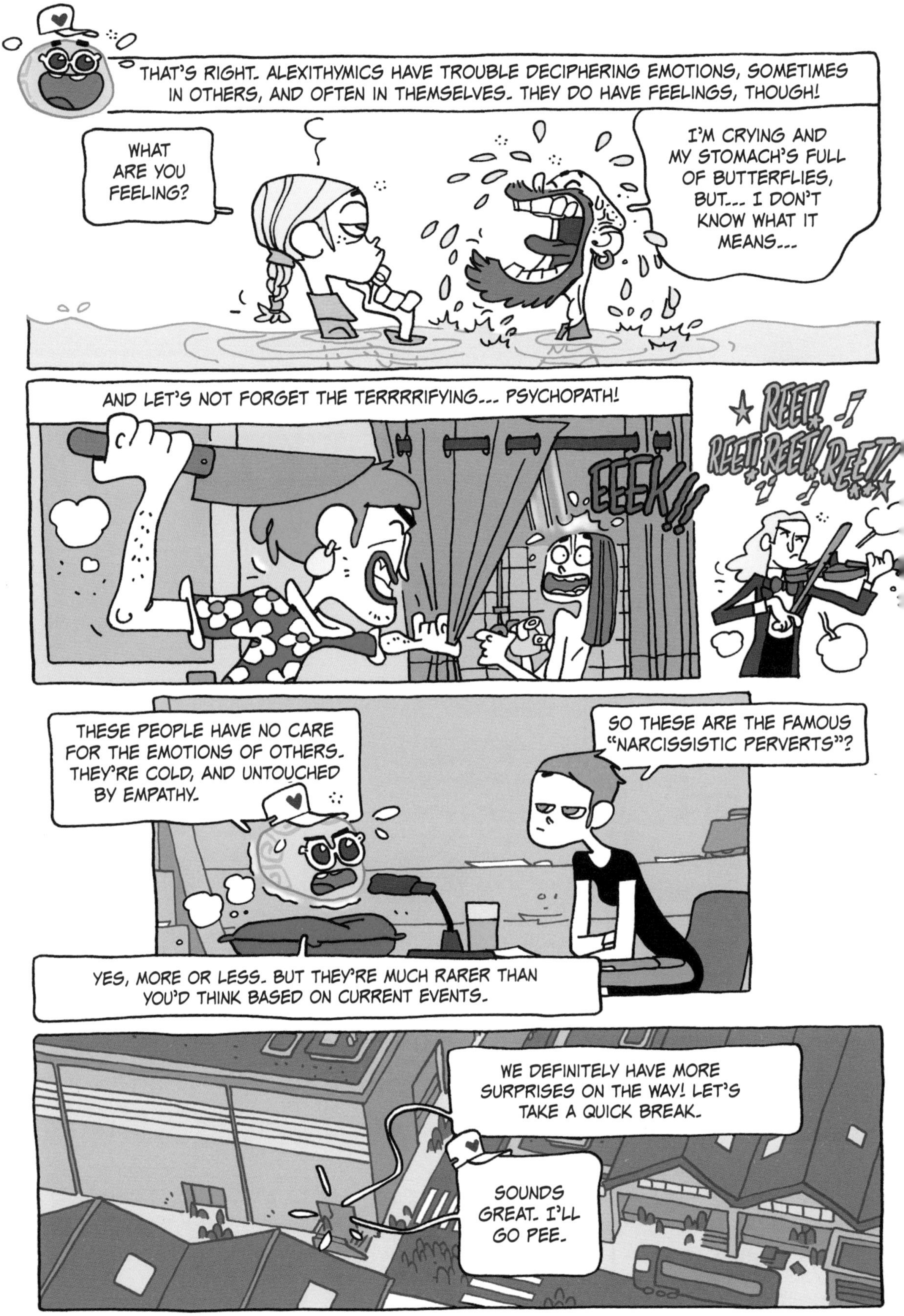
THAT'S RIGHT. ALEXITHYMICS HAVE TROUBLE DECIPHERING EMOTIONS, SOMETIMES IN OTHERS, AND OFTEN IN THEMSELVES. THEY DO HAVE FEELINGS, THOUGH!
WHAT ARE YOU FEELING?
I'M CRYING AND MY STOMACH'S FULL OF BUTTERFLIES, BUT... I DON'T KNOW WHAT IT MEANS...
AND LET'S NOT FORGET THE TERRRRIFYING... PSYCHOPATH!
EEEK!!!
REET! REET! REET! REET!
THESE PEOPLE HAVE NO CARE FOR THE EMOTIONS OF OTHERS. THEY'RE COLD, AND UNTOUCHED BY EMPATHY.
SO THESE ARE THE FAMOUS "NARCISSISTIC PERVERTS"?
YES, MORE OR LESS. BUT THEY'RE MUCH RARER THAN YOU'D THINK BASED ON CURRENT EVENTS.
WE DEFINITELY HAVE MORE SURPRISES ON THE WAY! LET'S TAKE A QUICK BREAK.
SOUNDS GREAT. I'LL GO PEE.

INTERLUDE

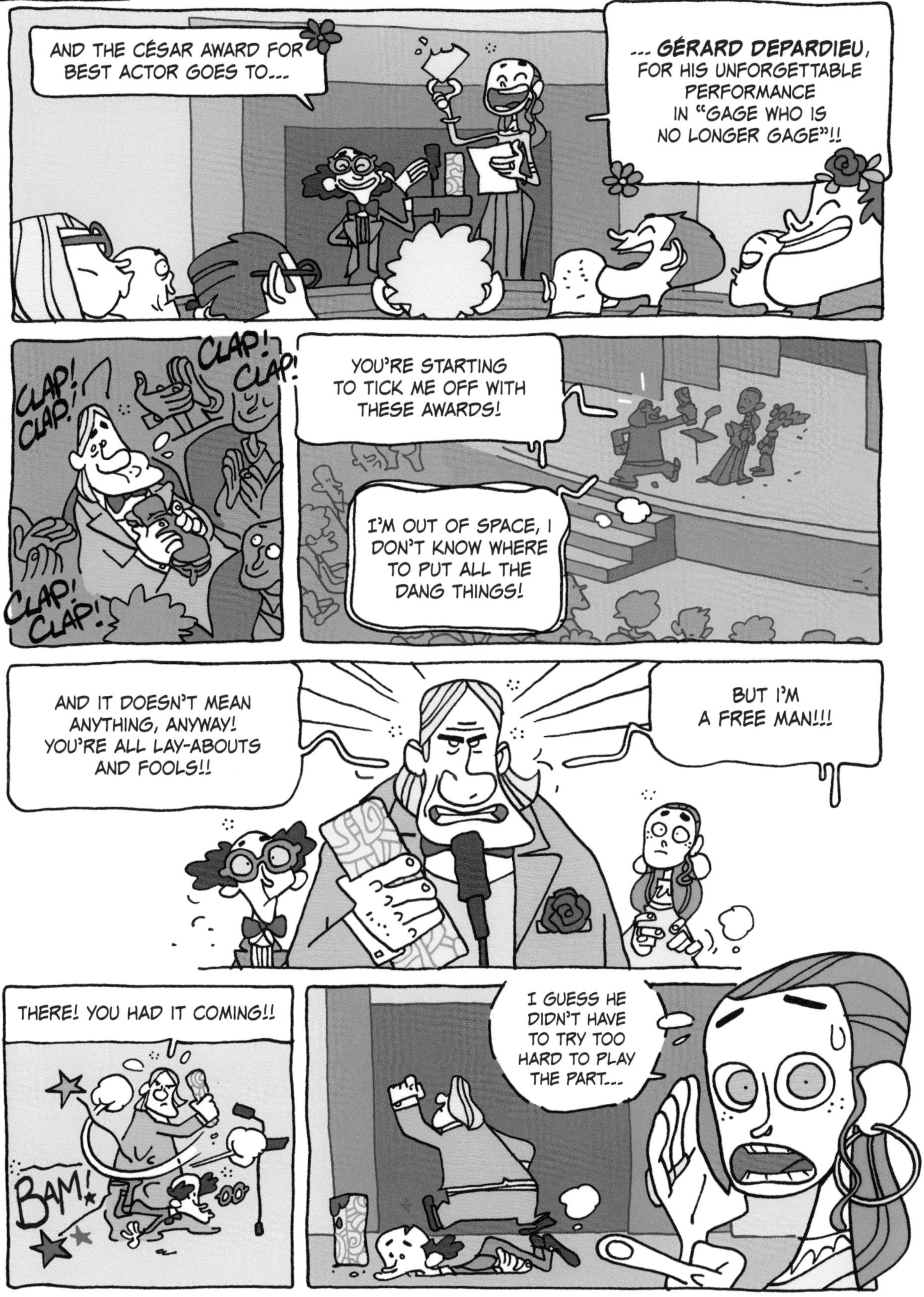

AND THE CÉSAR AWARD FOR BEST ACTOR GOES TO...
... GÉRARD DEPARDIEU, FOR HIS UNFORGETTABLE PERFORMANCE IN "GAGE WHO IS NO LONGER GAGE"!!
CLAP! CLAP!
CLAP! CLAP!
CLAP! CLAP!
YOU'RE STARTING TO TICK ME OFF WITH THESE AWARDS!
I'M OUT OF SPACE, I DON'T KNOW WHERE TO PUT ALL THE DANG THINGS!
AND IT DOESN'T MEAN ANYTHING, ANYWAY! YOU'RE ALL LAY-ABOUTS AND FOOLS!!
BUT I'M A FREE MAN!!!
THERE! YOU HAD IT COMING!!
BAM!
I GUESS HE DIDN'T HAVE TO TRY TOO HARD TO PLAY THE PART...

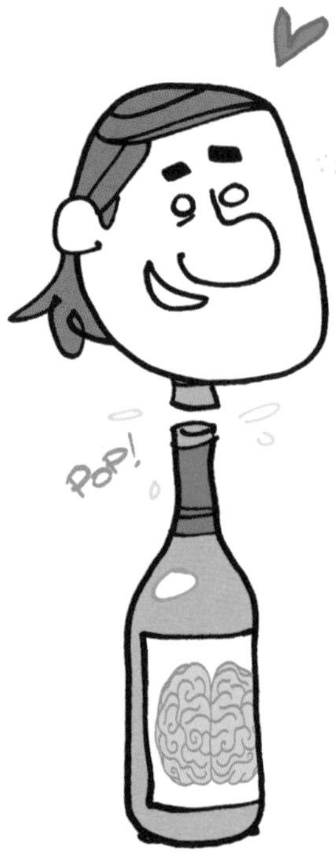
POP!

Chapter 5

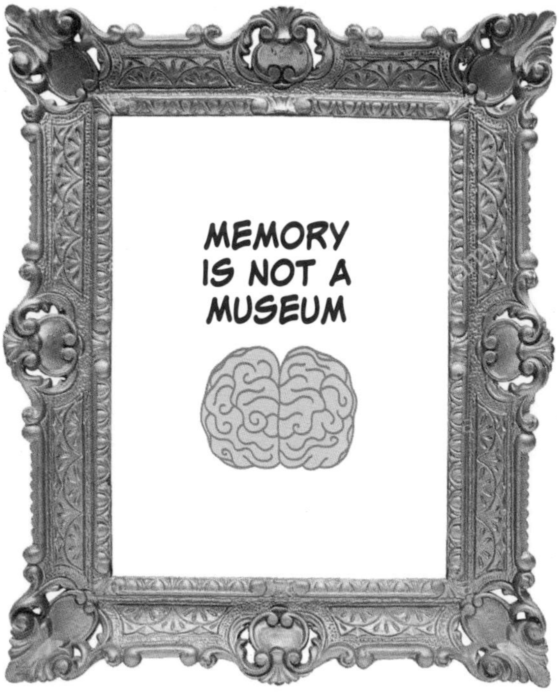
MEMORY
IS NOT A
MUSEUM

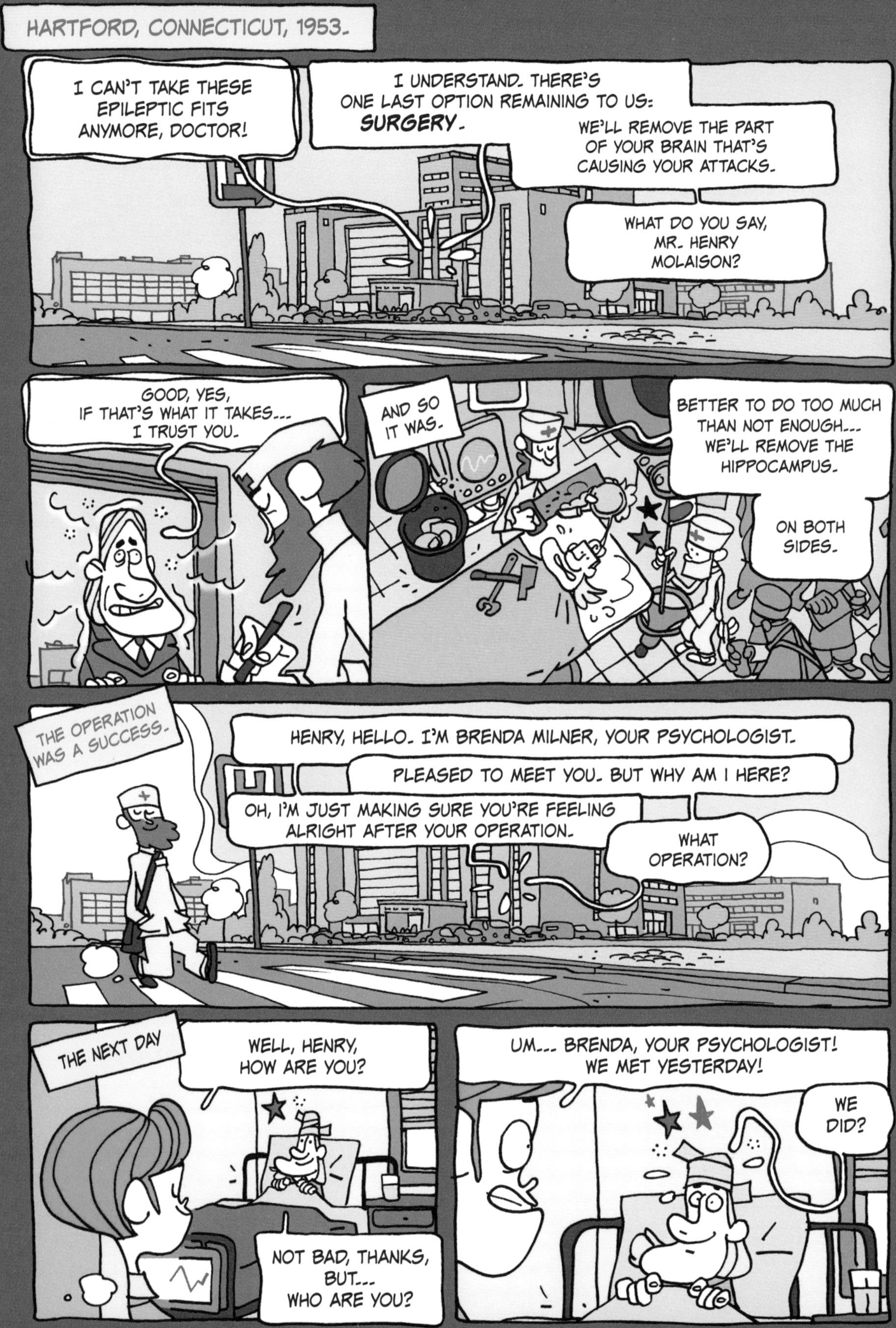
HARTFORD, CONNECTICUT, 1953.
I CAN'T TAKE THESE EPILEPTIC FITS ANYMORE, DOCTOR!
I UNDERSTAND. THERE'S ONE LAST OPTION REMAINING TO US: SURGERY.
WE'LL REMOVE THE PART OF YOUR BRAIN THAT'S CAUSING YOUR ATTACKS.
WHAT DO YOU SAY, MR. HENRY MOLAISON?
GOOD, YES, IF THAT'S WHAT IT TAKES... I TRUST YOU.
AND SO IT WAS.
BETTER TO DO TOO MUCH THAN NOT ENOUGH... WE'LL REMOVE THE HIPPOCAMPUS.
ON BOTH SIDES.
THE OPERATION WAS A SUCCESS.
HENRY, HELLO. I'M BRENDA MILNER, YOUR PSYCHOLOGIST.
PLEASED TO MEET YOU. BUT WHY AM I HERE?
OH, I'M JUST MAKING SURE YOU'RE FEELING ALRIGHT AFTER YOUR OPERATION.
WHAT OPERATION?
THE NEXT DAY
WELL, HENRY, HOW ARE YOU?
NOT BAD, THANKS, BUT... WHO ARE YOU?
UM... BRENDA, YOUR PSYCHOLOGIST! WE MET YESTERDAY!
WE DID?

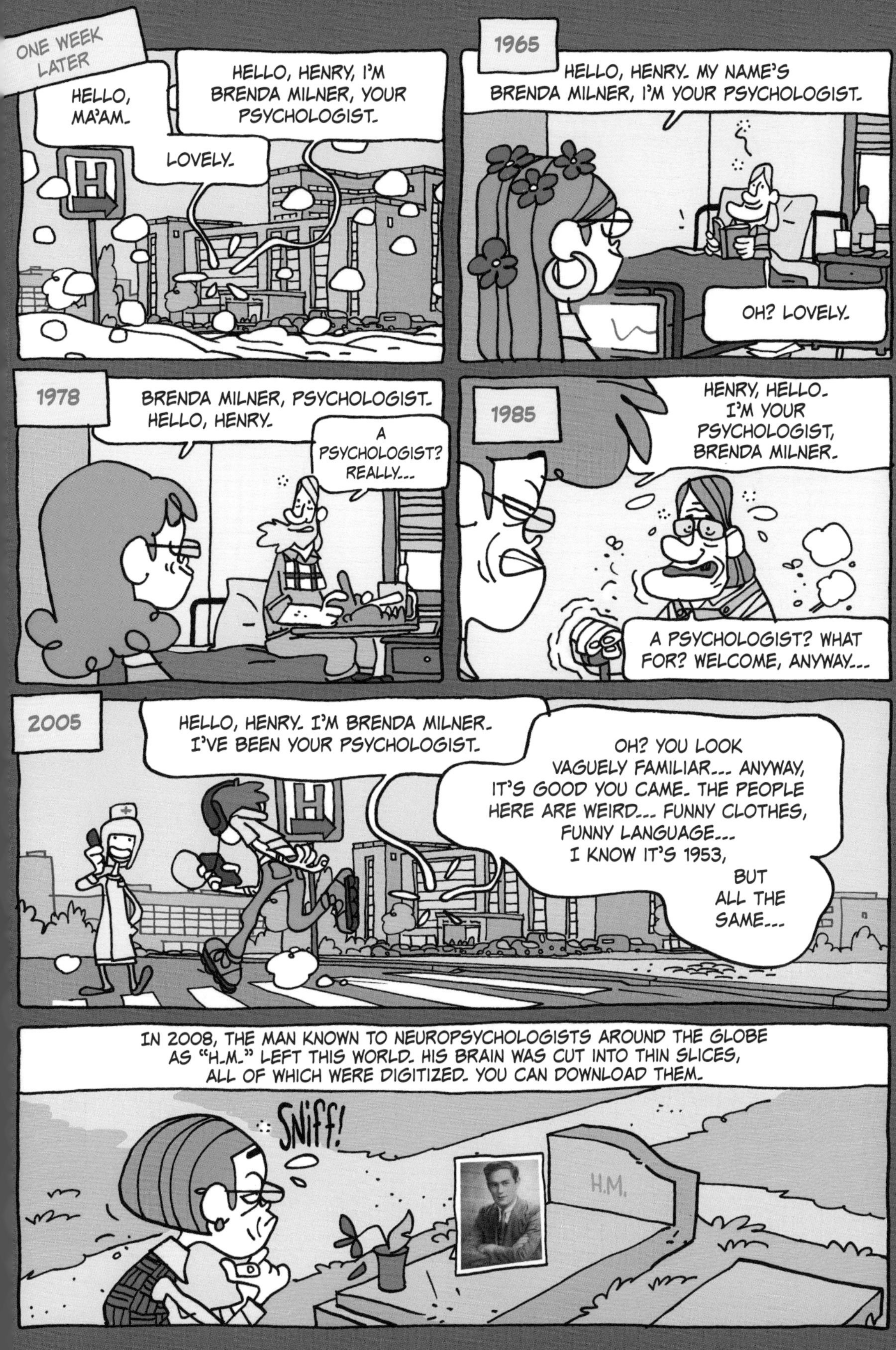
ONE WEEK LATER
HELLO, MA'AM.
HELLO, HENRY, I'M BRENDA MILNER, YOUR PSYCHOLOGIST.
LOVELY.
1965
HELLO, HENRY. MY NAME'S BRENDA MILNER, I'M YOUR PSYCHOLOGIST.
OH? LOVELY.
1978
BRENDA MILNER, PSYCHOLOGIST. HELLO, HENRY.
A PSYCHOLOGIST? REALLY...
1985
HENRY, HELLO. I'M YOUR PSYCHOLOGIST, BRENDA MILNER.
A PSYCHOLOGIST? WHAT FOR? WELCOME, ANYWAY...
2005
HELLO, HENRY. I'M BRENDA MILNER. I'VE BEEN YOUR PSYCHOLOGIST.
OH? YOU LOOK VAGUELY FAMILIAR... ANYWAY, IT'S GOOD YOU CAME. THE PEOPLE HERE ARE WEIRD... FUNNY CLOTHES, FUNNY LANGUAGE... I KNOW IT'S 1953,
BUT ALL THE SAME...
IN 2008, THE MAN KNOWN TO NEUROPSYCHOLOGISTS AROUND THE GLOBE AS "H.M." LEFT THIS WORLD. HIS BRAIN WAS CUT INTO THIN SLICES, ALL OF WHICH WERE DIGITIZED. YOU CAN DOWNLOAD THEM.
SNIFF!
H.M.

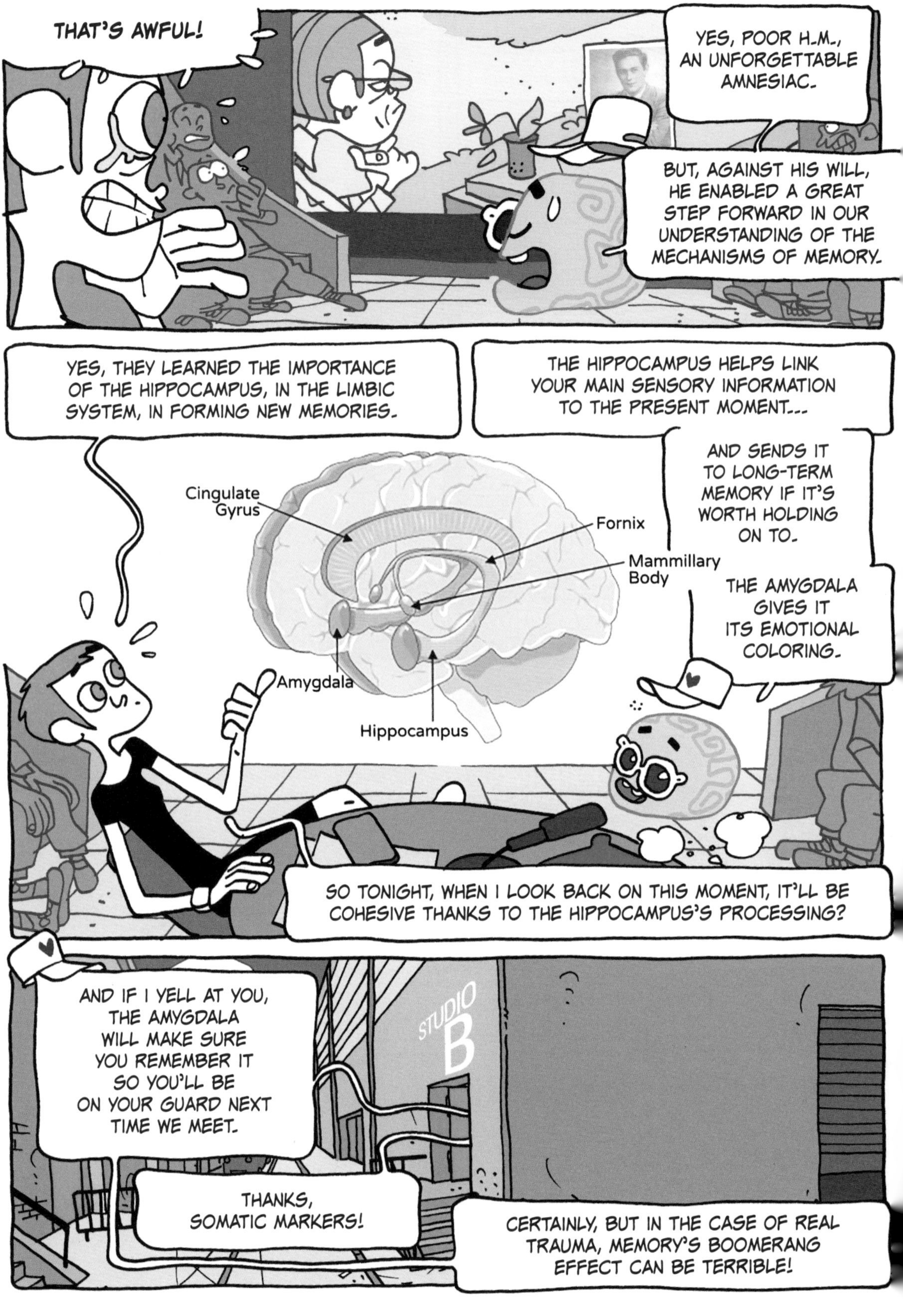

THAT'S AWFUL!
YES, POOR H.M., AN UNFORGETTABLE AMNESIAC.
BUT, AGAINST HIS WILL, HE ENABLED A GREAT STEP FORWARD IN OUR UNDERSTANDING OF THE MECHANISMS OF MEMORY.
YES, THEY LEARNED THE IMPORTANCE OF THE HIPPOCAMPUS, IN THE LIMBIC SYSTEM, IN FORMING NEW MEMORIES.
THE HIPPOCAMPUS HELPS LINK YOUR MAIN SENSORY INFORMATION TO THE PRESENT MOMENT...
AND SENDS IT TO LONG-TERM MEMORY IF IT'S WORTH HOLDING ON TO.
THE AMYGDALA GIVES IT ITS EMOTIONAL COLORING.
Cingulate Gyrus
Fornix
Mammillary Body
Amygdala
Hippocampus
SO TONIGHT, WHEN I LOOK BACK ON THIS MOMENT, IT'LL BE COHESIVE THANKS TO THE HIPPOCAMPUS'S PROCESSING?
AND IF I YELL AT YOU, THE AMYGDALA WILL MAKE SURE YOU REMEMBER IT SO YOU'LL BE ON YOUR GUARD NEXT TIME WE MEET.
STUDIO B
THANKS, SOMATIC MARKERS!
CERTAINLY, BUT IN THE CASE OF REAL TRAUMA, MEMORY'S BOOMERANG EFFECT CAN BE TERRIBLE!

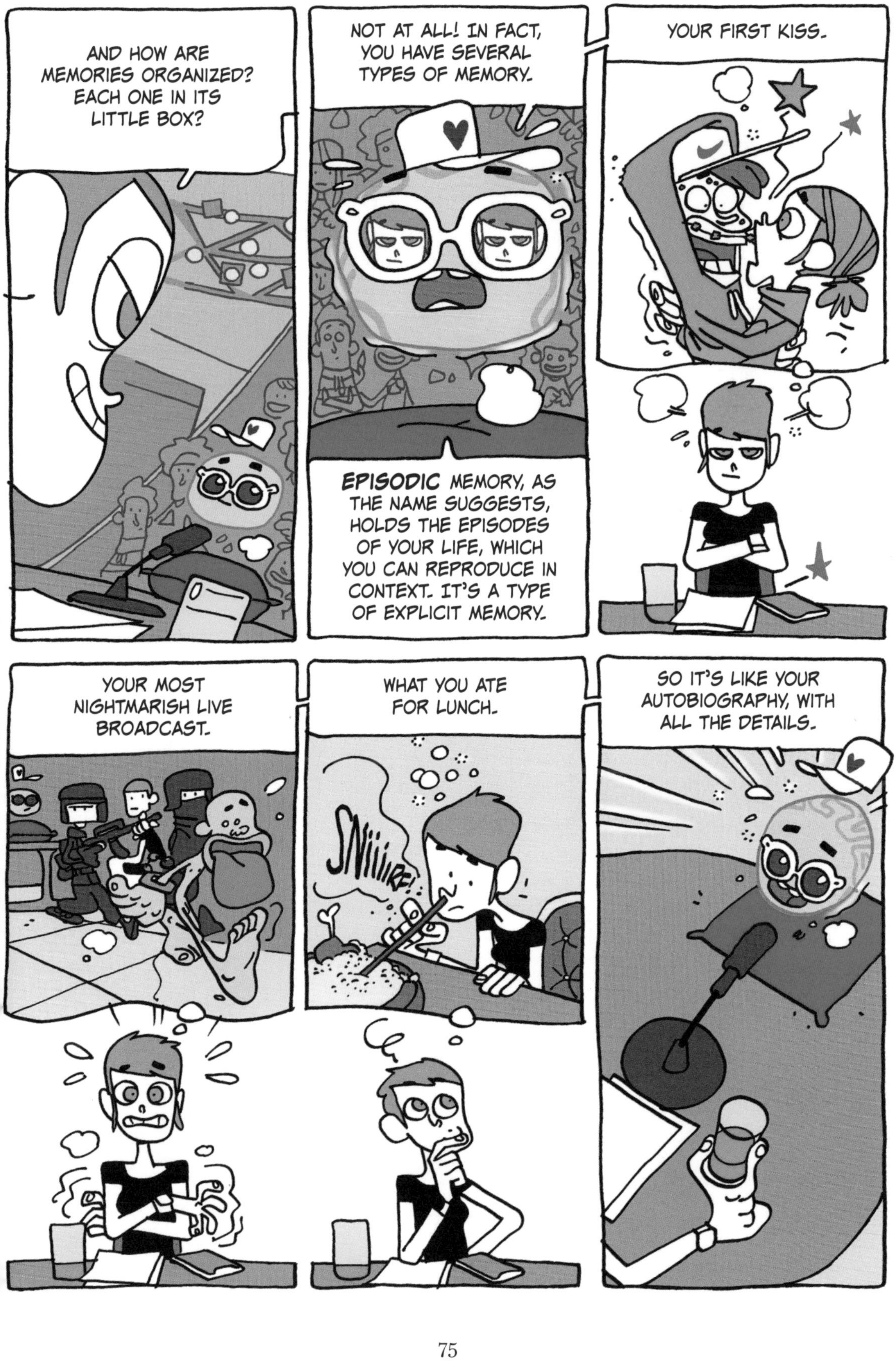
AND HOW ARE MEMORIES ORGANIZED? EACH ONE IN ITS LITTLE BOX?
NOT AT ALL! IN FACT, YOU HAVE SEVERAL TYPES OF MEMORY.
EPISODIC MEMORY, AS THE NAME SUGGESTS, HOLDS THE EPISODES OF YOUR LIFE, WHICH YOU CAN REPRODUCE IN CONTEXT. IT'S A TYPE OF EXPLICIT MEMORY.
YOUR FIRST KISS.
YOUR MOST NIGHTMARISH LIVE BROADCAST.
WHAT YOU ATE FOR LUNCH.
SNIIIIIRF!
SO IT'S LIKE YOUR AUTOBIOGRAPHY, WITH ALL THE DETAILS.

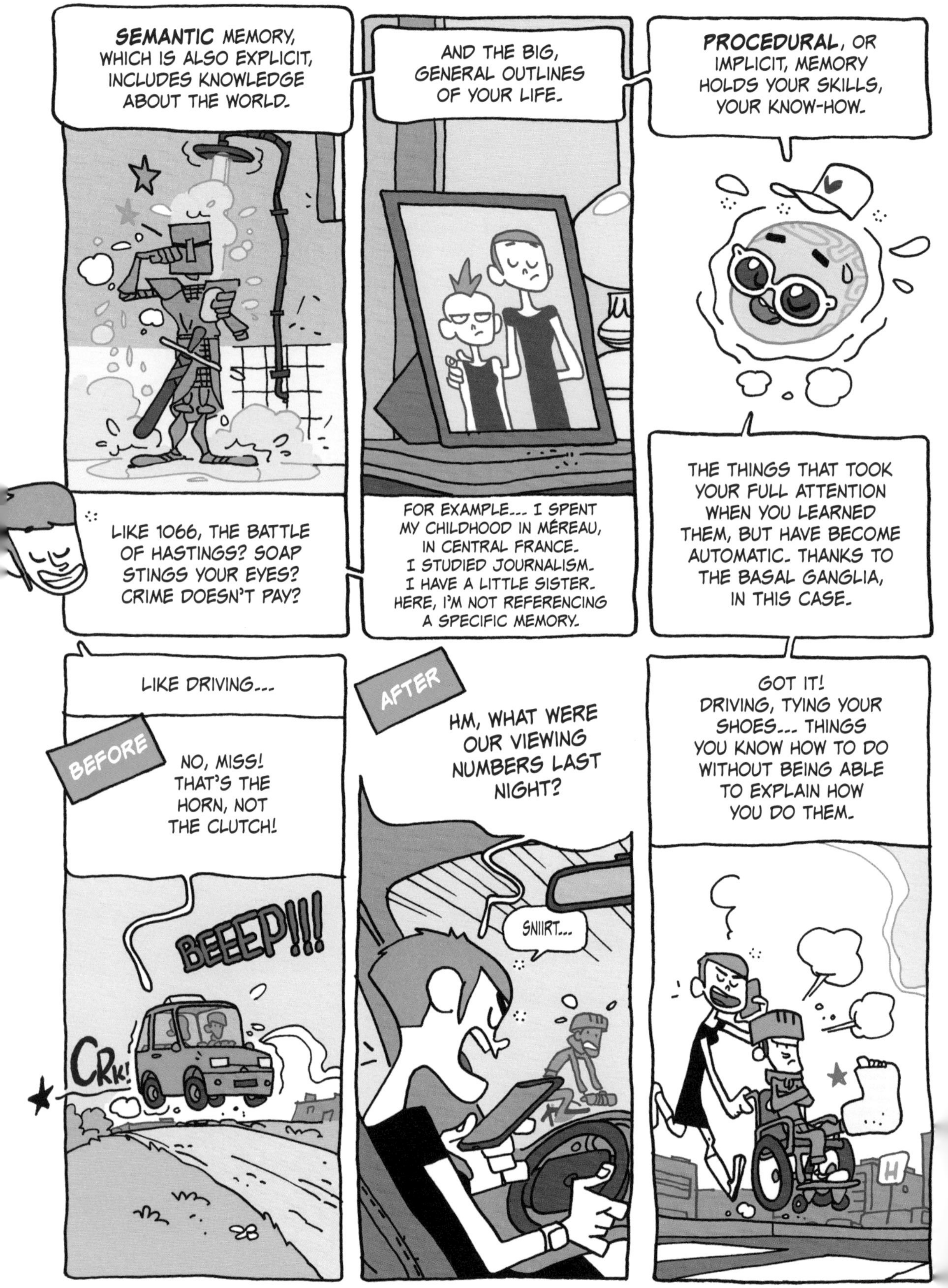
SEMANTIC MEMORY, WHICH IS ALSO EXPLICIT, INCLUDES KNOWLEDGE ABOUT THE WORLD.
AND THE BIG, GENERAL OUTLINES OF YOUR LIFE.
PROCEDURAL, OR IMPLICIT, MEMORY HOLDS YOUR SKILLS, YOUR KNOW-HOW.
LIKE 1066, THE BATTLE OF HASTINGS? SOAP STINGS YOUR EYES? CRIME DOESN'T PAY?
FOR EXAMPLE... I SPENT MY CHILDHOOD IN MÉREAU, IN CENTRAL FRANCE. I STUDIED JOURNALISM. I HAVE A LITTLE SISTER. HERE, I'M NOT REFERENCING A SPECIFIC MEMORY.
THE THINGS THAT TOOK YOUR FULL ATTENTION WHEN YOU LEARNED THEM, BUT HAVE BECOME AUTOMATIC. THANKS TO THE BASAL GANGLIA, IN THIS CASE.
LIKE DRIVING...
BEFORE
NO, MISS! THAT'S THE HORN, NOT THE CLUTCH!
BEEEP!!!
CRK!
AFTER
HM, WHAT WERE OUR VIEWING NUMBERS LAST NIGHT?
SNIIRT...
GOT IT! DRIVING, TYING YOUR SHOES... THINGS YOU KNOW HOW TO DO WITHOUT BEING ABLE TO EXPLAIN HOW YOU DO THEM.
H

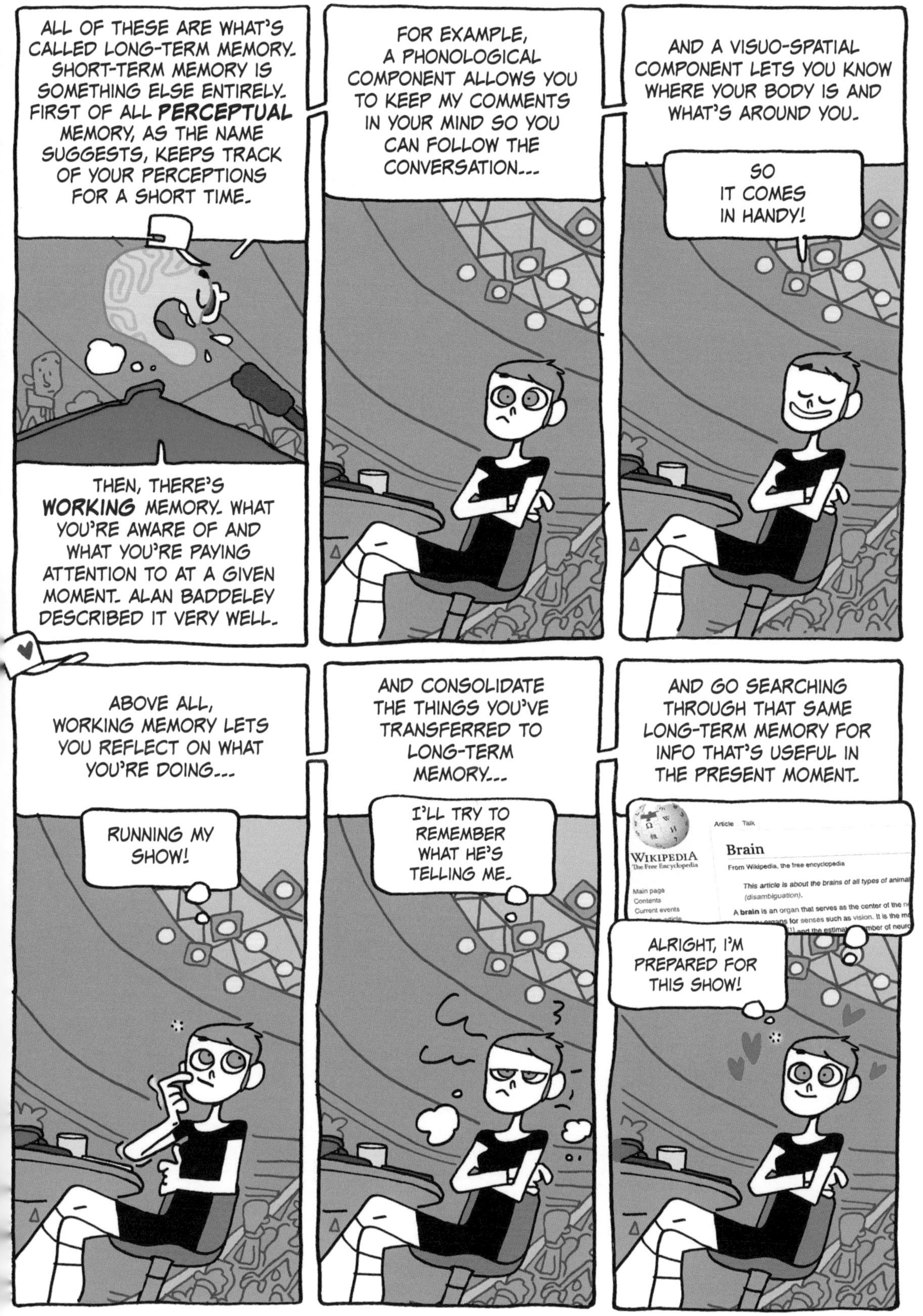

ALL OF THESE ARE WHAT'S CALLED LONG-TERM MEMORY. SHORT-TERM MEMORY IS SOMETHING ELSE ENTIRELY. FIRST OF ALL **PERCEPTUAL** MEMORY, AS THE NAME SUGGESTS, KEEPS TRACK OF YOUR PERCEPTIONS FOR A SHORT TIME.
THEN, THERE'S **WORKING** MEMORY. WHAT YOU'RE AWARE OF AND WHAT YOU'RE PAYING ATTENTION TO AT A GIVEN MOMENT. ALAN BADDELEY DESCRIBED IT VERY WELL.
FOR EXAMPLE, A PHONOLOGICAL COMPONENT ALLOWS YOU TO KEEP MY COMMENTS IN YOUR MIND SO YOU CAN FOLLOW THE CONVERSATION...
AND A VISUO-SPATIAL COMPONENT LETS YOU KNOW WHERE YOUR BODY IS AND WHAT'S AROUND YOU.
SO IT COMES IN HANDY!
ABOVE ALL, WORKING MEMORY LETS YOU REFLECT ON WHAT YOU'RE DOING...
RUNNING MY SHOW!
AND CONSOLIDATE THE THINGS YOU'VE TRANSFERRED TO LONG-TERM MEMORY...
I'LL TRY TO REMEMBER WHAT HE'S TELLING ME.
AND GO SEARCHING THROUGH THAT SAME LONG-TERM MEMORY FOR INFO THAT'S USEFUL IN THE PRESENT MOMENT.
WIKIPEDIA
The Free Encyclopedia
Article Talk
Brain
From Wikipedia, the free encyclopedia
Main page
Contents
Current events
ALRIGHT, I'M PREPARED FOR THIS SHOW!

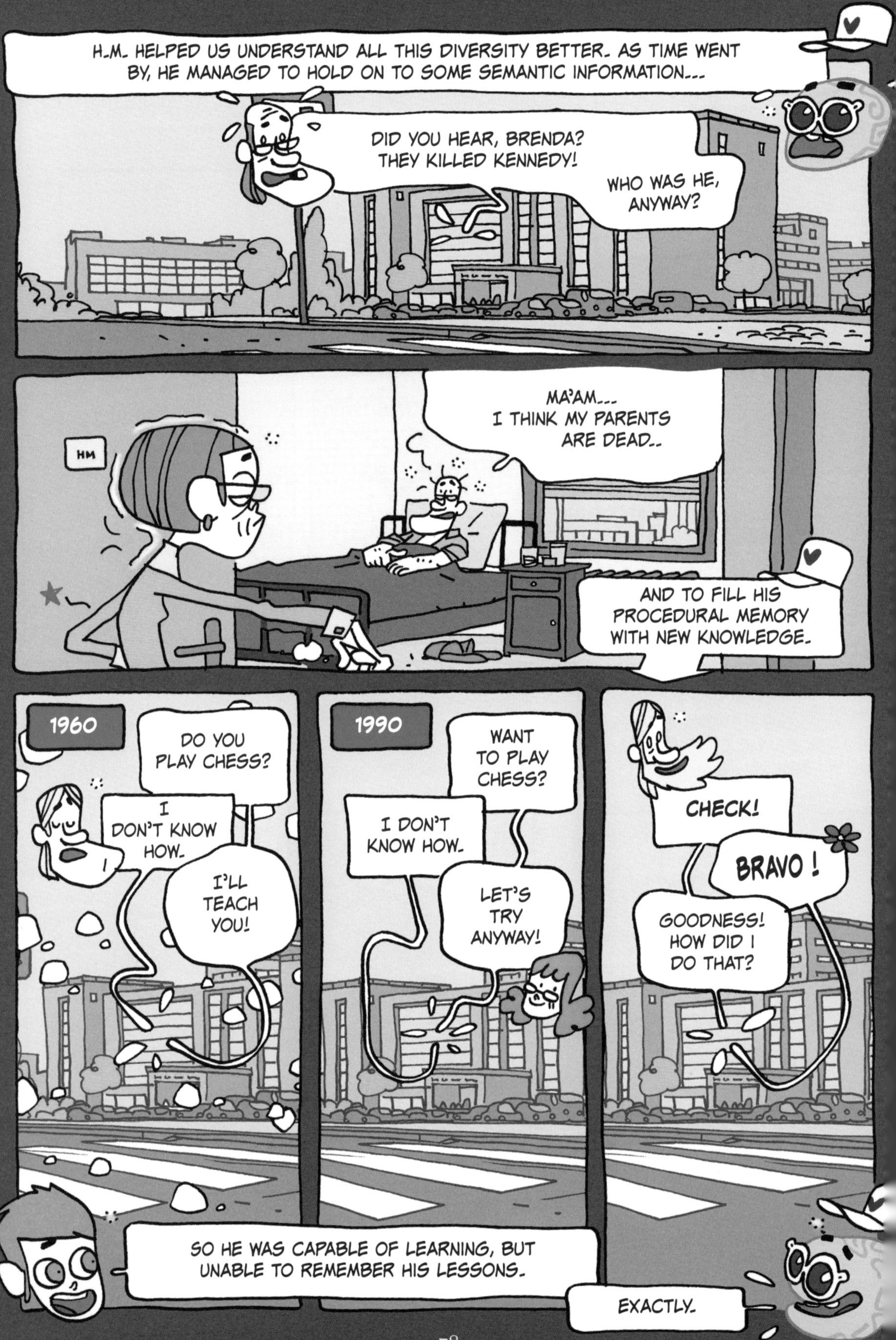
H.M. HELPED US UNDERSTAND ALL THIS DIVERSITY BETTER. AS TIME WENT BY, HE MANAGED TO HOLD ON TO SOME SEMANTIC INFORMATION...
DID YOU HEAR, BRENDA? THEY KILLED KENNEDY!
WHO WAS HE, ANYWAY?
HM
MA'AM... I THINK MY PARENTS ARE DEAD..
AND TO FILL HIS PROCEDURAL MEMORY WITH NEW KNOWLEDGE.
1960
DO YOU PLAY CHESS?
I DON'T KNOW HOW.
I'LL TEACH YOU!
1990
WANT TO PLAY CHESS?
I DON'T KNOW HOW.
LET'S TRY ANYWAY!
CHECK!
BRAVO !
GOODNESS! HOW DID I DO THAT?
SO HE WAS CAPABLE OF LEARNING, BUT UNABLE TO REMEMBER HIS LESSONS.
EXACTLY.

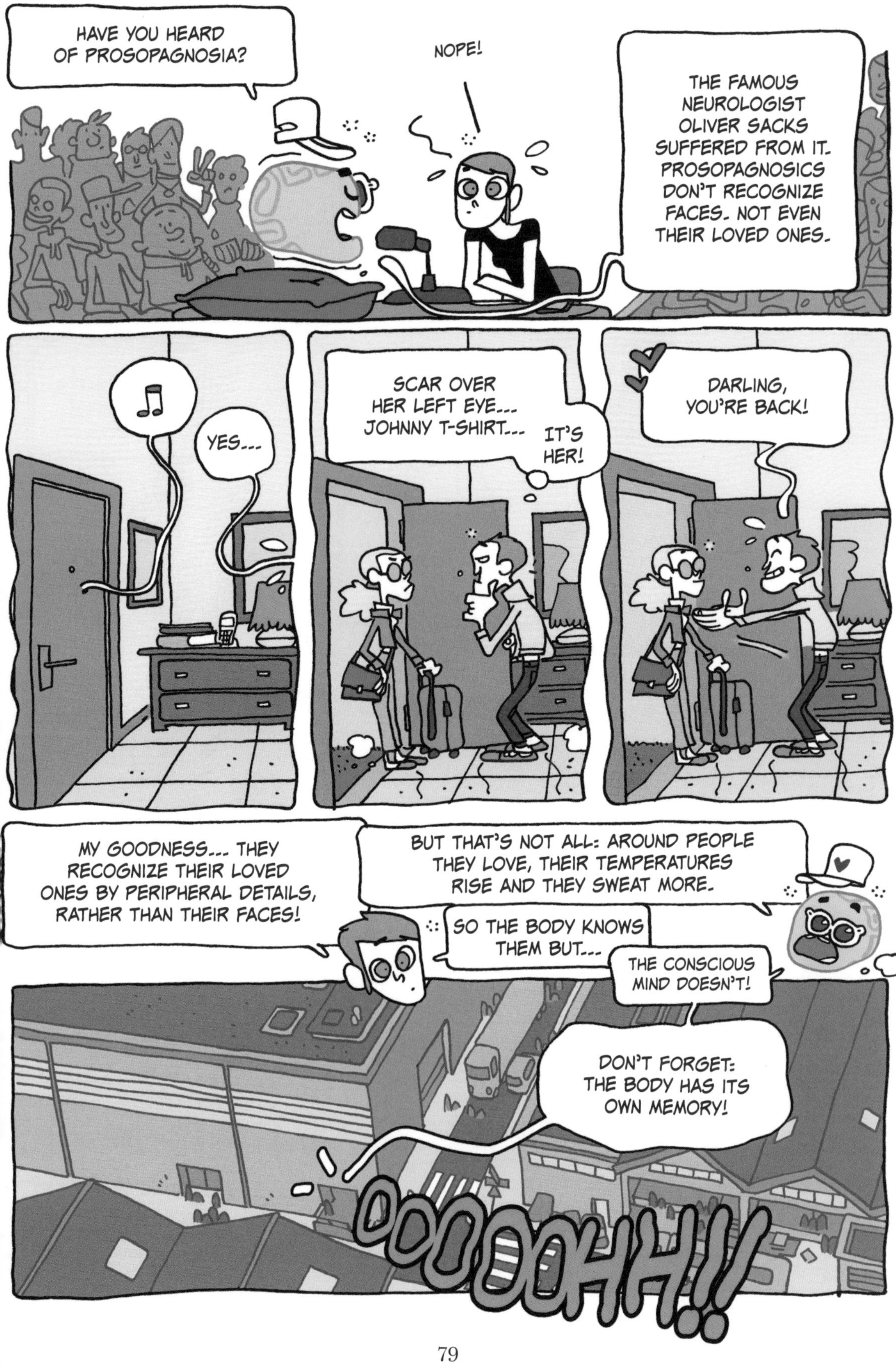
HAVE YOU HEARD OF PROSOPAGNOSIA?
NOPE!
THE FAMOUS NEUROLOGIST OLIVER SACKS SUFFERED FROM IT. PROSOPAGNOSICS DON'T RECOGNIZE FACES. NOT EVEN THEIR LOVED ONES.
YES...
SCAR OVER HER LEFT EYE... JOHNNY T-SHIRT...
IT'S HER!
DARLING, YOU'RE BACK!
MY GOODNESS... THEY RECOGNIZE THEIR LOVED ONES BY PERIPHERAL DETAILS, RATHER THAN THEIR FACES!
BUT THAT'S NOT ALL: AROUND PEOPLE THEY LOVE, THEIR TEMPERATURES RISE AND THEY SWEAT MORE.
SO THE BODY KNOWS THEM BUT...
THE CONSCIOUS MIND DOESN'T!
DON'T FORGET: THE BODY HAS ITS OWN MEMORY!
OOOOOHH!!

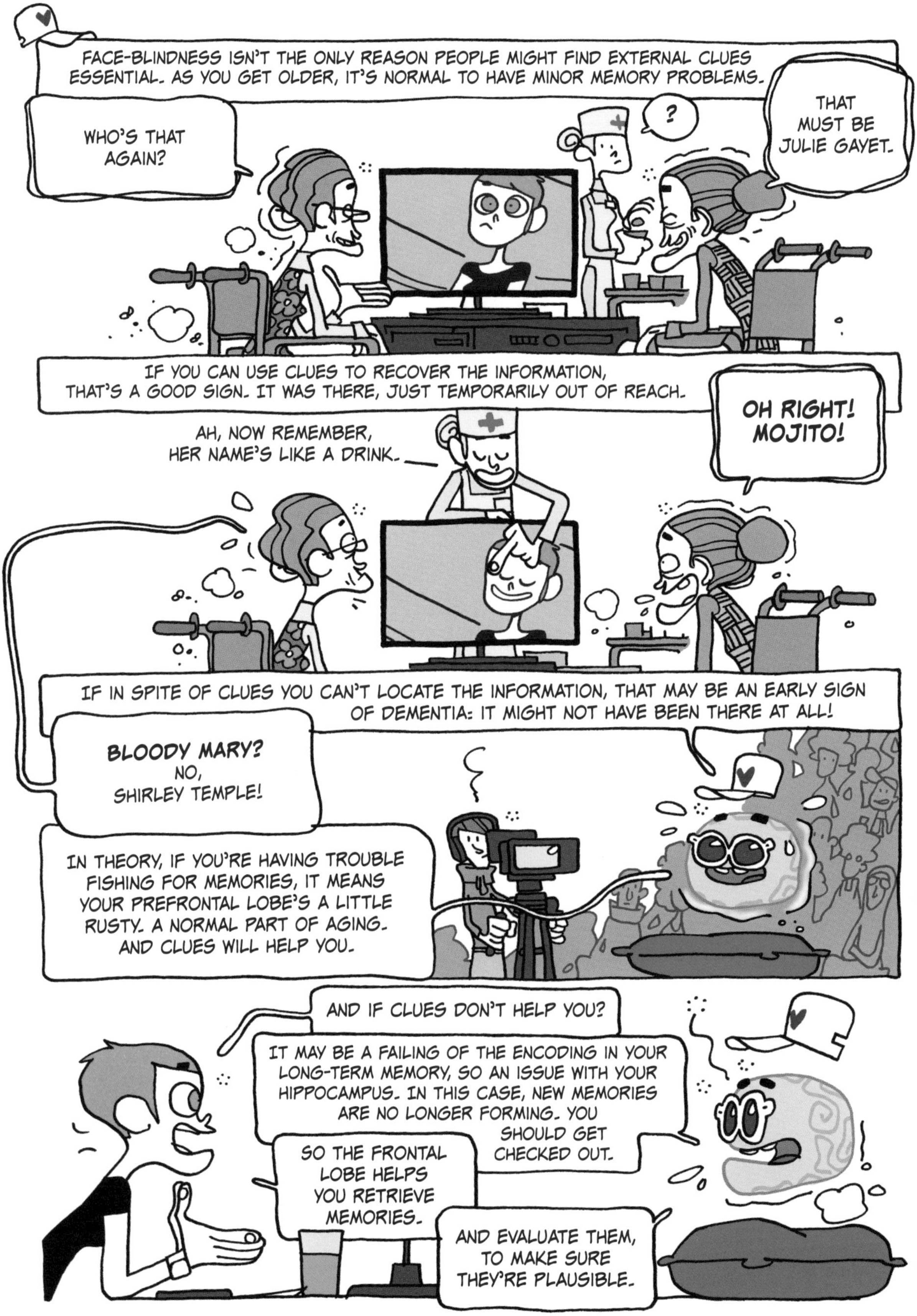
FACE-BLINDNESS ISN'T THE ONLY REASON PEOPLE MIGHT FIND EXTERNAL CLUES ESSENTIAL. AS YOU GET OLDER, IT'S NORMAL TO HAVE MINOR MEMORY PROBLEMS.
WHO'S THAT AGAIN?
?
THAT MUST BE JULIE GAYET.
IF YOU CAN USE CLUES TO RECOVER THE INFORMATION, THAT'S A GOOD SIGN. IT WAS THERE, JUST TEMPORARILY OUT OF REACH.
AH, NOW REMEMBER, HER NAME'S LIKE A DRINK.
OH RIGHT! MOJITO!
IF IN SPITE OF CLUES YOU CAN'T LOCATE THE INFORMATION, THAT MAY BE AN EARLY SIGN OF DEMENTIA: IT MIGHT NOT HAVE BEEN THERE AT ALL!
BLOODY MARY? NO, SHIRLEY TEMPLE!
IN THEORY, IF YOU'RE HAVING TROUBLE FISHING FOR MEMORIES, IT MEANS YOUR PREFRONTAL LOBE'S A LITTLE RUSTY. A NORMAL PART OF AGING. AND CLUES WILL HELP YOU.
AND IF CLUES DON'T HELP YOU?
IT MAY BE A FAILING OF THE ENCODING IN YOUR LONG-TERM MEMORY, SO AN ISSUE WITH YOUR HIPPOCAMPUS. IN THIS CASE, NEW MEMORIES ARE NO LONGER FORMING. YOU SHOULD GET CHECKED OUT.
SO THE FRONTAL LOBE HELPS YOU RETRIEVE MEMORIES.
AND EVALUATE THEM, TO MAKE SURE THEY'RE PLAUSIBLE.

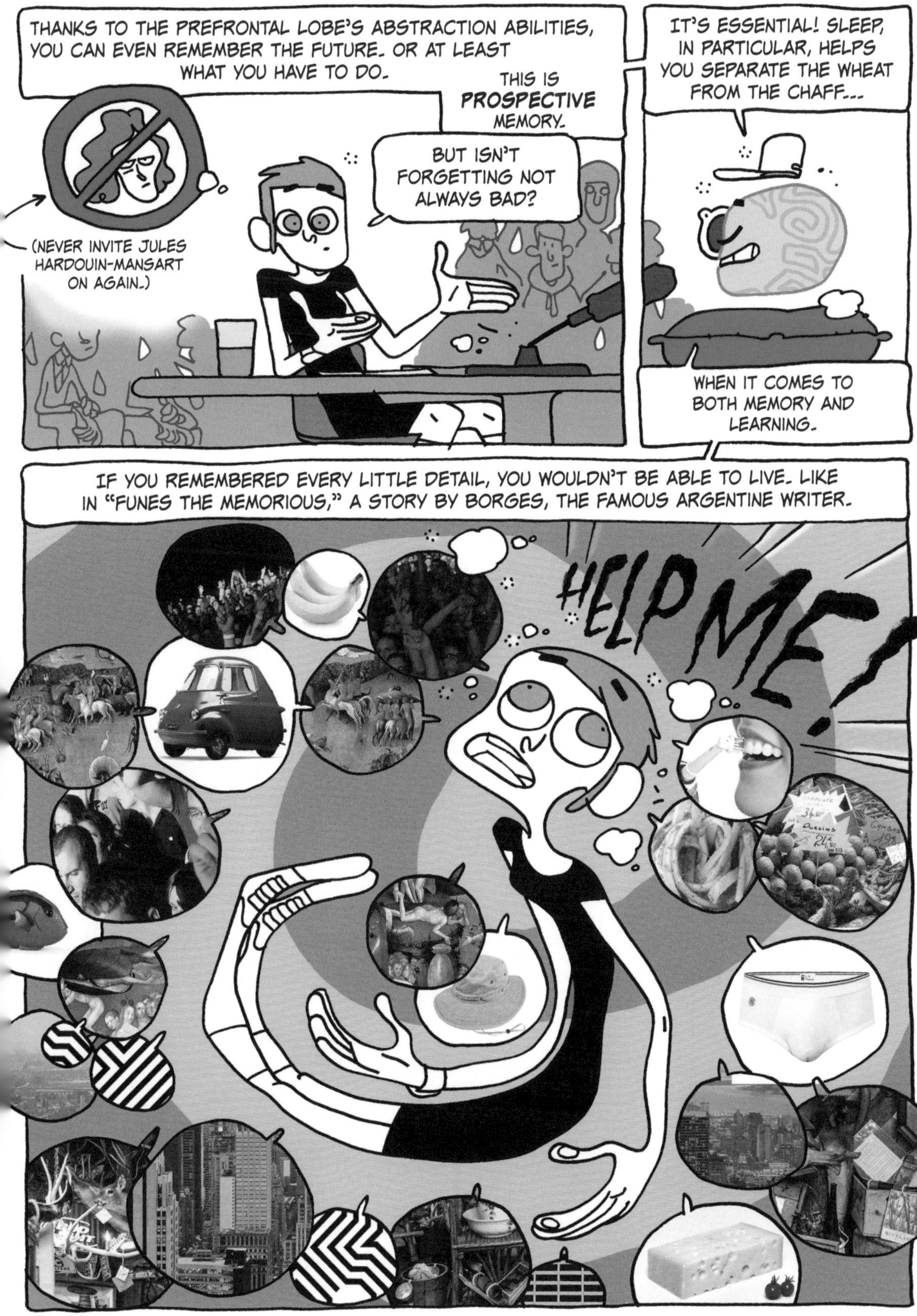

THANKS TO THE PREFRONTAL LOBE'S ABSTRACTION ABILITIES, YOU CAN EVEN REMEMBER THE FUTURE. OR AT LEAST WHAT YOU HAVE TO DO.
THIS IS **PROSPECTIVE** MEMORY.
(NEVER INVITE JULES HARDOUIN-MANSART ON AGAIN.)
BUT ISN'T FORGETTING NOT ALWAYS BAD?
IT'S ESSENTIAL! SLEEP, IN PARTICULAR, HELPS YOU SEPARATE THE WHEAT FROM THE CHAFF...
WHEN IT COMES TO BOTH MEMORY AND LEARNING.
IF YOU REMEMBERED EVERY LITTLE DETAIL, YOU WOULDN'T BE ABLE TO LIVE. LIKE IN "FUNES THE MEMORIOUS," A STORY BY BORGES, THE FAMOUS ARGENTINE WRITER.
HELP ME!

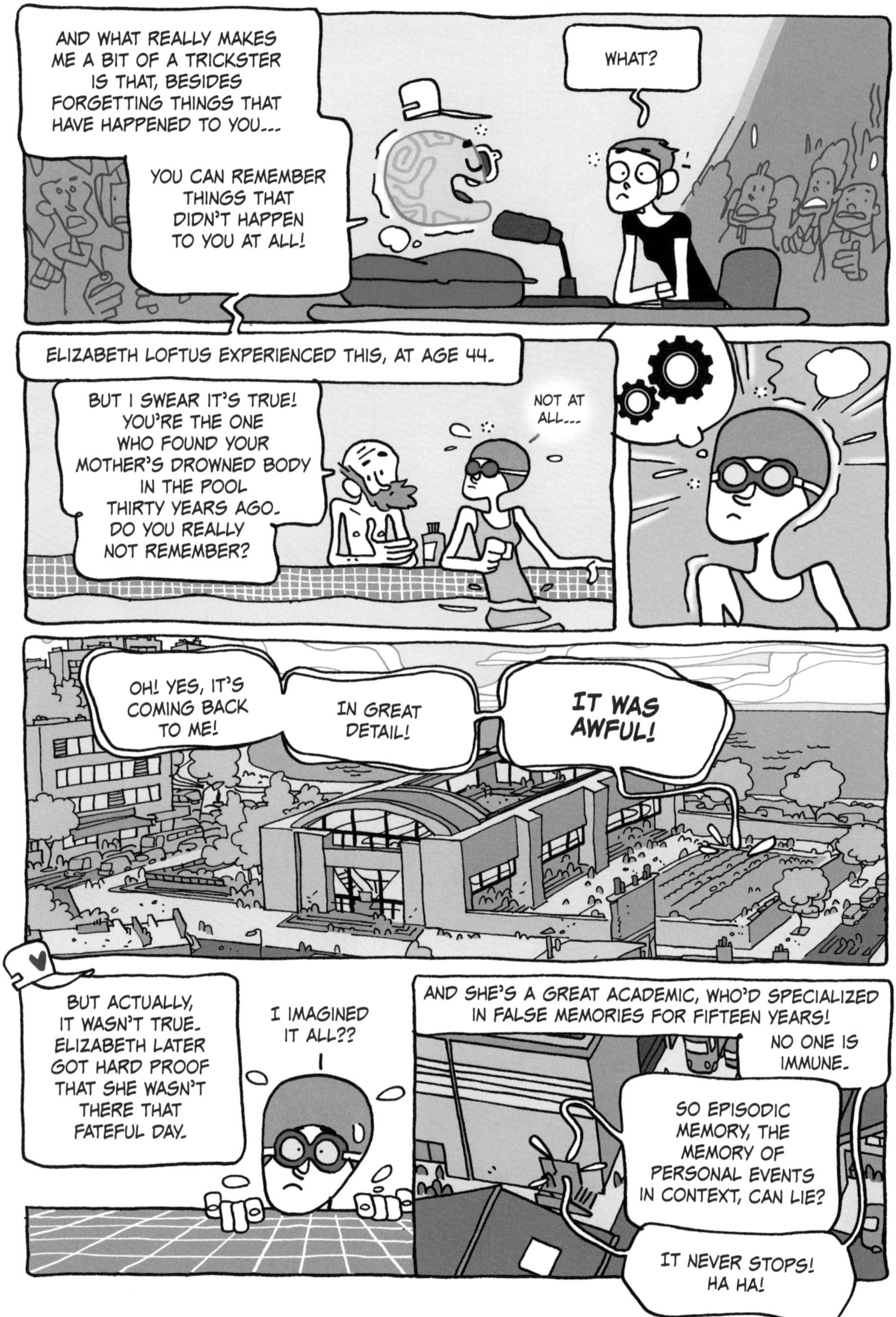
AND WHAT REALLY MAKES ME A BIT OF A TRICKSTER IS THAT, BESIDES FORGETTING THINGS THAT HAVE HAPPENED TO YOU...
YOU CAN REMEMBER THINGS THAT DIDN'T HAPPEN TO YOU AT ALL!
WHAT?
ELIZABETH LOFTUS EXPERIENCED THIS, AT AGE 44.
BUT I SWEAR IT'S TRUE! YOU'RE THE ONE WHO FOUND YOUR MOTHER'S DROWNED BODY IN THE POOL THIRTY YEARS AGO. DO YOU REALLY NOT REMEMBER?
NOT AT ALL...
OH! YES, IT'S COMING BACK TO ME!
IN GREAT DETAIL!
IT WAS AWFUL!
BUT ACTUALLY, IT WASN'T TRUE. ELIZABETH LATER GOT HARD PROOF THAT SHE WASN'T THERE THAT FATEFUL DAY.
I IMAGINED IT ALL??
AND SHE'S A GREAT ACADEMIC, WHO'D SPECIALIZED IN FALSE MEMORIES FOR FIFTEEN YEARS!
NO ONE IS IMMUNE.
SO EPISODIC MEMORY, THE MEMORY OF PERSONAL EVENTS IN CONTEXT, CAN LIE?
IT NEVER STOPS! HA HA!

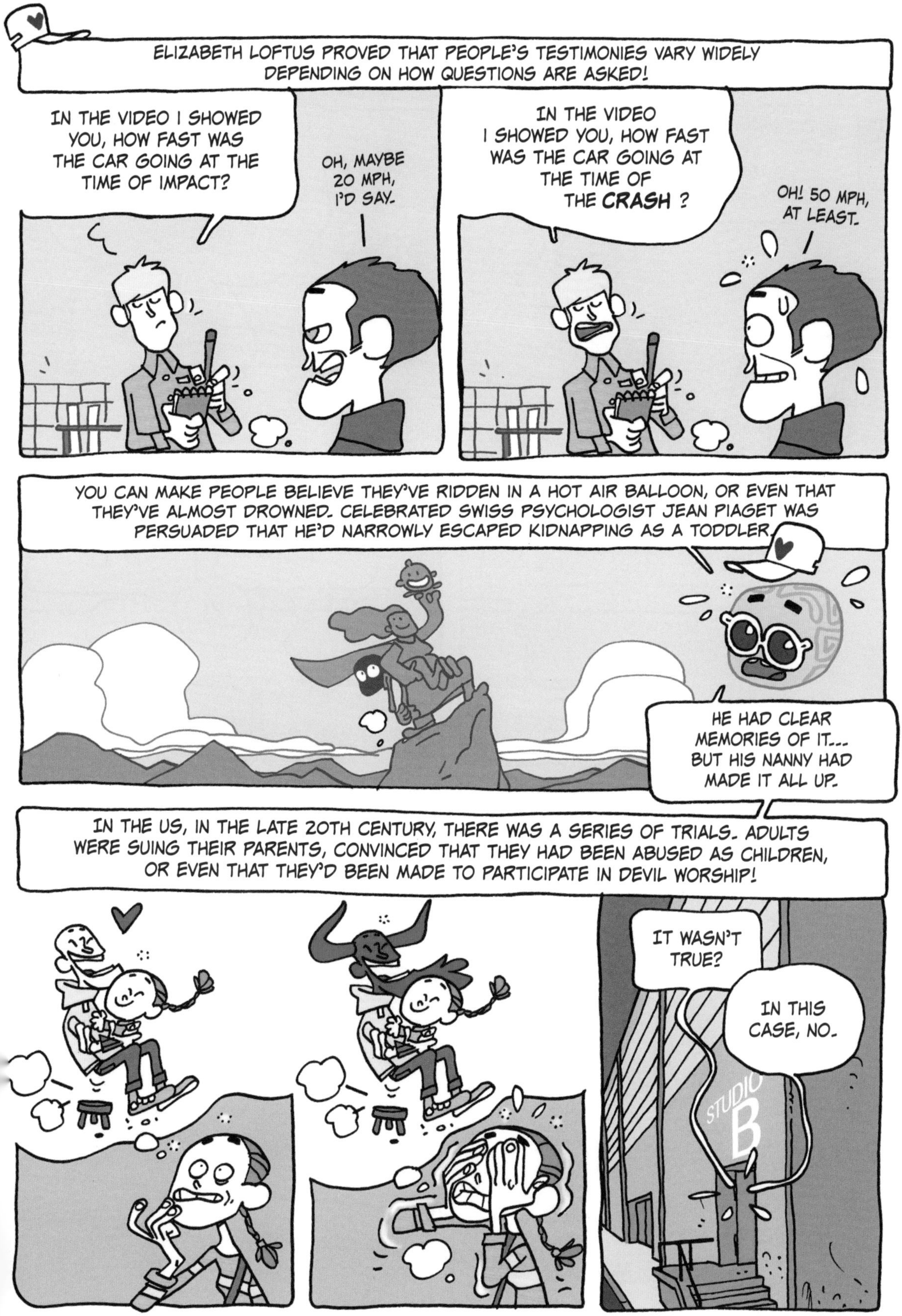
ELIZABETH LOFTUS PROVED THAT PEOPLE'S TESTIMONIES VARY WIDELY DEPENDING ON HOW QUESTIONS ARE ASKED!
IN THE VIDEO I SHOWED YOU, HOW FAST WAS THE CAR GOING AT THE TIME OF IMPACT?
OH, MAYBE 20 MPH, I'D SAY.
IN THE VIDEO I SHOWED YOU, HOW FAST WAS THE CAR GOING AT THE TIME OF THE **CRASH** ?
OH! 50 MPH, AT LEAST.
YOU CAN MAKE PEOPLE BELIEVE THEY'VE RIDDEN IN A HOT AIR BALLOON, OR EVEN THAT THEY'VE ALMOST DROWNED. CELEBRATED SWISS PSYCHOLOGIST JEAN PIAGET WAS PERSUADED THAT HE'D NARROWLY ESCAPED KIDNAPPING AS A TODDLER.
HE HAD CLEAR MEMORIES OF IT... BUT HIS NANNY HAD MADE IT ALL UP.
IN THE US, IN THE LATE 20TH CENTURY, THERE WAS A SERIES OF TRIALS. ADULTS WERE SUING THEIR PARENTS, CONVINCED THAT THEY HAD BEEN ABUSED AS CHILDREN, OR EVEN THAT THEY'D BEEN MADE TO PARTICIPATE IN DEVIL WORSHIP!
IT WASN'T TRUE?
IN THIS CASE, NO.
STUDIO B

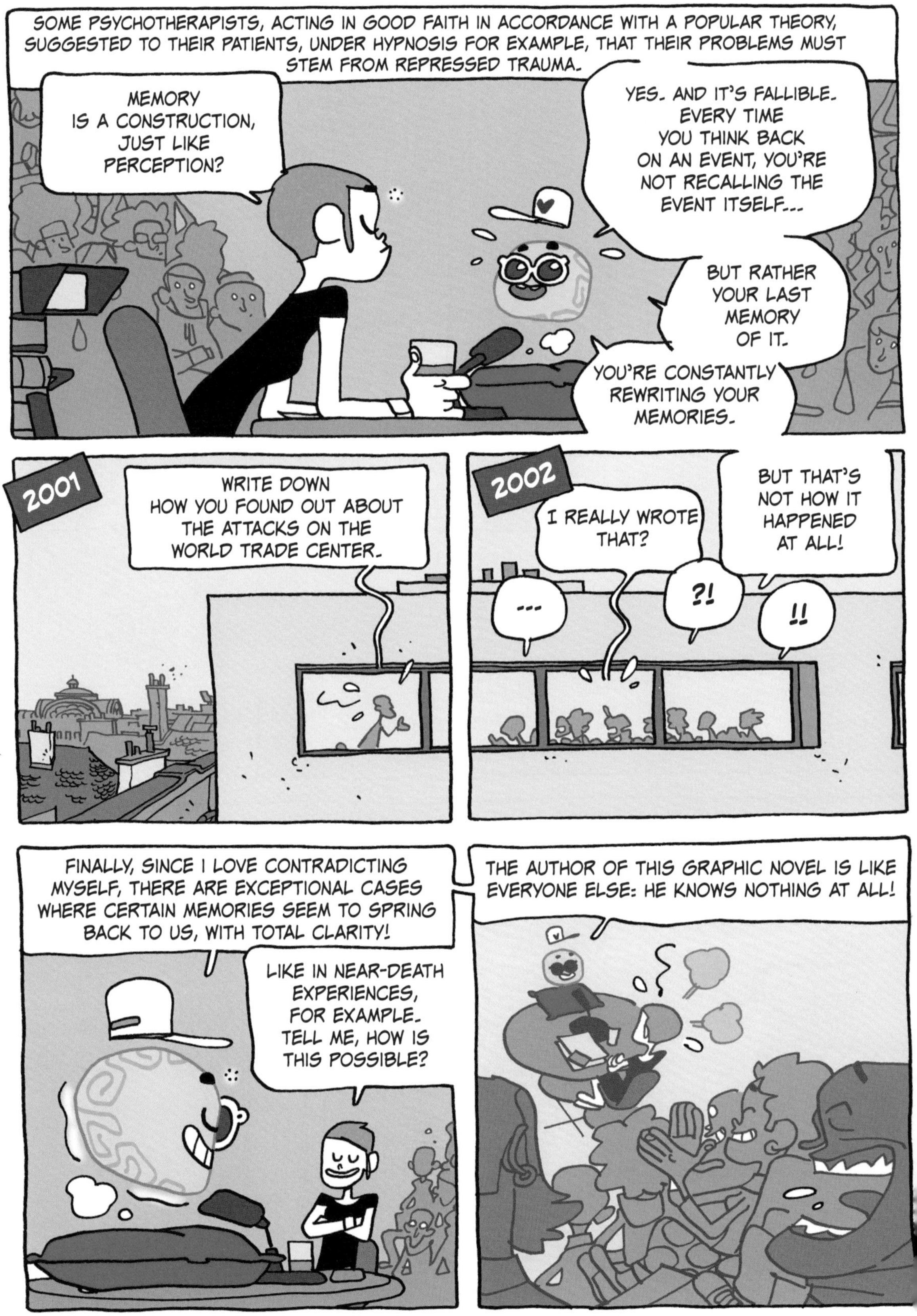

SOME PSYCHOTHERAPISTS, ACTING IN GOOD FAITH IN ACCORDANCE WITH A POPULAR THEORY, SUGGESTED TO THEIR PATIENTS, UNDER HYPNOSIS FOR EXAMPLE, THAT THEIR PROBLEMS MUST STEM FROM REPRESSED TRAUMA.
MEMORY IS A CONSTRUCTION, JUST LIKE PERCEPTION?
YES. AND IT'S FALLIBLE. EVERY TIME YOU THINK BACK ON AN EVENT, YOU'RE NOT RECALLING THE EVENT ITSELF...
BUT RATHER YOUR LAST MEMORY OF IT.
YOU'RE CONSTANTLY REWRITING YOUR MEMORIES.
2001
WRITE DOWN HOW YOU FOUND OUT ABOUT THE ATTACKS ON THE WORLD TRADE CENTER.
2002
I REALLY WROTE THAT?
BUT THAT'S NOT HOW IT HAPPENED AT ALL!
...
?!
!!
FINALLY, SINCE I LOVE CONTRADICTING MYSELF, THERE ARE EXCEPTIONAL CASES WHERE CERTAIN MEMORIES SEEM TO SPRING BACK TO US, WITH TOTAL CLARITY!
LIKE IN NEAR-DEATH EXPERIENCES, FOR EXAMPLE. TELL ME, HOW IS THIS POSSIBLE?
THE AUTHOR OF THIS GRAPHIC NOVEL IS LIKE EVERYONE ELSE: HE KNOWS NOTHING AT ALL!

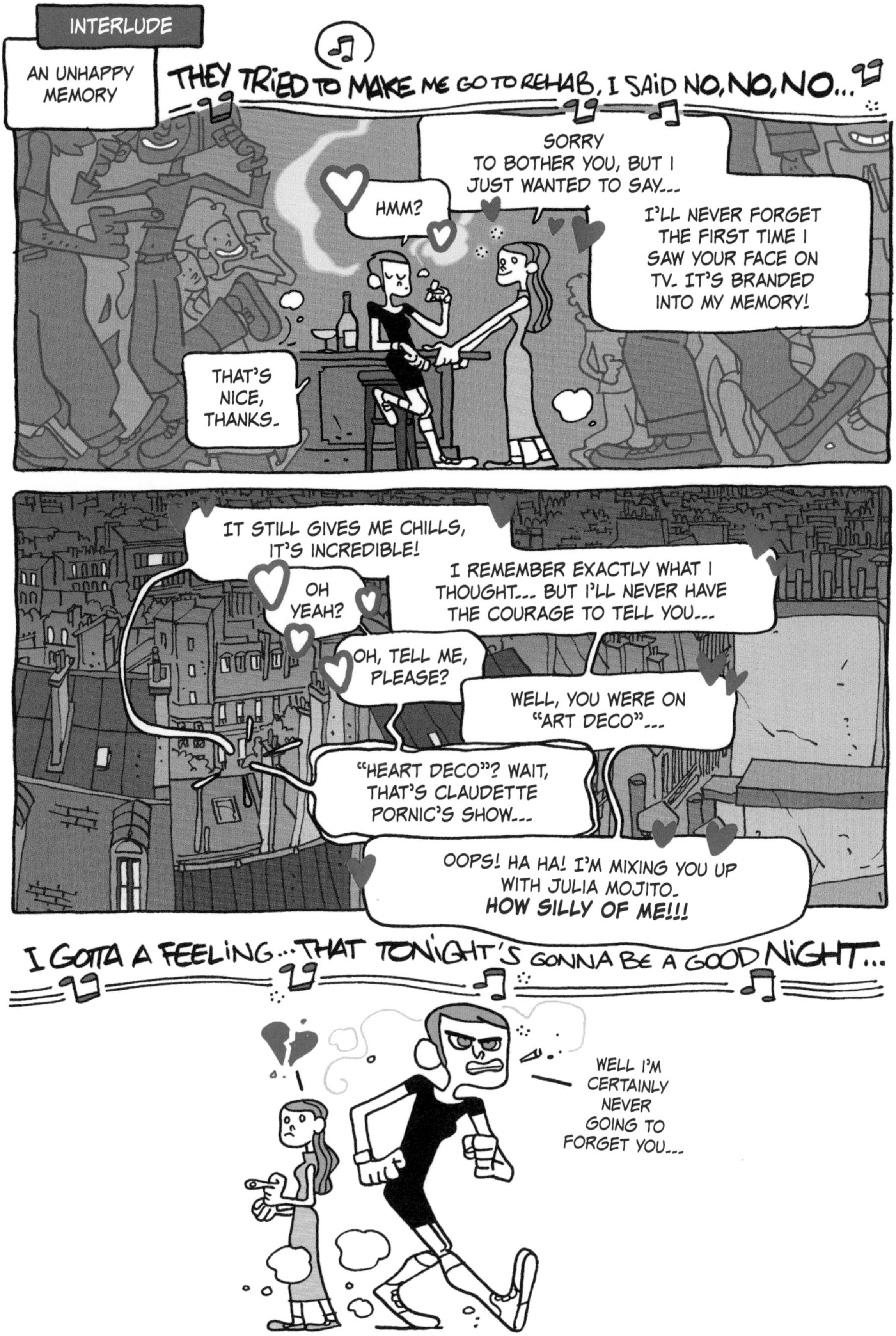
INTERLUDE
AN UNHAPPY MEMORY
THEY TRIED TO MAKE ME GO TO REHAB, I SAID NO, NO, NO...
SORRY TO BOTHER YOU, BUT I JUST WANTED TO SAY...
HMM?
I'LL NEVER FORGET THE FIRST TIME I SAW YOUR FACE ON TV. IT'S BRANDED INTO MY MEMORY!
THAT'S NICE, THANKS.
IT STILL GIVES ME CHILLS, IT'S INCREDIBLE!
OH YEAH?
I REMEMBER EXACTLY WHAT I THOUGHT... BUT I'LL NEVER HAVE THE COURAGE TO TELL YOU...
OH, TELL ME, PLEASE?
WELL, YOU WERE ON "ART DECO"...
"HEART DECO"? WAIT, THAT'S CLAUDETTE PORNIC'S SHOW...
OOPS! HA HA! I'M MIXING YOU UP WITH JULIA MOJITO. HOW SILLY OF ME!!!
I GOTTA A FEELING...THAT TONIGHT'S GONNA BE A GOOD NIGHT...
WELL I'M CERTAINLY NEVER GOING TO FORGET YOU...

Chapter 6

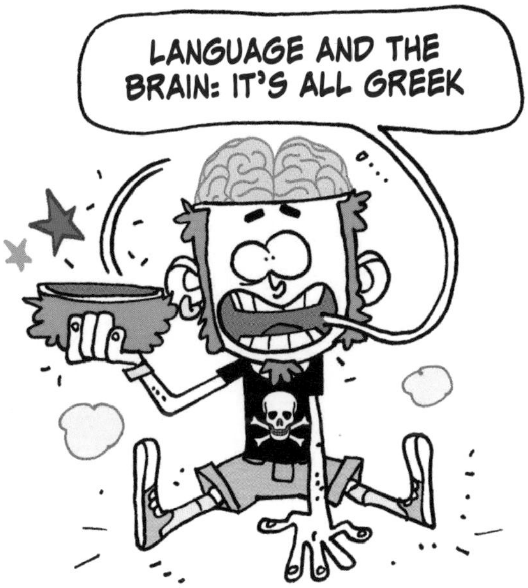
LANGUAGE AND THE BRAIN: IT'S ALL GREEK

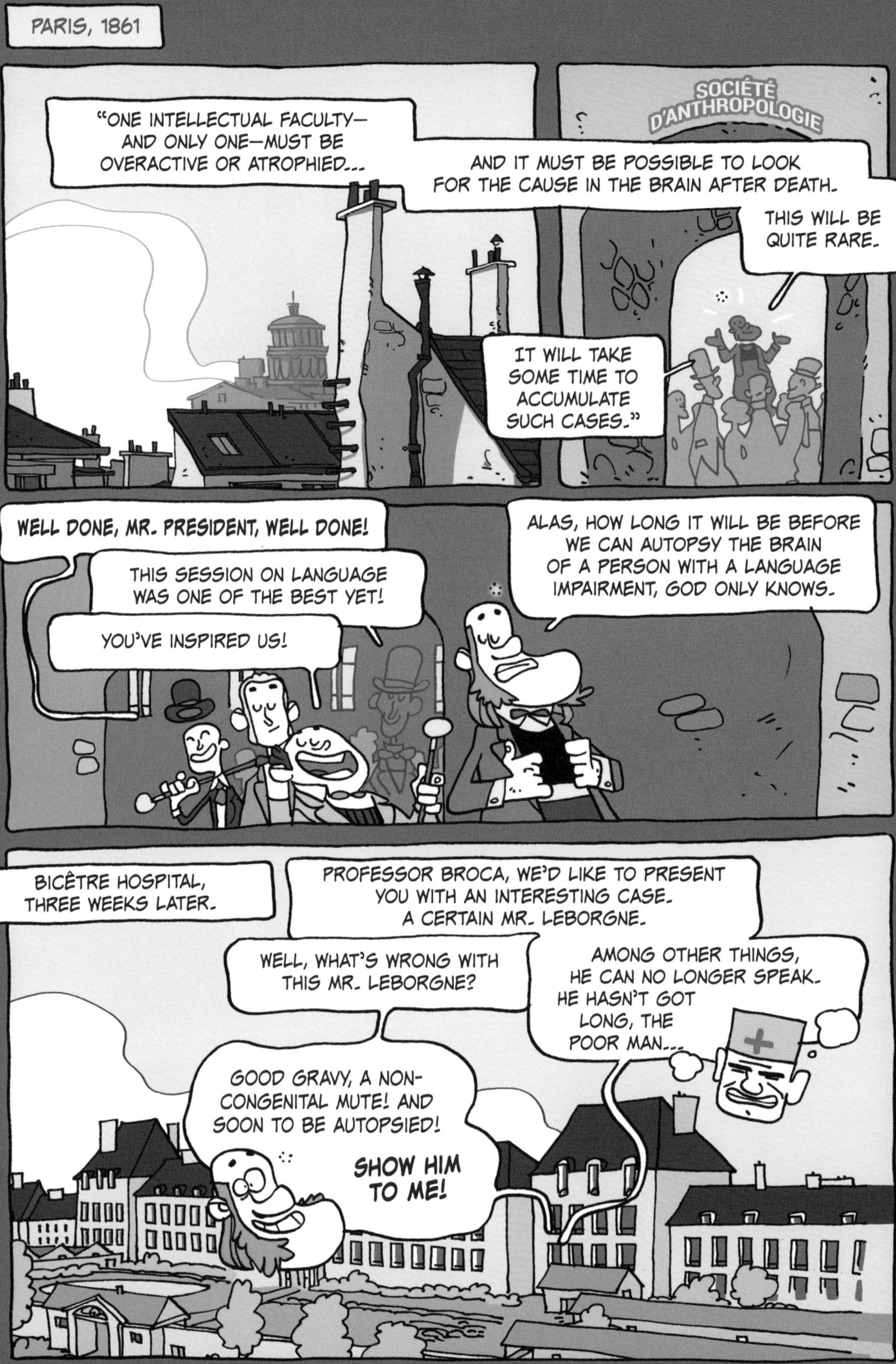
PARIS, 1861
"ONE INTELLECTUAL FACULTY— AND ONLY ONE—MUST BE OVERACTIVE OR ATROPHIED...
AND IT MUST BE POSSIBLE TO LOOK FOR THE CAUSE IN THE BRAIN AFTER DEATH.
SOCIÉTÉ D'ANTHROPOLOGIE
THIS WILL BE QUITE RARE.
IT WILL TAKE SOME TIME TO ACCUMULATE SUCH CASES."
WELL DONE, MR. PRESIDENT, WELL DONE!
THIS SESSION ON LANGUAGE WAS ONE OF THE BEST YET!
YOU'VE INSPIRED US!
ALAS, HOW LONG IT WILL BE BEFORE WE CAN AUTOPSY THE BRAIN OF A PERSON WITH A LANGUAGE IMPAIRMENT, GOD ONLY KNOWS.
BICÊTRE HOSPITAL, THREE WEEKS LATER.
PROFESSOR BROCA, WE'D LIKE TO PRESENT YOU WITH AN INTERESTING CASE. A CERTAIN MR. LEBORGNE.
WELL, WHAT'S WRONG WITH THIS MR. LEBORGNE?
AMONG OTHER THINGS, HE CAN NO LONGER SPEAK. HE HASN'T GOT LONG, THE POOR MAN...
GOOD GRAVY, A NON-CONGENITAL MUTE! AND SOON TO BE AUTOPSIED!
SHOW HIM TO ME!

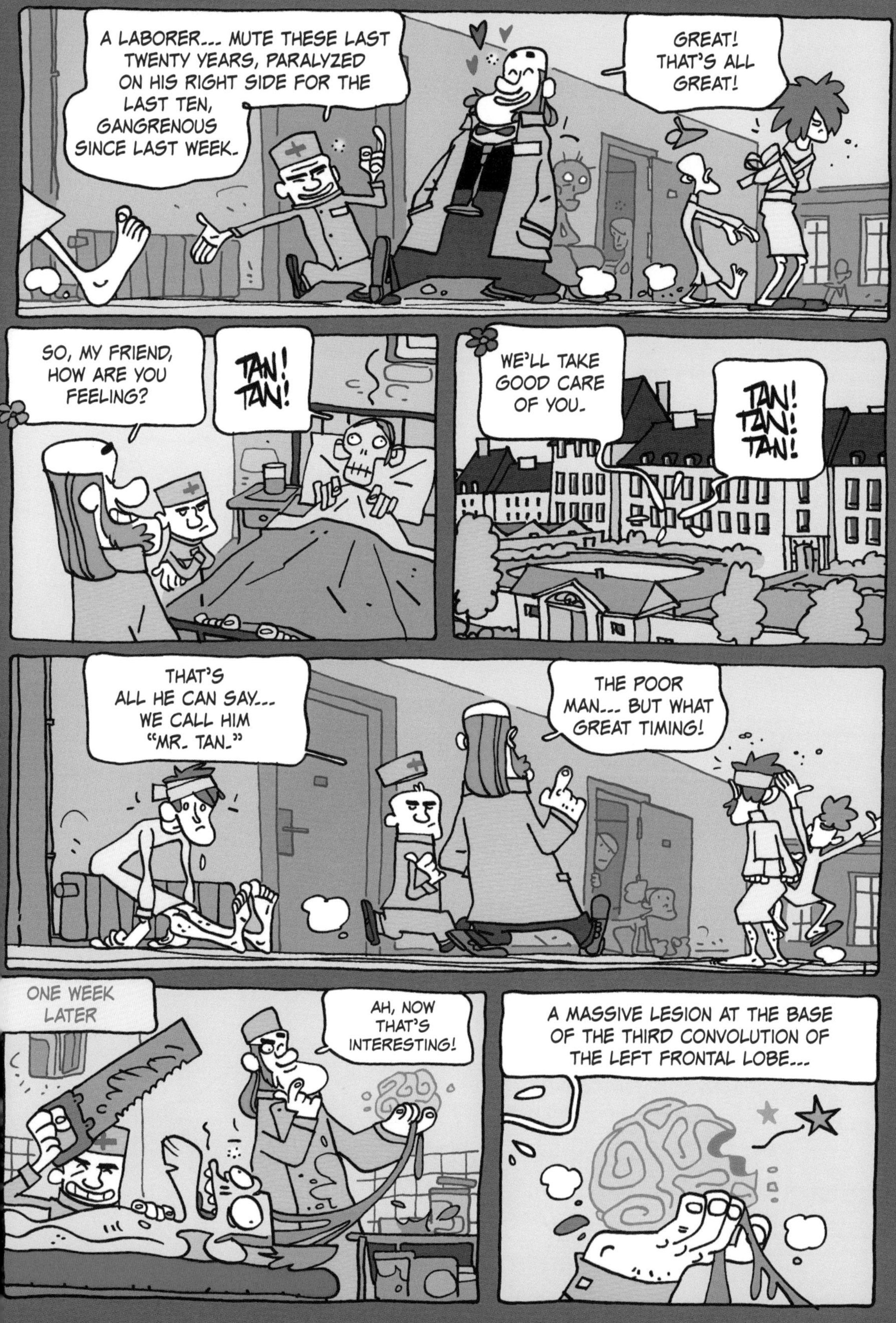

A LABORER... MUTE THESE LAST TWENTY YEARS, PARALYZED ON HIS RIGHT SIDE FOR THE LAST TEN, GANGRENOUS SINCE LAST WEEK.
GREAT! THAT'S ALL GREAT!
SO, MY FRIEND, HOW ARE YOU FEELING?
TAN! TAN!
WE'LL TAKE GOOD CARE OF YOU.
TAN! TAN! TAN!
THAT'S ALL HE CAN SAY... WE CALL HIM "MR. TAN."
THE POOR MAN... BUT WHAT GREAT TIMING!
ONE WEEK LATER
AH, NOW THAT'S INTERESTING!
A MASSIVE LESION AT THE BASE OF THE THIRD CONVOLUTION OF THE LEFT FRONTAL LOBE...

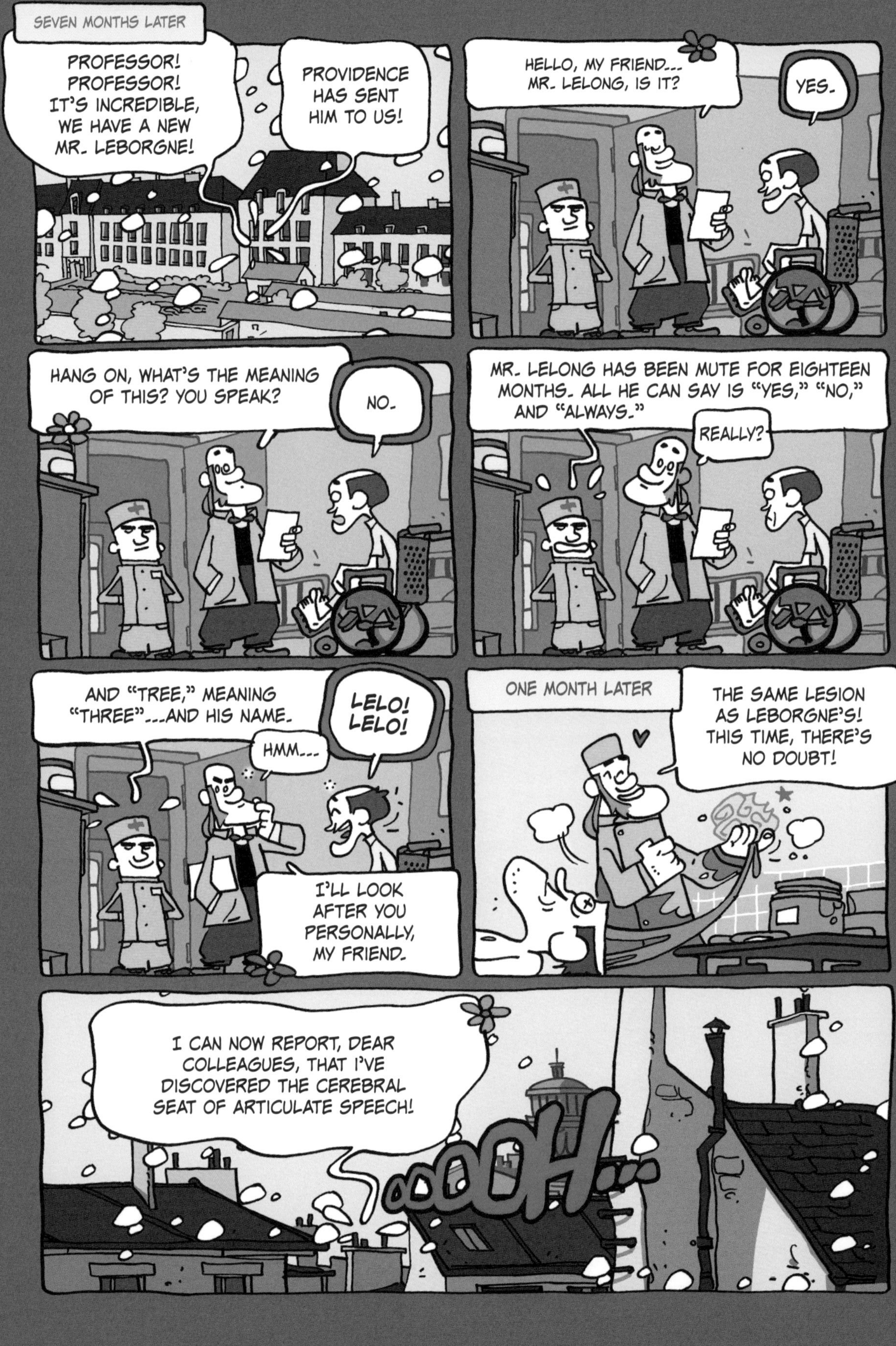
SEVEN MONTHS LATER
PROFESSOR! PROFESSOR! IT'S INCREDIBLE, WE HAVE A NEW MR. LEBORGNE!
PROVIDENCE HAS SENT HIM TO US!
HELLO, MY FRIEND... MR. LELONG, IS IT?
YES.
HANG ON, WHAT'S THE MEANING OF THIS? YOU SPEAK?
NO.
MR. LELONG HAS BEEN MUTE FOR EIGHTEEN MONTHS. ALL HE CAN SAY IS "YES," "NO," AND "ALWAYS."
REALLY?
AND "TREE," MEANING "THREE"...AND HIS NAME.
LELO! LELO!
HMM...
I'LL LOOK AFTER YOU PERSONALLY, MY FRIEND.
ONE MONTH LATER
THE SAME LESION AS LEBORGNE'S! THIS TIME, THERE'S NO DOUBT!
I CAN NOW REPORT, DEAR COLLEAGUES, THAT I'VE DISCOVERED THE CEREBRAL SEAT OF ARTICULATE SPEECH!
OOOH...

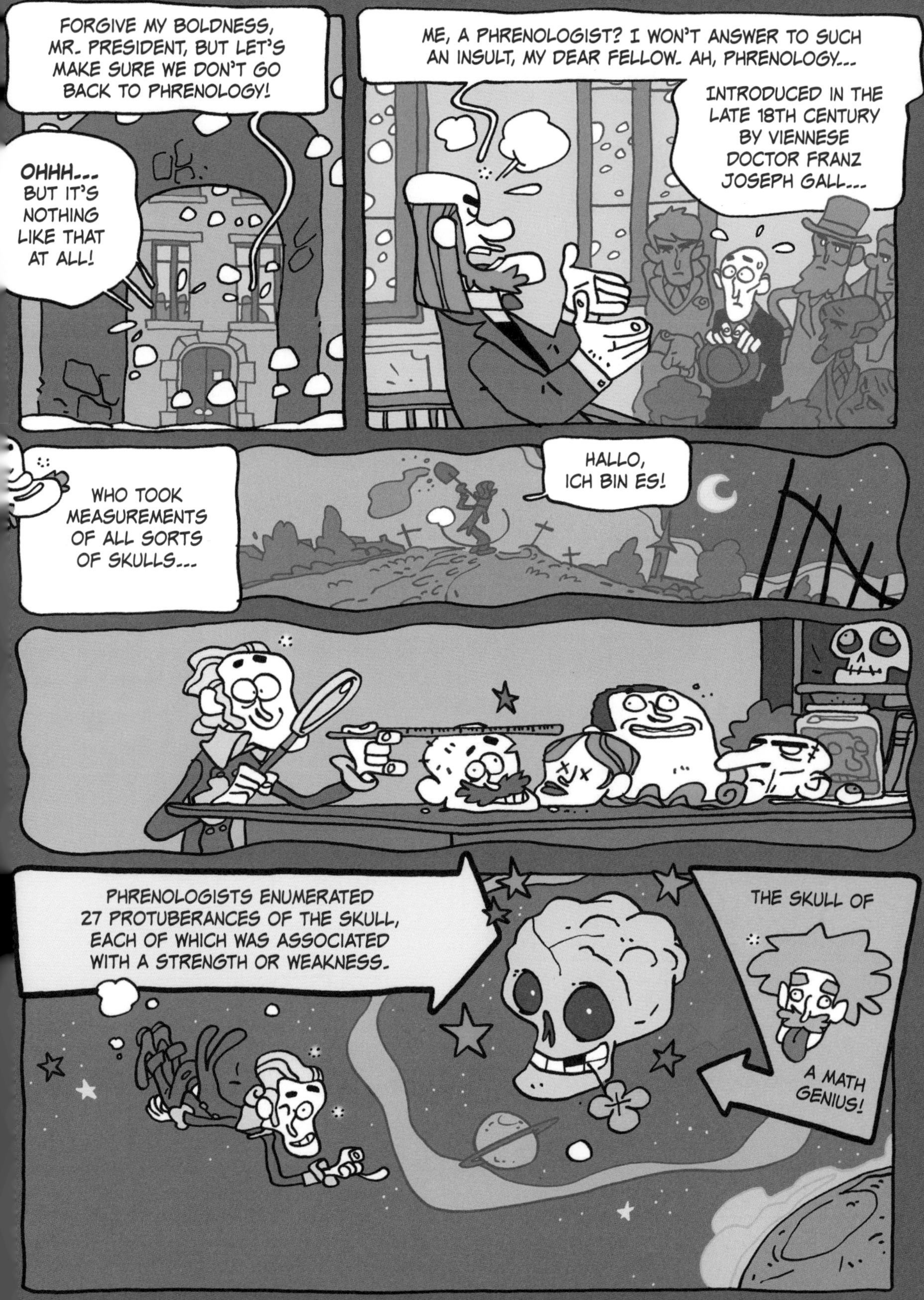

FORGIVE MY BOLDNESS, MR. PRESIDENT, BUT LET'S MAKE SURE WE DON'T GO BACK TO PHRENOLOGY!
OHHH... BUT IT'S NOTHING LIKE THAT AT ALL!
ME, A PHRENOLOGIST? I WON'T ANSWER TO SUCH AN INSULT, MY DEAR FELLOW. AH, PHRENOLOGY...
INTRODUCED IN THE LATE 18TH CENTURY BY VIENNESE DOCTOR FRANZ JOSEPH GALL...
WHO TOOK MEASUREMENTS OF ALL SORTS OF SKULLS...
HALLO, ICH BIN ES!
PHRENOLOGISTS ENUMERATED 27 PROTUBERANCES OF THE SKULL, EACH OF WHICH WAS ASSOCIATED WITH A STRENGTH OR WEAKNESS.
THE SKULL OF
A MATH GENIUS!

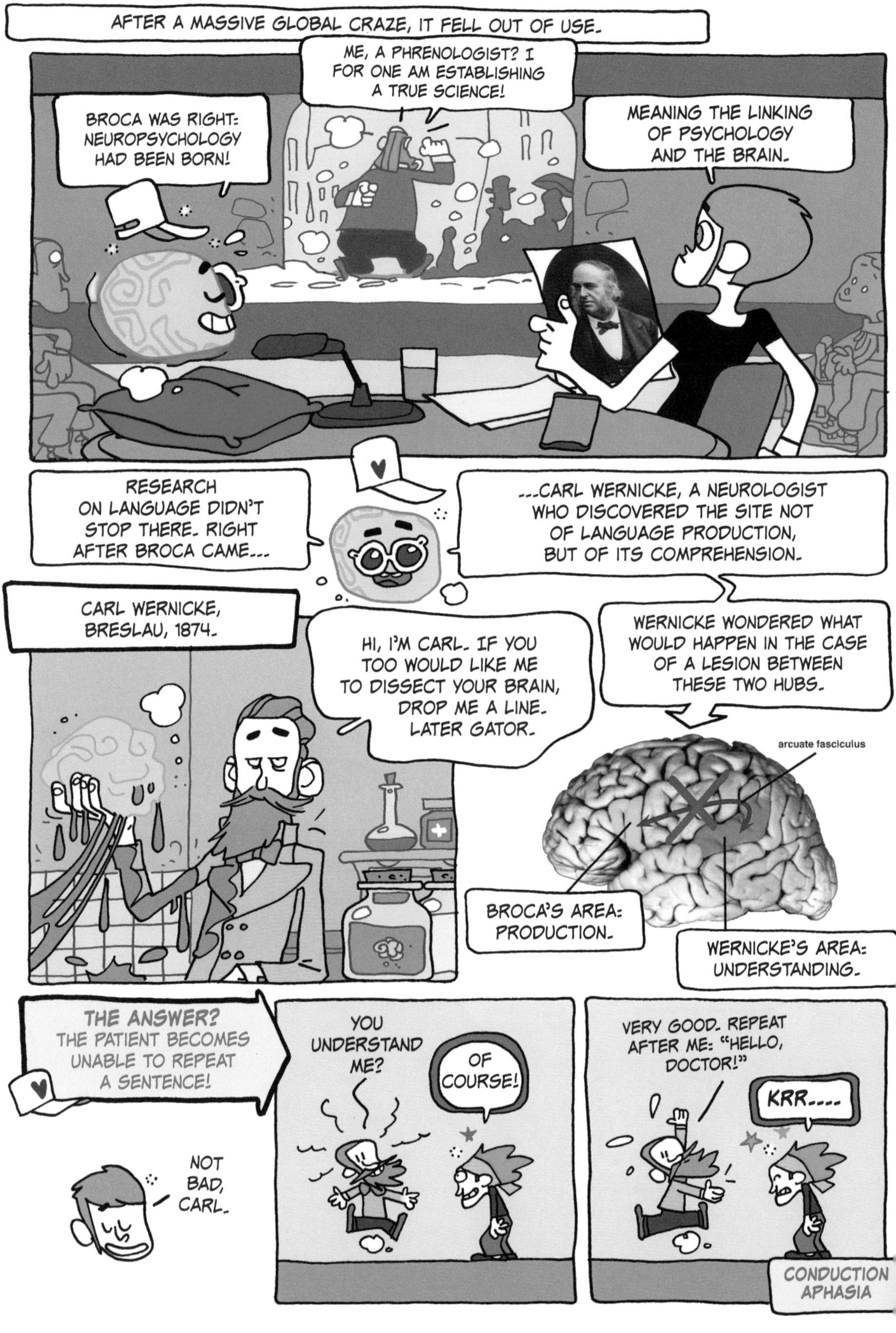

AFTER A MASSIVE GLOBAL CRAZE, IT FELL OUT OF USE.
BROCA WAS RIGHT: NEUROPSYCHOLOGY HAD BEEN BORN!
ME, A PHRENOLOGIST? I FOR ONE AM ESTABLISHING A TRUE SCIENCE!
MEANING THE LINKING OF PSYCHOLOGY AND THE BRAIN.
RESEARCH ON LANGUAGE DIDN'T STOP THERE. RIGHT AFTER BROCA CAME...
...CARL WERNICKE, A NEUROLOGIST WHO DISCOVERED THE SITE NOT OF LANGUAGE PRODUCTION, BUT OF ITS COMPREHENSION.
CARL WERNICKE, BRESLAU, 1874.
HI, I'M CARL. IF YOU TOO WOULD LIKE ME TO DISSECT YOUR BRAIN, DROP ME A LINE. LATER GATOR.
WERNICKE WONDERED WHAT WOULD HAPPEN IN THE CASE OF A LESION BETWEEN THESE TWO HUBS.
arcuate fasciculus
BROCA'S AREA: PRODUCTION.
WERNICKE'S AREA: UNDERSTANDING.
THE ANSWER? THE PATIENT BECOMES UNABLE TO REPEAT A SENTENCE!
NOT BAD, CARL.
YOU UNDERSTAND ME?
OF COURSE!
VERY GOOD. REPEAT AFTER ME: "HELLO, DOCTOR!"
KRR....
CONDUCTION APHASIA

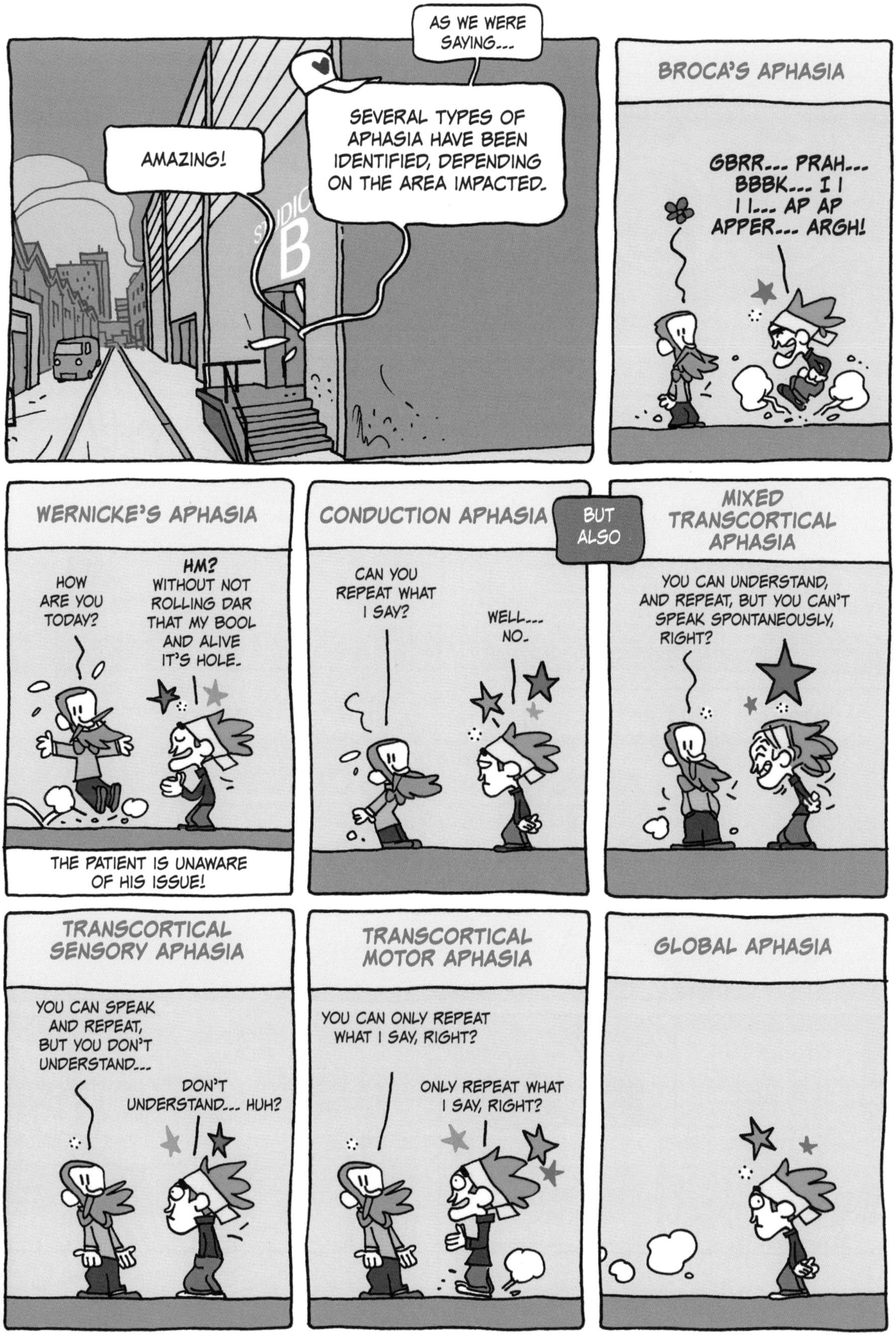
AS WE WERE SAYING...
AMAZING!
SEVERAL TYPES OF APHASIA HAVE BEEN IDENTIFIED, DEPENDING ON THE AREA IMPACTED.
STUDIO B
BROCA'S APHASIA
GBRR... PRAH... BBBK... I I I I... AP AP APPER... ARGH!
WERNICKE'S APHASIA
HOW ARE YOU TODAY?
HM? WITHOUT NOT ROLLING DAR THAT MY BOOL AND ALIVE IT'S HOLE.
THE PATIENT IS UNAWARE OF HIS ISSUE!
CONDUCTION APHASIA
CAN YOU REPEAT WHAT I SAY?
WELL... NO.
BUT ALSO
MIXED TRANSCORTICAL APHASIA
YOU CAN UNDERSTAND, AND REPEAT, BUT YOU CAN'T SPEAK SPONTANEOUSLY, RIGHT?
TRANSCORTICAL SENSORY APHASIA
YOU CAN SPEAK AND REPEAT, BUT YOU DON'T UNDERSTAND...
DON'T UNDERSTAND... HUH?
TRANSCORTICAL MOTOR APHASIA
YOU CAN ONLY REPEAT WHAT I SAY, RIGHT?
ONLY REPEAT WHAT I SAY, RIGHT?
GLOBAL APHASIA

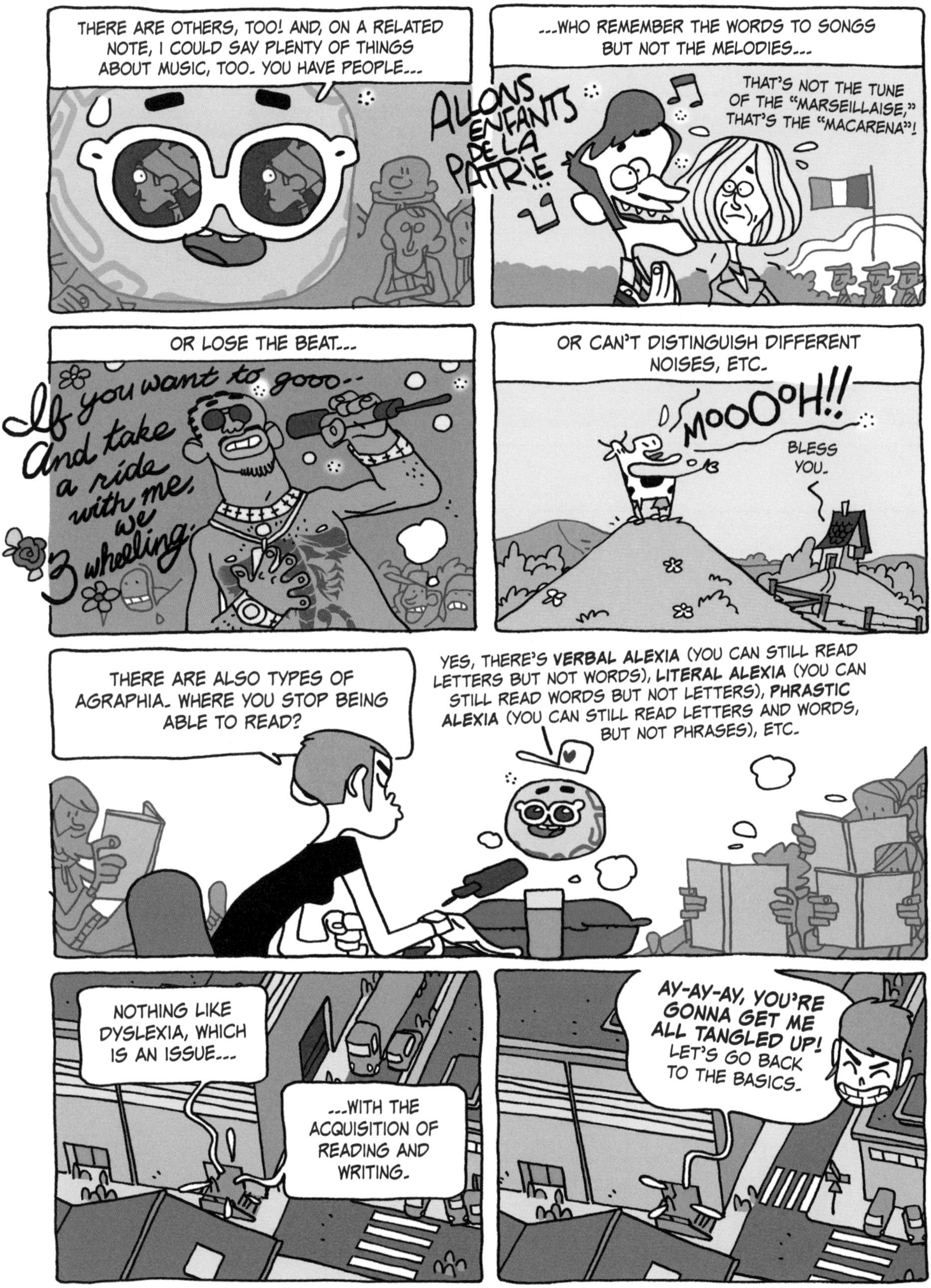

THERE ARE OTHERS, TOO! AND, ON A RELATED NOTE, I COULD SAY PLENTY OF THINGS ABOUT MUSIC, TOO. YOU HAVE PEOPLE...
...WHO REMEMBER THE WORDS TO SONGS BUT NOT THE MELODIES...
ALLONS ENFANTS DE LA PATRIE...
THAT'S NOT THE TUNE OF THE "MARSEILLAISE," THAT'S THE "MACARENA"!
OR LOSE THE BEAT...
If you want to gooo... And take a ride with me, we 3 wheeling.
OR CAN'T DISTINGUISH DIFFERENT NOISES, ETC.
MOOOOH!!
BLESS YOU.
THERE ARE ALSO TYPES OF AGRAPHIA. WHERE YOU STOP BEING ABLE TO READ?
YES, THERE'S VERBAL ALEXIA (YOU CAN STILL READ LETTERS BUT NOT WORDS), LITERAL ALEXIA (YOU CAN STILL READ WORDS BUT NOT LETTERS), PHRASTIC ALEXIA (YOU CAN STILL READ LETTERS AND WORDS, BUT NOT PHRASES), ETC.
NOTHING LIKE DYSLEXIA, WHICH IS AN ISSUE...
...WITH THE ACQUISITION OF READING AND WRITING.
AY-AY-AY, YOU'RE GONNA GET ME ALL TANGLED UP! LET'S GO BACK TO THE BASICS.

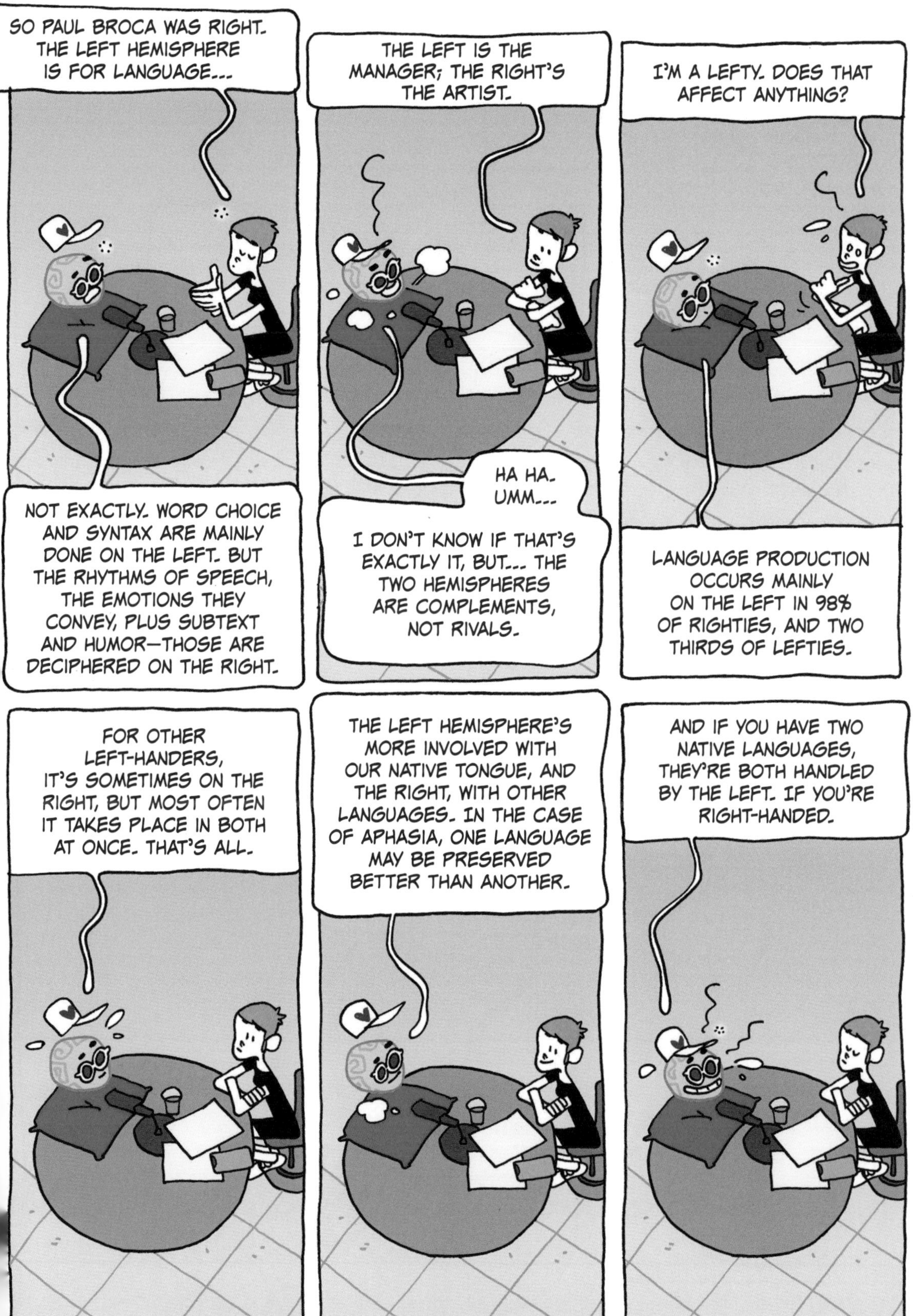
SO PAUL BROCA WAS RIGHT. THE LEFT HEMISPHERE IS FOR LANGUAGE...
NOT EXACTLY. WORD CHOICE AND SYNTAX ARE MAINLY DONE ON THE LEFT. BUT THE RHYTHMS OF SPEECH, THE EMOTIONS THEY CONVEY, PLUS SUBTEXT AND HUMOR—THOSE ARE DECIPHERED ON THE RIGHT.
THE LEFT IS THE MANAGER; THE RIGHT'S THE ARTIST.
HA HA. UMM...
I DON'T KNOW IF THAT'S EXACTLY IT, BUT... THE TWO HEMISPHERES ARE COMPLEMENTS, NOT RIVALS.
I'M A LEFTY. DOES THAT AFFECT ANYTHING?
LANGUAGE PRODUCTION OCCURS MAINLY ON THE LEFT IN 98% OF RIGHTIES, AND TWO THIRDS OF LEFTIES.
FOR OTHER LEFT-HANDERS, IT'S SOMETIMES ON THE RIGHT, BUT MOST OFTEN IT TAKES PLACE IN BOTH AT ONCE. THAT'S ALL.
THE LEFT HEMISPHERE'S MORE INVOLVED WITH OUR NATIVE TONGUE, AND THE RIGHT, WITH OTHER LANGUAGES. IN THE CASE OF APHASIA, ONE LANGUAGE MAY BE PRESERVED BETTER THAN ANOTHER.
AND IF YOU HAVE TWO NATIVE LANGUAGES, THEY'RE BOTH HANDLED BY THE LEFT. IF YOU'RE RIGHT-HANDED.

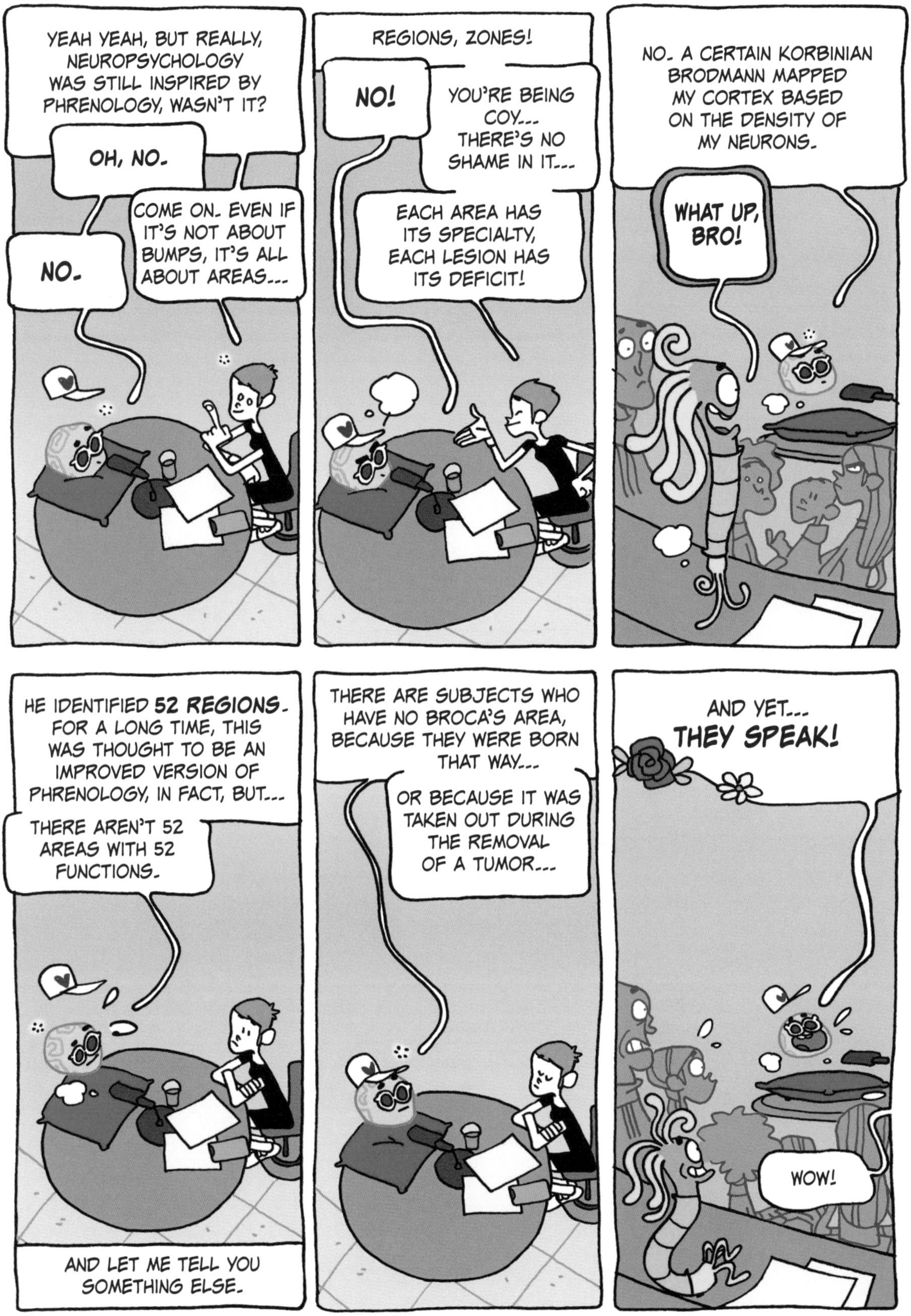
YEAH YEAH, BUT REALLY, NEUROPSYCHOLOGY WAS STILL INSPIRED BY PHRENOLOGY, WASN'T IT?
OH, NO.
NO.
COME ON. EVEN IF IT'S NOT ABOUT BUMPS, IT'S ALL ABOUT AREAS...
REGIONS, ZONES!
NO!
YOU'RE BEING COY... THERE'S NO SHAME IN IT...
EACH AREA HAS ITS SPECIALTY, EACH LESION HAS ITS DEFICIT!
NO. A CERTAIN KORBINIAN BRODMANN MAPPED MY CORTEX BASED ON THE DENSITY OF MY NEURONS.
WHAT UP, BRO!
HE IDENTIFIED **52 REGIONS**. FOR A LONG TIME, THIS WAS THOUGHT TO BE AN IMPROVED VERSION OF PHRENOLOGY, IN FACT, BUT...
THERE AREN'T 52 AREAS WITH 52 FUNCTIONS.
AND LET ME TELL YOU SOMETHING ELSE.
THERE ARE SUBJECTS WHO HAVE NO BROCA'S AREA, BECAUSE THEY WERE BORN THAT WAY...
OR BECAUSE IT WAS TAKEN OUT DURING THE REMOVAL OF A TUMOR...
AND YET... **THEY SPEAK!**
WOW!

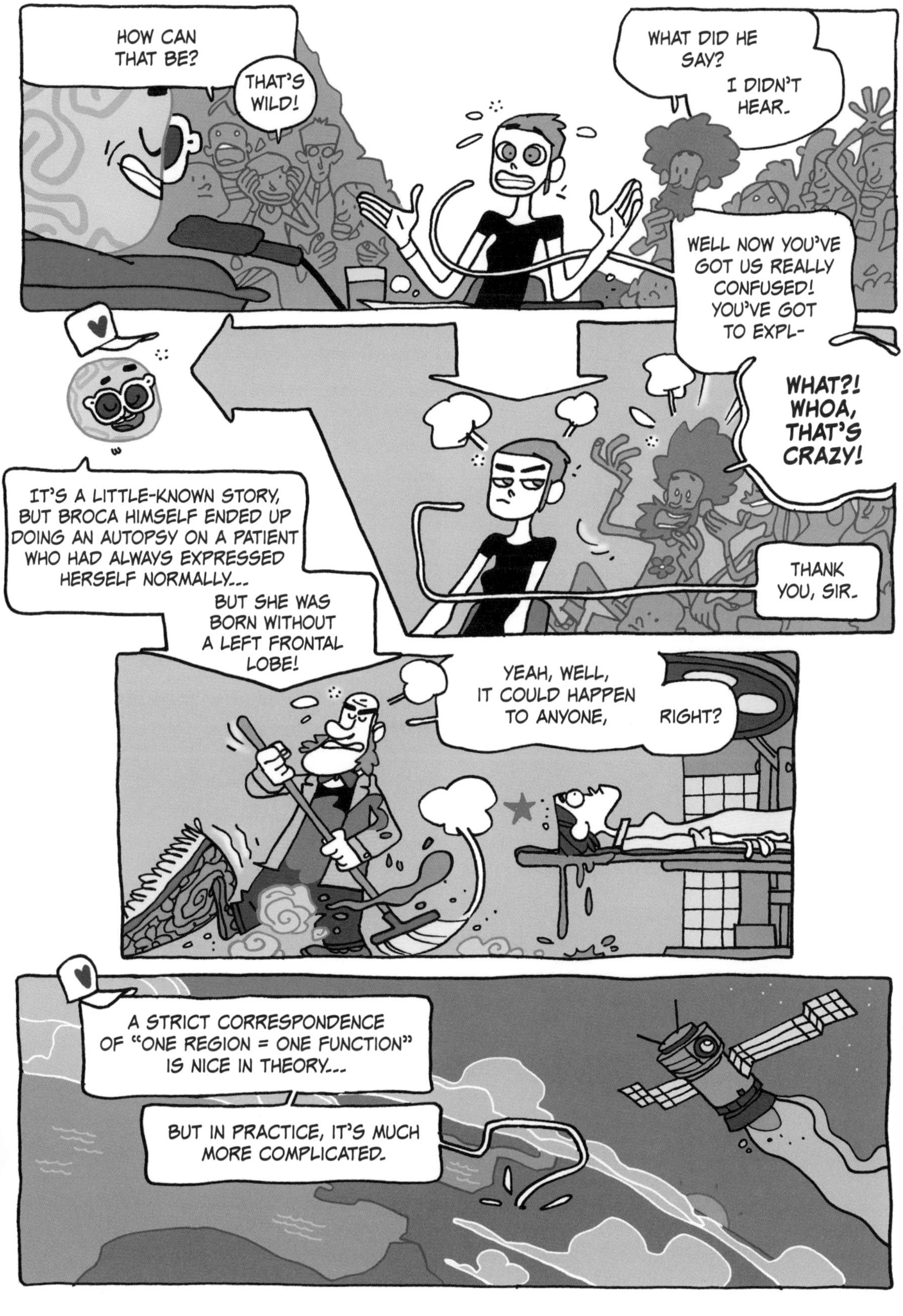

HOW CAN THAT BE?
THAT'S WILD!
WHAT DID HE SAY?
I DIDN'T HEAR.
WELL NOW YOU'VE GOT US REALLY CONFUSED! YOU'VE GOT TO EXPL-
WHAT?! WHOA, THAT'S CRAZY!
THANK YOU, SIR.
IT'S A LITTLE-KNOWN STORY, BUT BROCA HIMSELF ENDED UP DOING AN AUTOPSY ON A PATIENT WHO HAD ALWAYS EXPRESSED HERSELF NORMALLY...
BUT SHE WAS BORN WITHOUT A LEFT FRONTAL LOBE!
YEAH, WELL, IT COULD HAPPEN TO ANYONE,
RIGHT?
A STRICT CORRESPONDENCE OF "ONE REGION = ONE FUNCTION" IS NICE IN THEORY...
BUT IN PRACTICE, IT'S MUCH MORE COMPLICATED.

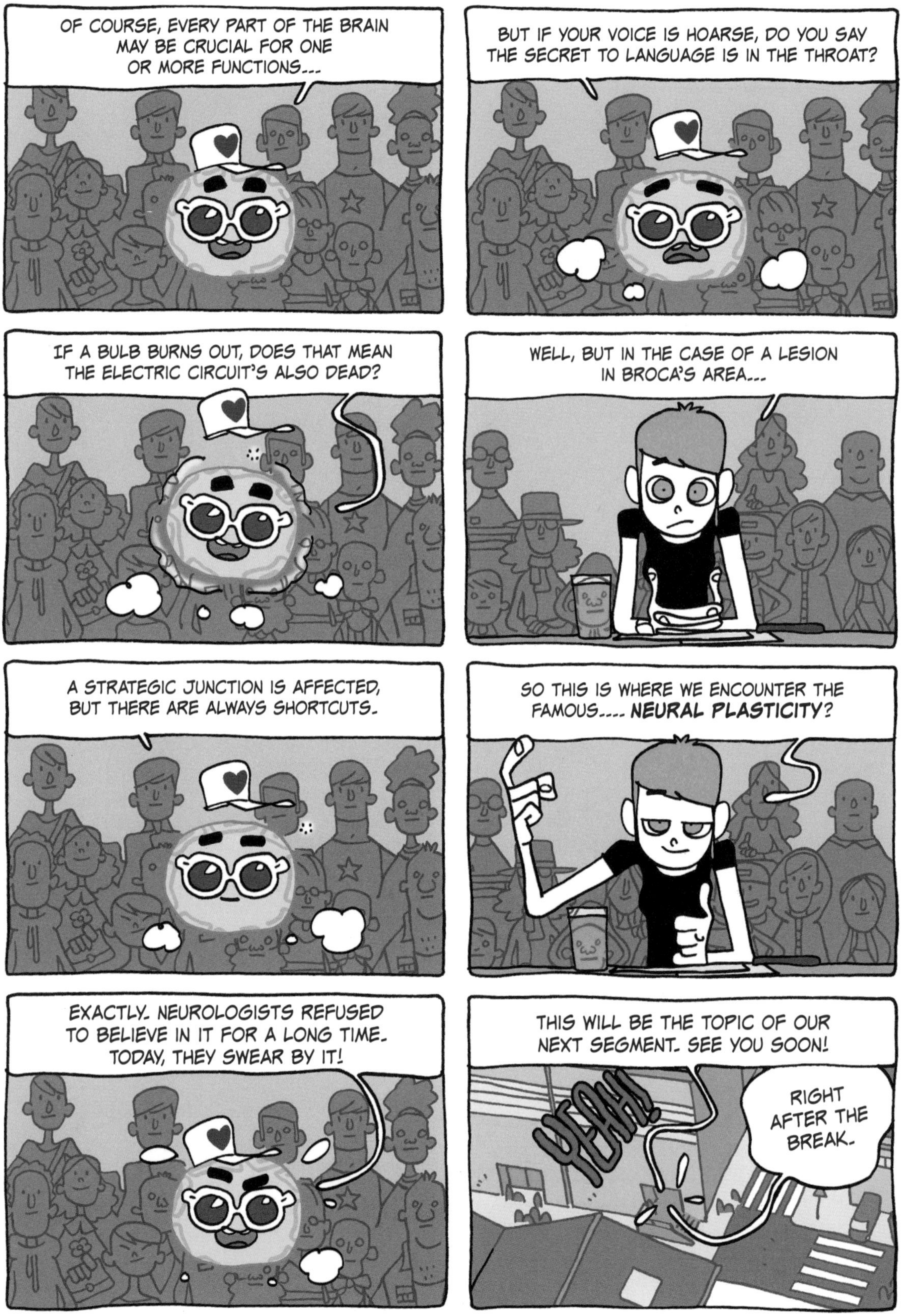

OF COURSE, EVERY PART OF THE BRAIN MAY BE CRUCIAL FOR ONE OR MORE FUNCTIONS...
BUT IF YOUR VOICE IS HOARSE, DO YOU SAY THE SECRET TO LANGUAGE IS IN THE THROAT?
IF A BULB BURNS OUT, DOES THAT MEAN THE ELECTRIC CIRCUIT'S ALSO DEAD?
WELL, BUT IN THE CASE OF A LESION IN BROCA'S AREA...
A STRATEGIC JUNCTION IS AFFECTED, BUT THERE ARE ALWAYS SHORTCUTS.
SO THIS IS WHERE WE ENCOUNTER THE FAMOUS.... **NEURAL PLASTICITY**?
EXACTLY. NEUROLOGISTS REFUSED TO BELIEVE IN IT FOR A LONG TIME. TODAY, THEY SWEAR BY IT!
THIS WILL BE THE TOPIC OF OUR NEXT SEGMENT. SEE YOU SOON!
YEAH!
RIGHT AFTER THE BREAK.

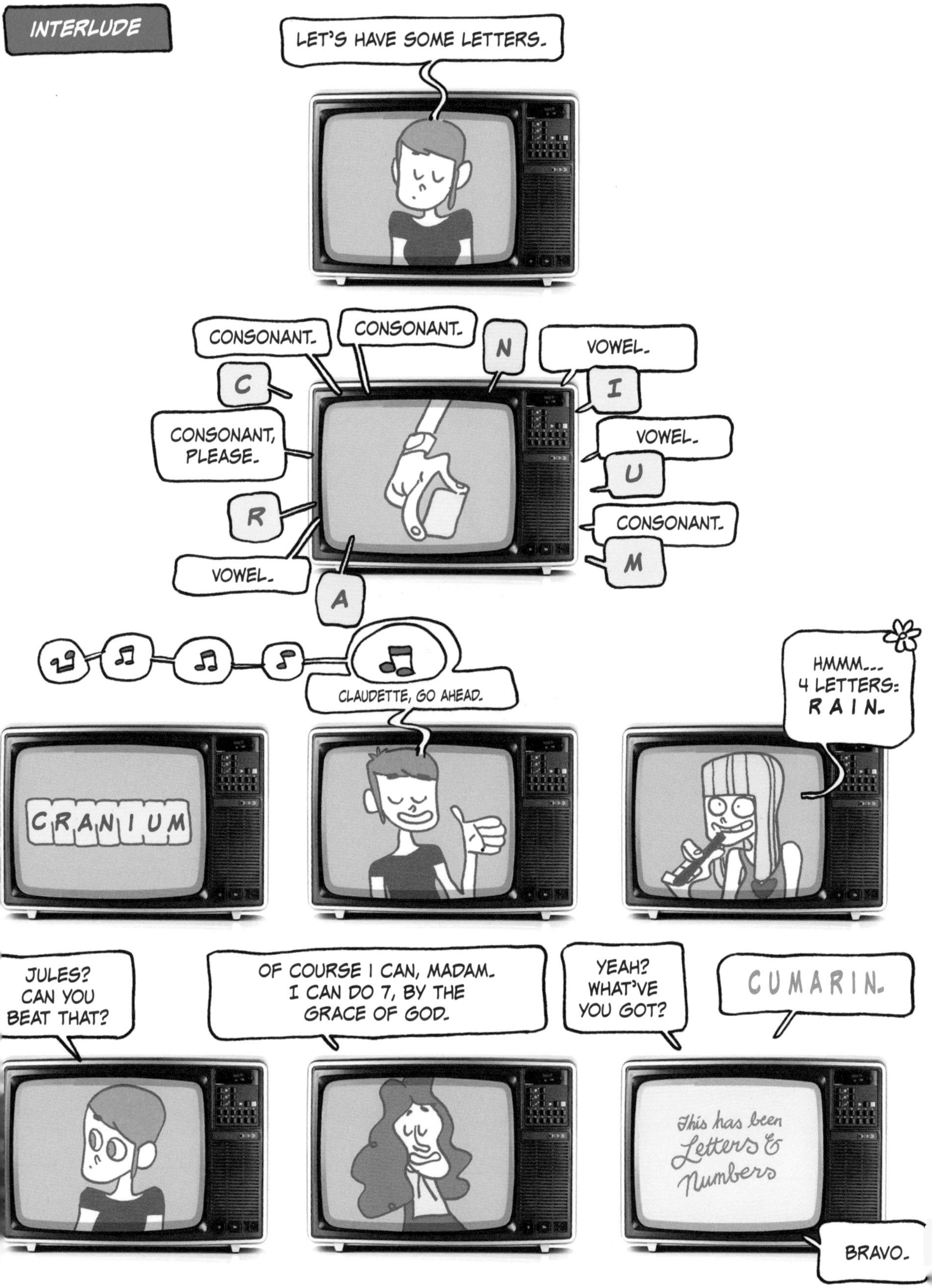
INTERLUDE
LET'S HAVE SOME LETTERS.
CONSONANT.
CONSONANT.
N
VOWEL.
C
I
CONSONANT, PLEASE.
VOWEL.
U
R
CONSONANT.
M
VOWEL.
A
CLAUDETTE, GO AHEAD.
HMMM... 4 LETTERS: R A I N.
CRANIUM
JULES? CAN YOU BEAT THAT?
OF COURSE I CAN, MADAM. I CAN DO 7, BY THE GRACE OF GOD.
YEAH? WHAT'VE YOU GOT?
CUMARIN.
This has been Letters & Numbers
BRAVO.

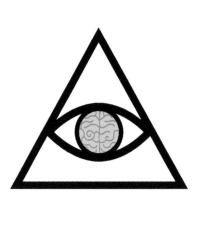

Chapter 7

LIFE, DEATH, AND THE BRAIN RESURRECTED

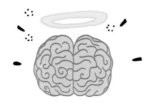

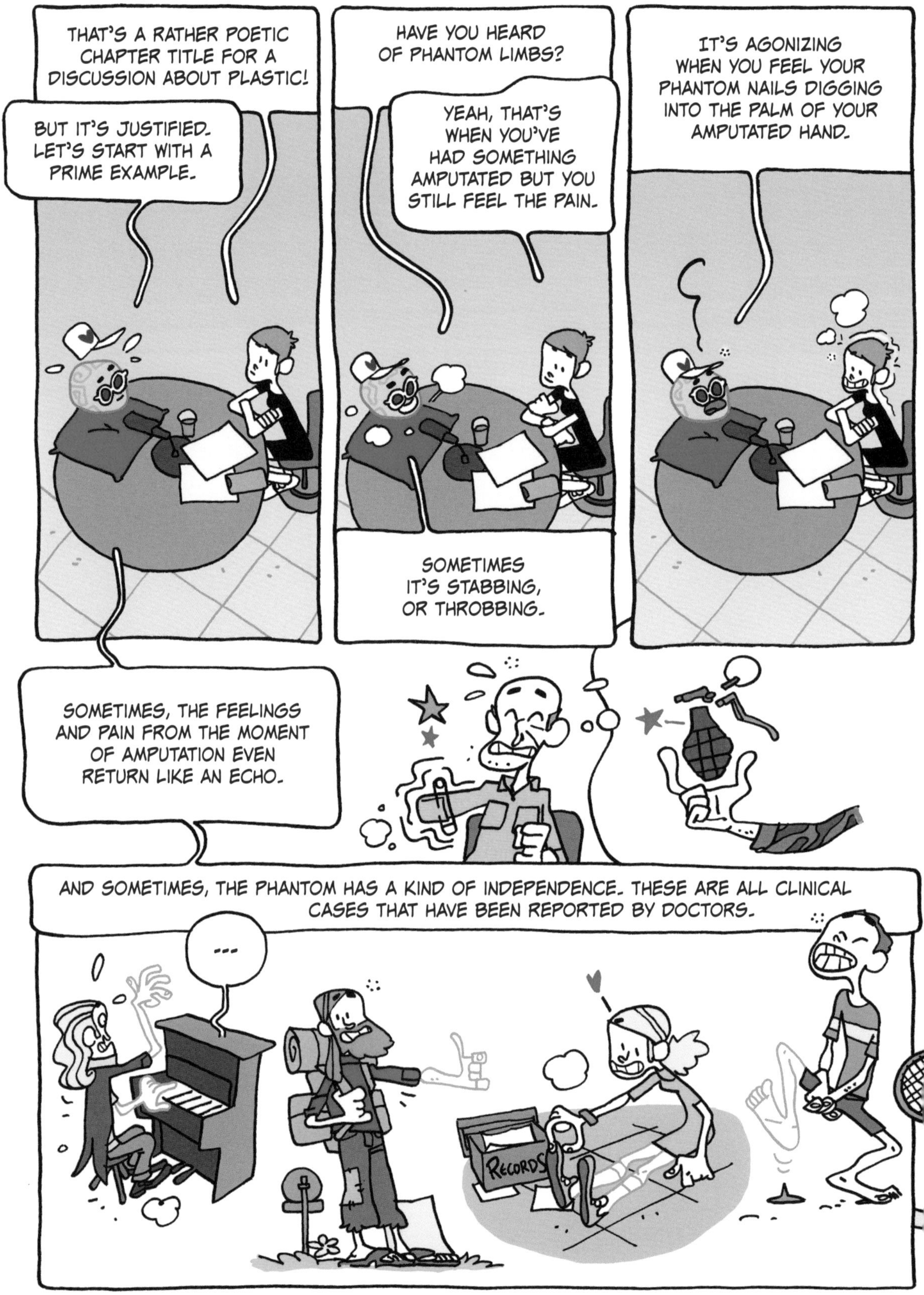

THAT'S A RATHER POETIC CHAPTER TITLE FOR A DISCUSSION ABOUT PLASTIC!
BUT IT'S JUSTIFIED. LET'S START WITH A PRIME EXAMPLE.
HAVE YOU HEARD OF PHANTOM LIMBS?
YEAH, THAT'S WHEN YOU'VE HAD SOMETHING AMPUTATED BUT YOU STILL FEEL THE PAIN.
SOMETIMES IT'S STABBING, OR THROBBING.
IT'S AGONIZING WHEN YOU FEEL YOUR PHANTOM NAILS DIGGING INTO THE PALM OF YOUR AMPUTATED HAND.
SOMETIMES, THE FEELINGS AND PAIN FROM THE MOMENT OF AMPUTATION EVEN RETURN LIKE AN ECHO.
AND SOMETIMES, THE PHANTOM HAS A KIND OF INDEPENDENCE. THESE ARE ALL CLINICAL CASES THAT HAVE BEEN REPORTED BY DOCTORS.
...
RECORDS

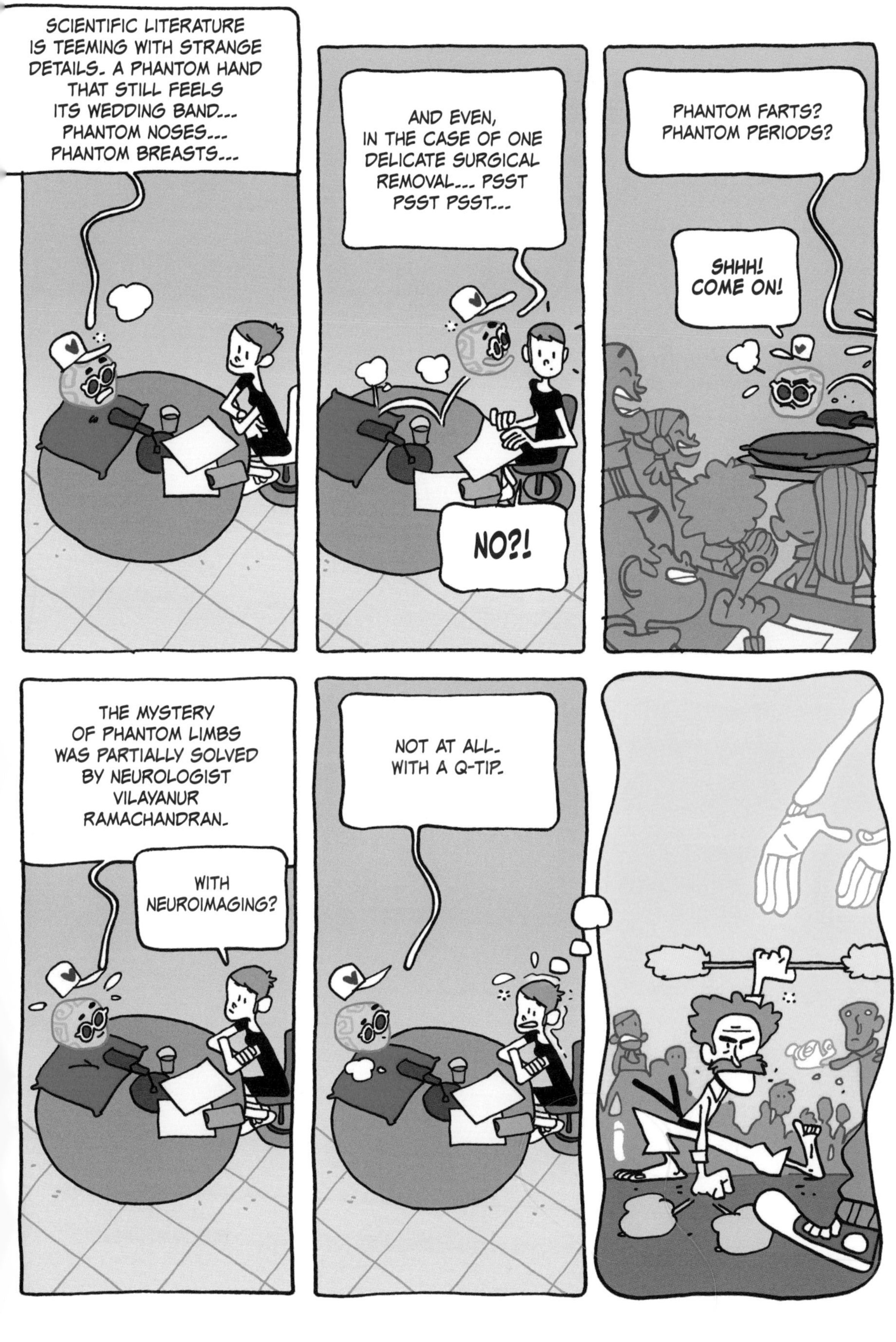
SCIENTIFIC LITERATURE IS TEEMING WITH STRANGE DETAILS. A PHANTOM HAND THAT STILL FEELS ITS WEDDING BAND... PHANTOM NOSES... PHANTOM BREASTS...
AND EVEN, IN THE CASE OF ONE DELICATE SURGICAL REMOVAL... PSST PSST PSST...
NO?!
PHANTOM FARTS? PHANTOM PERIODS?
SHHH! COME ON!
THE MYSTERY OF PHANTOM LIMBS WAS PARTIALLY SOLVED BY NEUROLOGIST VILAYANUR RAMACHANDRAN.
WITH NEUROIMAGING?
NOT AT ALL. WITH A Q-TIP.

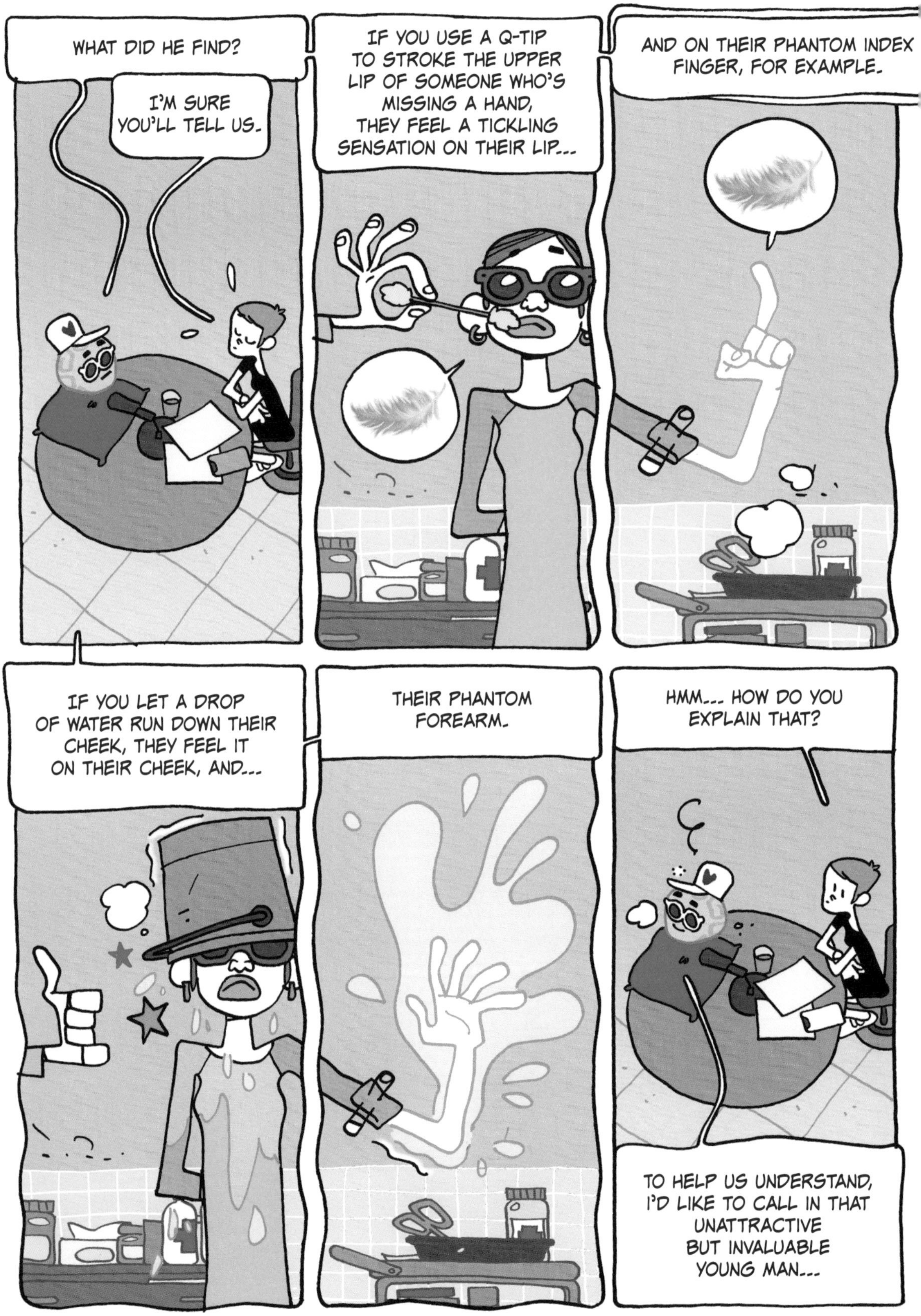
WHAT DID HE FIND?
I'M SURE YOU'LL TELL US.
IF YOU USE A Q-TIP TO STROKE THE UPPER LIP OF SOMEONE WHO'S MISSING A HAND, THEY FEEL A TICKLING SENSATION ON THEIR LIP...
AND ON THEIR PHANTOM INDEX FINGER, FOR EXAMPLE.
IF YOU LET A DROP OF WATER RUN DOWN THEIR CHEEK, THEY FEEL IT ON THEIR CHEEK, AND...
THEIR PHANTOM FOREARM.
HMM... HOW DO YOU EXPLAIN THAT?
TO HELP US UNDERSTAND, I'D LIKE TO CALL IN THAT UNATTRACTIVE BUT INVALUABLE YOUNG MAN...

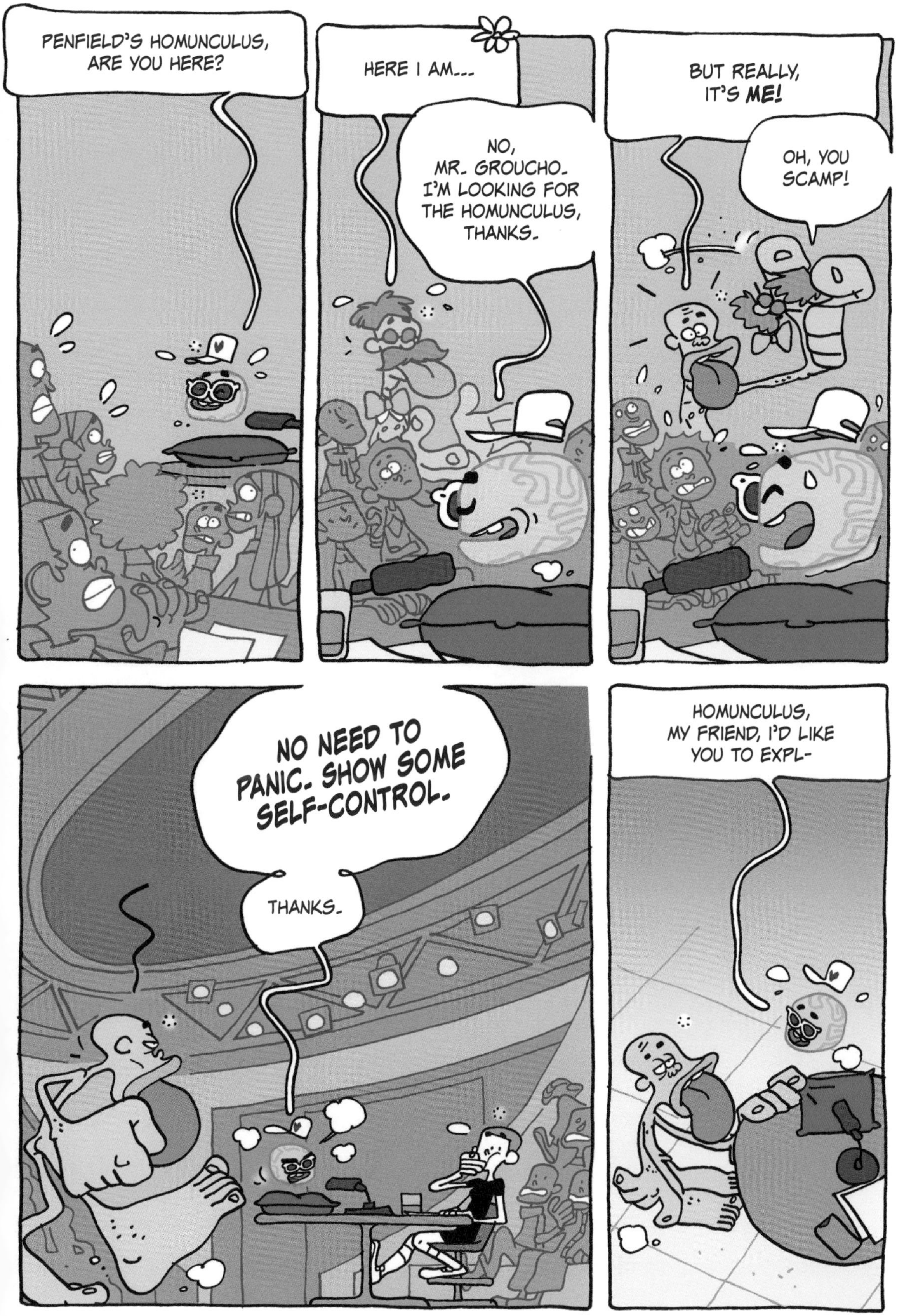
PENFIELD'S HOMUNCULUS, ARE YOU HERE?
HERE I AM...
NO, MR. GROUCHO. I'M LOOKING FOR THE HOMUNCULUS, THANKS.
BUT REALLY, IT'S **ME!**
OH, YOU SCAMP!
NO NEED TO PANIC. SHOW SOME SELF-CONTROL.
THANKS.
HOMUNCULUS, MY FRIEND, I'D LIKE YOU TO EXPL-

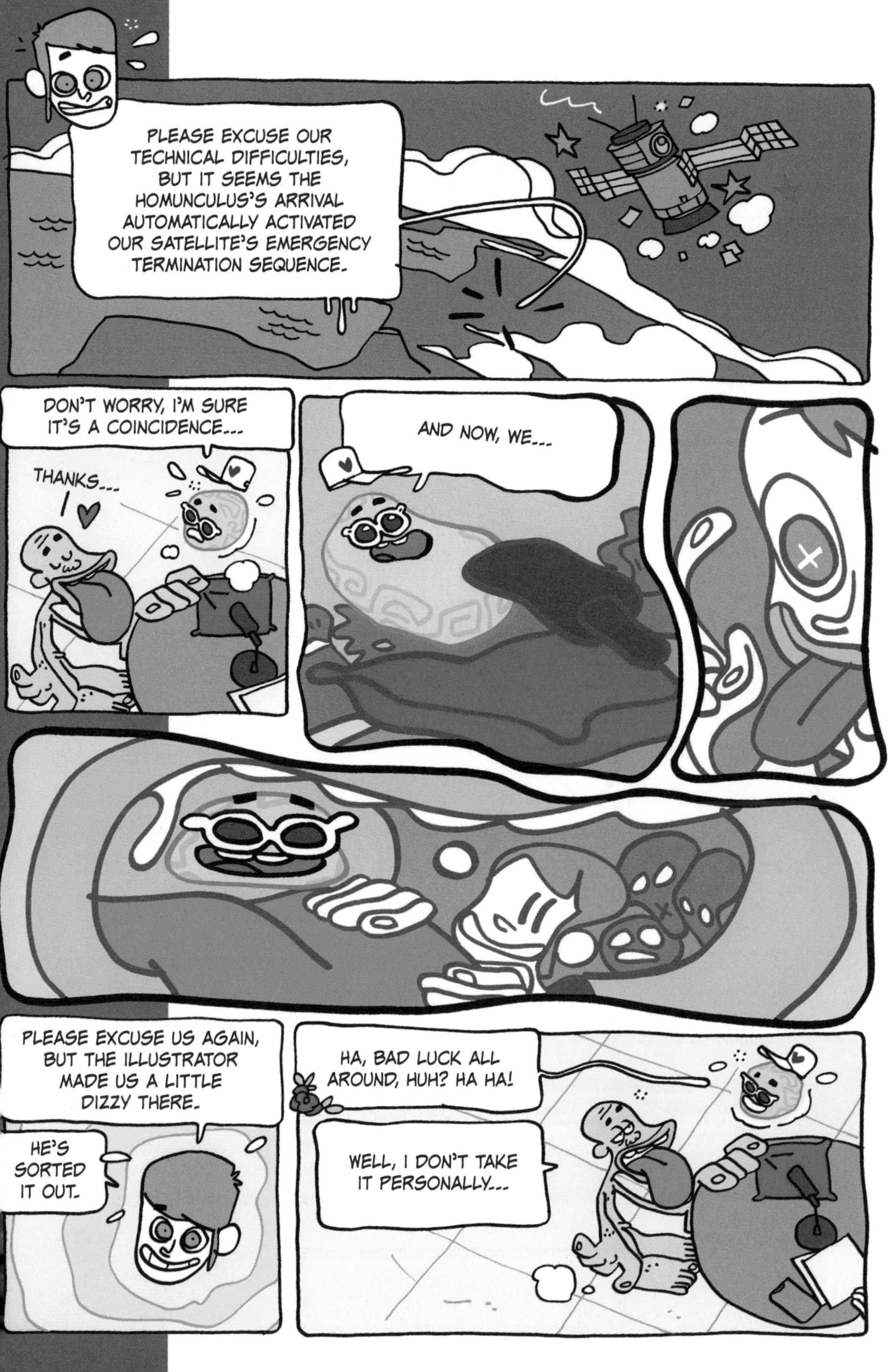
PLEASE EXCUSE OUR TECHNICAL DIFFICULTIES, BUT IT SEEMS THE HOMUNCULUS'S ARRIVAL AUTOMATICALLY ACTIVATED OUR SATELLITE'S EMERGENCY TERMINATION SEQUENCE.
DON'T WORRY, I'M SURE IT'S A COINCIDENCE...
THANKS...
AND NOW, WE...
PLEASE EXCUSE US AGAIN, BUT THE ILLUSTRATOR MADE US A LITTLE DIZZY THERE.
HE'S SORTED IT OUT.
HA, BAD LUCK ALL AROUND, HUH? HA HA!
WELL, I DON'T TAKE IT PERSONALLY...

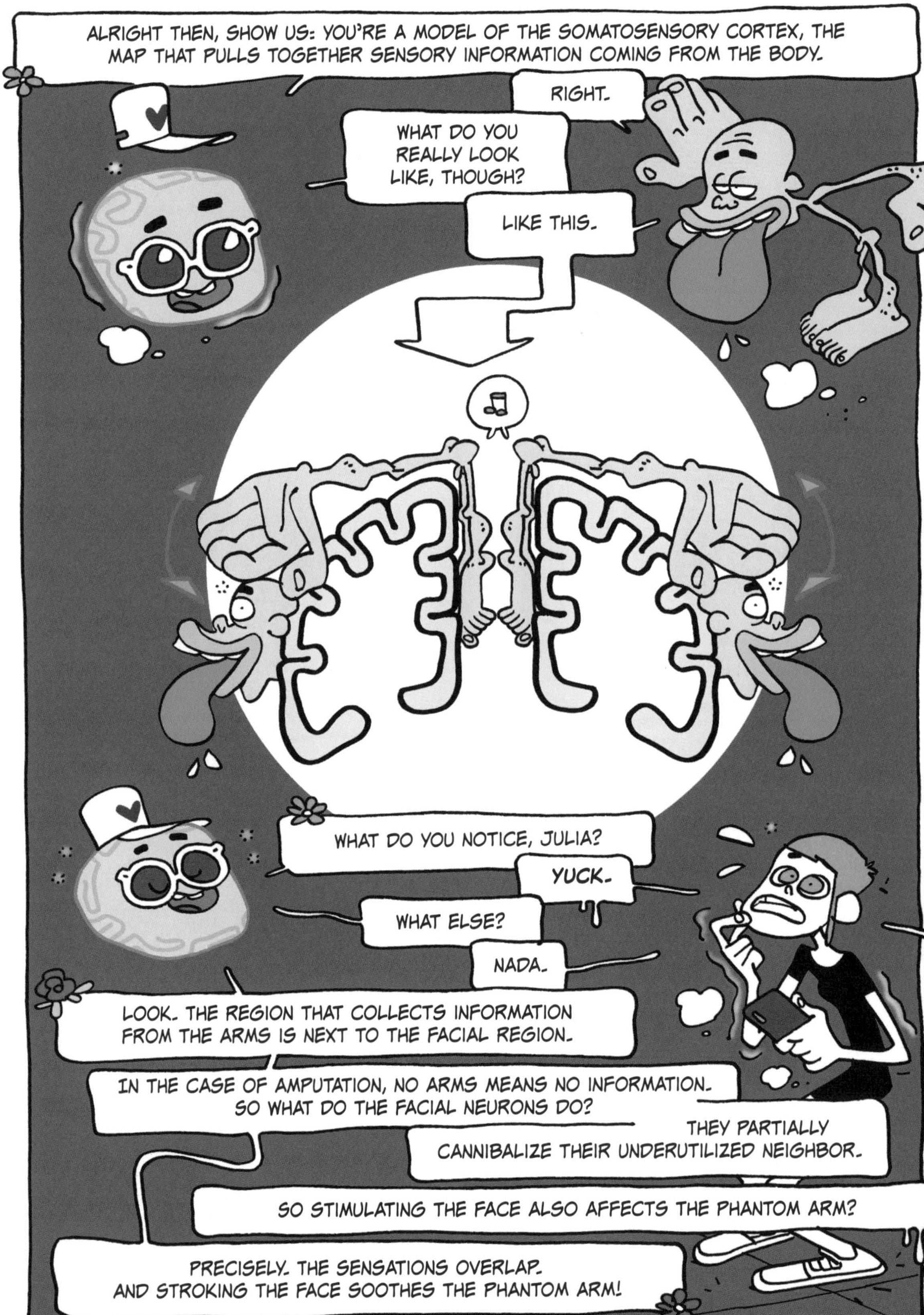
ALRIGHT THEN, SHOW US: YOU'RE A MODEL OF THE SOMATOSENSORY CORTEX, THE MAP THAT PULLS TOGETHER SENSORY INFORMATION COMING FROM THE BODY.
RIGHT.
WHAT DO YOU REALLY LOOK LIKE, THOUGH?
LIKE THIS.
WHAT DO YOU NOTICE, JULIA?
YUCK.
WHAT ELSE?
NADA.
LOOK. THE REGION THAT COLLECTS INFORMATION FROM THE ARMS IS NEXT TO THE FACIAL REGION.
IN THE CASE OF AMPUTATION, NO ARMS MEANS NO INFORMATION. SO WHAT DO THE FACIAL NEURONS DO?
THEY PARTIALLY CANNIBALIZE THEIR UNDERUTILIZED NEIGHBOR.
SO STIMULATING THE FACE ALSO AFFECTS THE PHANTOM ARM?
PRECISELY. THE SENSATIONS OVERLAP. AND STROKING THE FACE SOOTHES THE PHANTOM ARM!

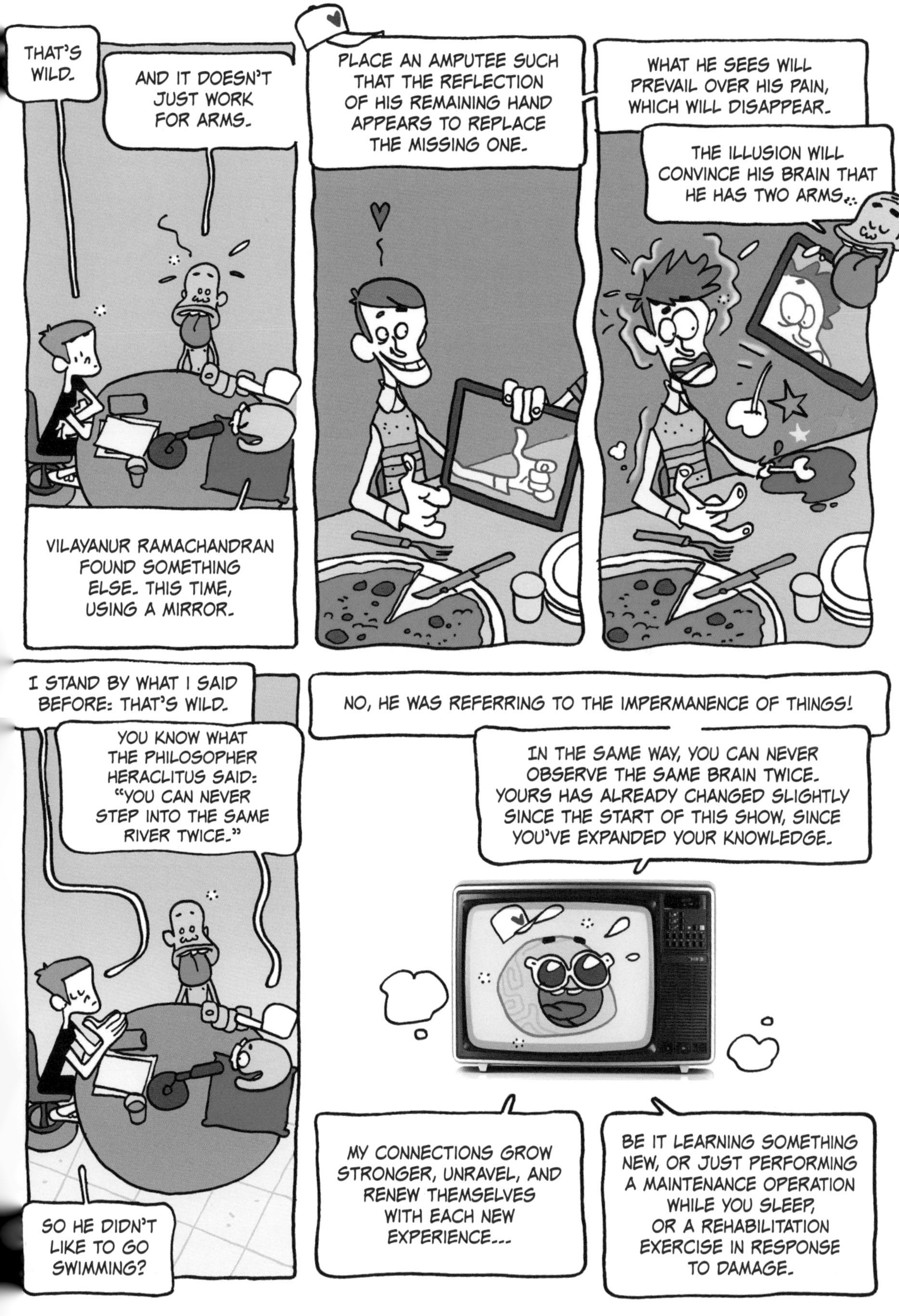
THAT'S WILD.
AND IT DOESN'T JUST WORK FOR ARMS.
VILAYANUR RAMACHANDRAN FOUND SOMETHING ELSE. THIS TIME, USING A MIRROR.
PLACE AN AMPUTEE SUCH THAT THE REFLECTION OF HIS REMAINING HAND APPEARS TO REPLACE THE MISSING ONE.
WHAT HE SEES WILL PREVAIL OVER HIS PAIN, WHICH WILL DISAPPEAR.
THE ILLUSION WILL CONVINCE HIS BRAIN THAT HE HAS TWO ARMS.
I STAND BY WHAT I SAID BEFORE: THAT'S WILD.
YOU KNOW WHAT THE PHILOSOPHER HERACLITUS SAID: "YOU CAN NEVER STEP INTO THE SAME RIVER TWICE."
SO HE DIDN'T LIKE TO GO SWIMMING?
NO, HE WAS REFERRING TO THE IMPERMANENCE OF THINGS!
IN THE SAME WAY, YOU CAN NEVER OBSERVE THE SAME BRAIN TWICE. YOURS HAS ALREADY CHANGED SLIGHTLY SINCE THE START OF THIS SHOW, SINCE YOU'VE EXPANDED YOUR KNOWLEDGE.
MY CONNECTIONS GROW STRONGER, UNRAVEL, AND RENEW THEMSELVES WITH EACH NEW EXPERIENCE...
BE IT LEARNING SOMETHING NEW, OR JUST PERFORMING A MAINTENANCE OPERATION WHILE YOU SLEEP, OR A REHABILITATION EXERCISE IN RESPONSE TO DAMAGE.

THIS EVERYDAY PLASTICITY BEGINS IN UTERO AND CONTINUES TO YOUR DYING BREATH...

IN THE WOMB THERE IS A VERITABLE BIG BANG OF NEURONS: UP TO 250,000 NEW NEURONS ARE PRODUCED EVERY MINUTE!

YEAH! WICKED!

AT BIRTH, I HAVE NEARLY ALL MY NEURONS. BUT I'LL CONSTANTLY CREATE NEW CONNECTIONS AMONG THEM...

THROUGH THE EXPERIENCES OF CHILDHOOD.

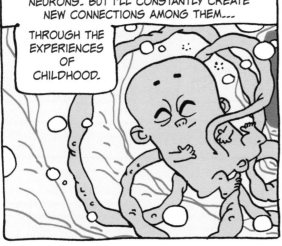

SOME UNNECESSARY ONES WILL DISAPPEAR, AND THE FREED-UP NEURONS WILL BE FITTED INTO OTHER CIRCUITS.

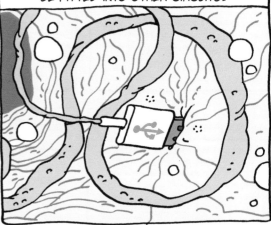

I'M CONSTANTLY CUTTING, RECONNECTING, PRUNING, IMPROVING...

THIS METHOD IS APPROVED BY JULES HARDOUIN-MANSART.

NOBEL LAUREATE GERALD EDELMAN DESCRIBED THIS PERMANENT ADAPTATION AS "NEURAL DARWINISM." THAT'S ONE WAY OF LOOKING AT IT.

THE HUMAN BRAIN, IT'S AWESOME.

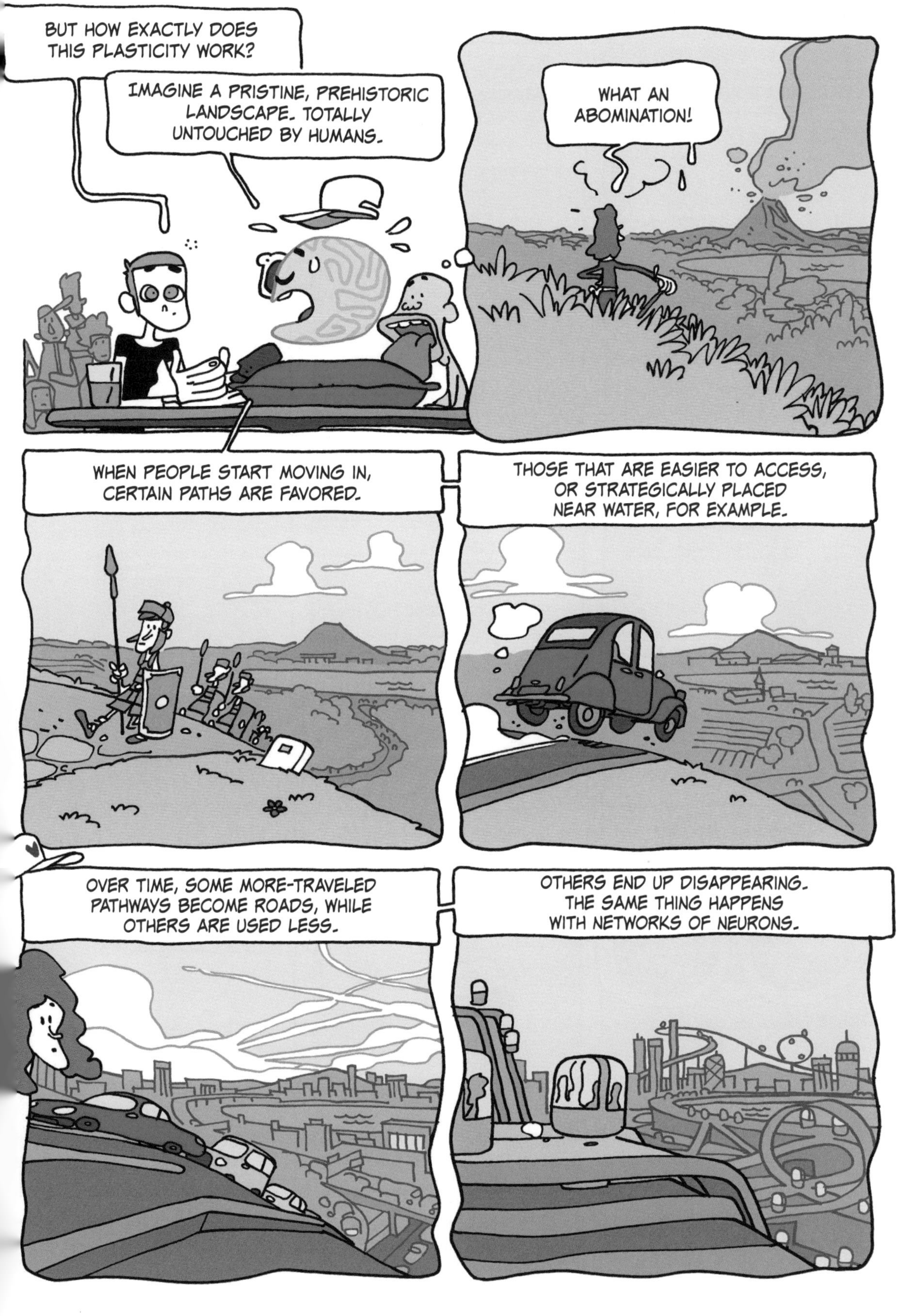
BUT HOW EXACTLY DOES THIS PLASTICITY WORK?
IMAGINE A PRISTINE, PREHISTORIC LANDSCAPE. TOTALLY UNTOUCHED BY HUMANS.
WHAT AN ABOMINATION!
WHEN PEOPLE START MOVING IN, CERTAIN PATHS ARE FAVORED.
THOSE THAT ARE EASIER TO ACCESS, OR STRATEGICALLY PLACED NEAR WATER, FOR EXAMPLE.
OVER TIME, SOME MORE-TRAVELED PATHWAYS BECOME ROADS, WHILE OTHERS ARE USED LESS.
OTHERS END UP DISAPPEARING. THE SAME THING HAPPENS WITH NETWORKS OF NEURONS.

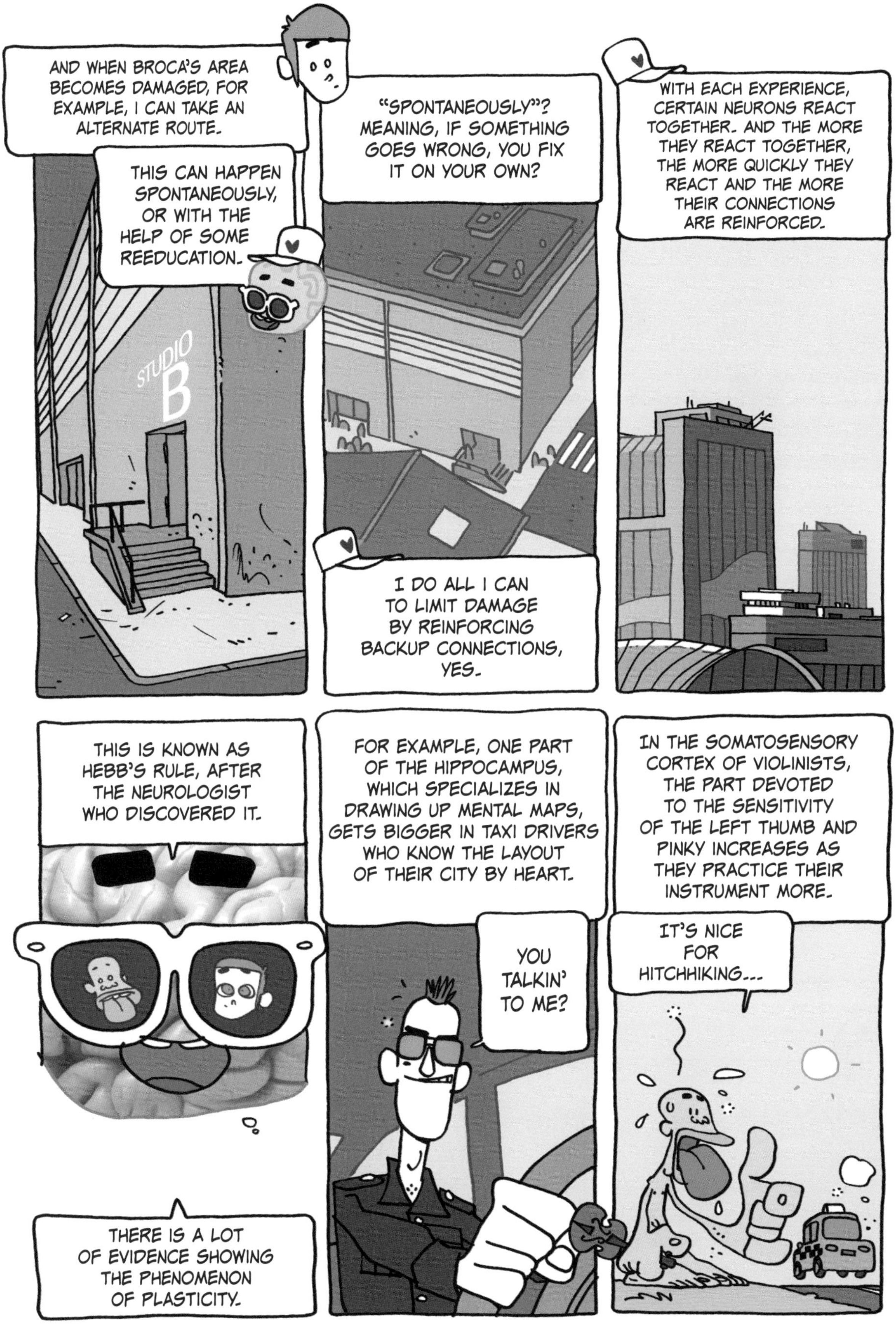

AND WHEN BROCA'S AREA BECOMES DAMAGED, FOR EXAMPLE, I CAN TAKE AN ALTERNATE ROUTE.
THIS CAN HAPPEN SPONTANEOUSLY, OR WITH THE HELP OF SOME REEDUCATION.
STUDIO B
"SPONTANEOUSLY"? MEANING, IF SOMETHING GOES WRONG, YOU FIX IT ON YOUR OWN?
I DO ALL I CAN TO LIMIT DAMAGE BY REINFORCING BACKUP CONNECTIONS, YES.
WITH EACH EXPERIENCE, CERTAIN NEURONS REACT TOGETHER. AND THE MORE THEY REACT TOGETHER, THE MORE QUICKLY THEY REACT AND THE MORE THEIR CONNECTIONS ARE REINFORCED.
THIS IS KNOWN AS HEBB'S RULE, AFTER THE NEUROLOGIST WHO DISCOVERED IT.
THERE IS A LOT OF EVIDENCE SHOWING THE PHENOMENON OF PLASTICITY.
FOR EXAMPLE, ONE PART OF THE HIPPOCAMPUS, WHICH SPECIALIZES IN DRAWING UP MENTAL MAPS, GETS BIGGER IN TAXI DRIVERS WHO KNOW THE LAYOUT OF THEIR CITY BY HEART.
YOU TALKIN' TO ME?
IN THE SOMATOSENSORY CORTEX OF VIOLINISTS, THE PART DEVOTED TO THE SENSITIVITY OF THE LEFT THUMB AND PINKY INCREASES AS THEY PRACTICE THEIR INSTRUMENT MORE.
IT'S NICE FOR HITCHHIKING...

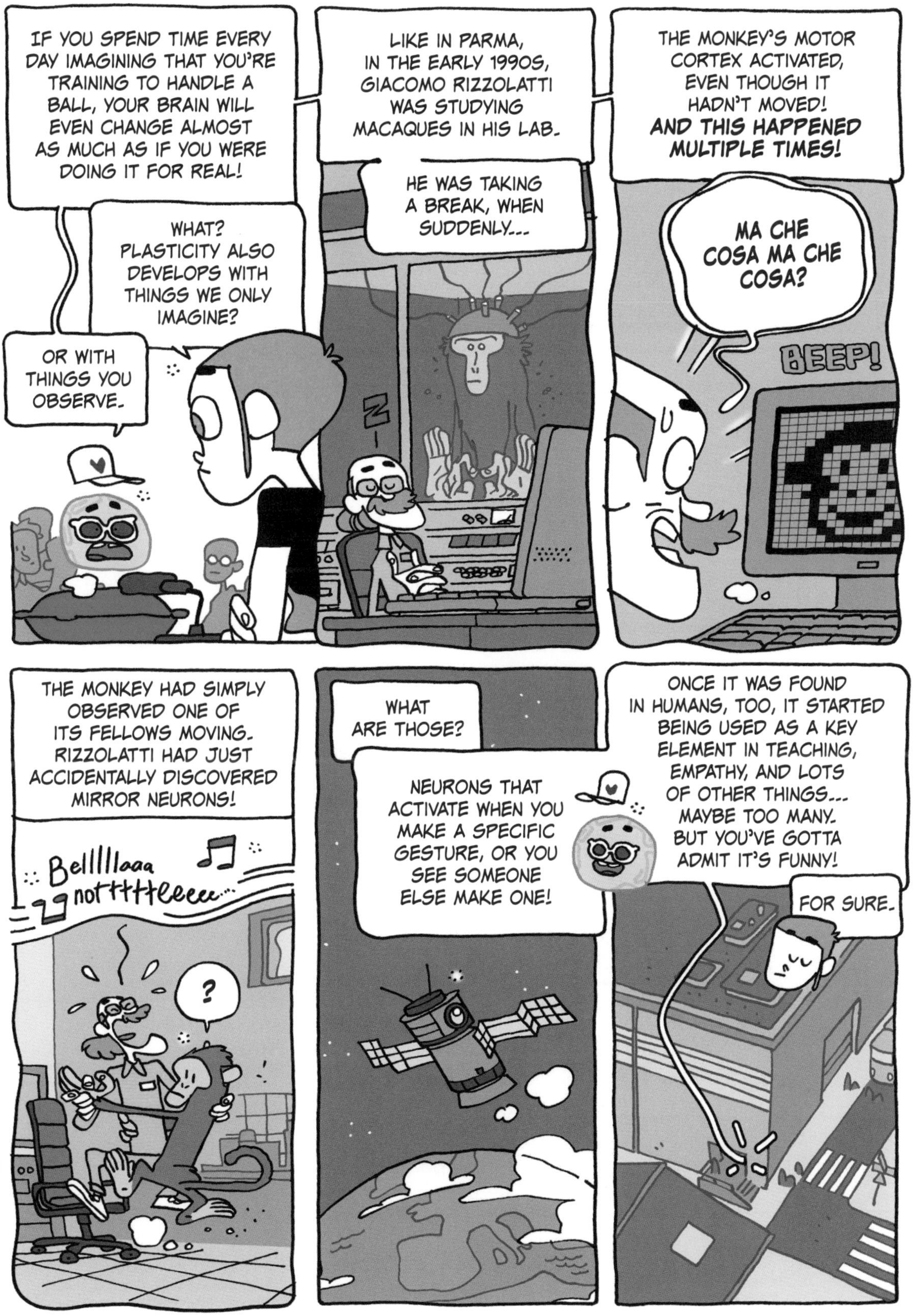
IF YOU SPEND TIME EVERY DAY IMAGINING THAT YOU'RE TRAINING TO HANDLE A BALL, YOUR BRAIN WILL EVEN CHANGE ALMOST AS MUCH AS IF YOU WERE DOING IT FOR REAL!
WHAT? PLASTICITY ALSO DEVELOPS WITH THINGS WE ONLY IMAGINE?
OR WITH THINGS YOU OBSERVE.
LIKE IN PARMA, IN THE EARLY 1990S, GIACOMO RIZZOLATTI WAS STUDYING MACAQUES IN HIS LAB.
HE WAS TAKING A BREAK, WHEN SUDDENLY...
THE MONKEY'S MOTOR CORTEX ACTIVATED, EVEN THOUGH IT HADN'T MOVED! **AND THIS HAPPENED MULTIPLE TIMES!**
MA CHE COSA MA CHE COSA?
BEEP!
THE MONKEY HAD SIMPLY OBSERVED ONE OF ITS FELLOWS MOVING. RIZZOLATTI HAD JUST ACCIDENTALLY DISCOVERED MIRROR NEURONS!
Belllllaaa notttttteeee...
?
WHAT ARE THOSE?
NEURONS THAT ACTIVATE WHEN YOU MAKE A SPECIFIC GESTURE, OR YOU SEE SOMEONE ELSE MAKE ONE!
ONCE IT WAS FOUND IN HUMANS, TOO, IT STARTED BEING USED AS A KEY ELEMENT IN TEACHING, EMPATHY, AND LOTS OF OTHER THINGS... MAYBE TOO MANY. BUT YOU'VE GOTTA ADMIT IT'S FUNNY!
FOR SURE.

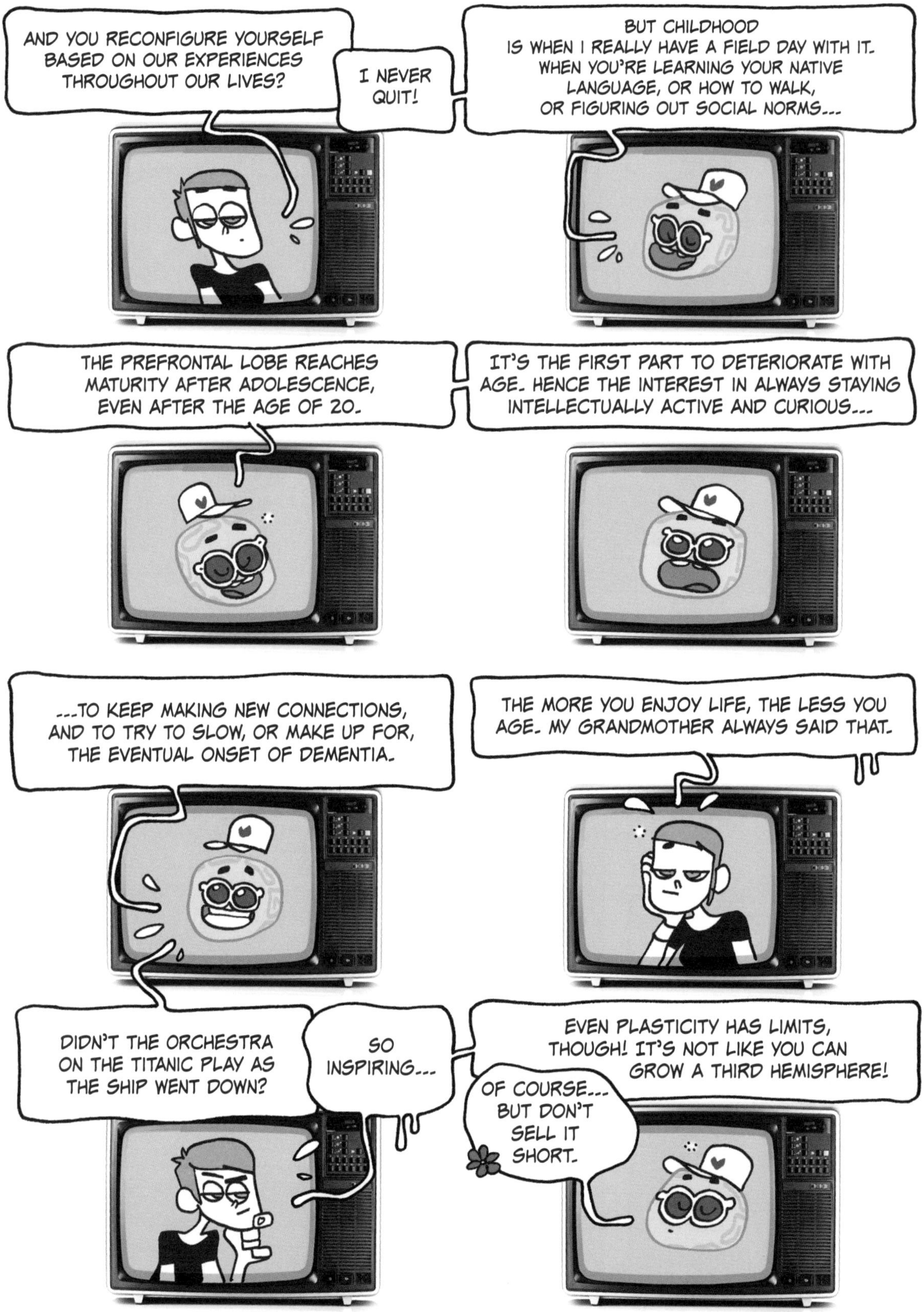
AND YOU RECONFIGURE YOURSELF BASED ON OUR EXPERIENCES THROUGHOUT OUR LIVES?
I NEVER QUIT!
BUT CHILDHOOD IS WHEN I REALLY HAVE A FIELD DAY WITH IT. WHEN YOU'RE LEARNING YOUR NATIVE LANGUAGE, OR HOW TO WALK, OR FIGURING OUT SOCIAL NORMS...
THE PREFRONTAL LOBE REACHES MATURITY AFTER ADOLESCENCE, EVEN AFTER THE AGE OF 20.
IT'S THE FIRST PART TO DETERIORATE WITH AGE. HENCE THE INTEREST IN ALWAYS STAYING INTELLECTUALLY ACTIVE AND CURIOUS...
...TO KEEP MAKING NEW CONNECTIONS, AND TO TRY TO SLOW, OR MAKE UP FOR, THE EVENTUAL ONSET OF DEMENTIA.
THE MORE YOU ENJOY LIFE, THE LESS YOU AGE. MY GRANDMOTHER ALWAYS SAID THAT.
DIDN'T THE ORCHESTRA ON THE TITANIC PLAY AS THE SHIP WENT DOWN?
SO INSPIRING...
EVEN PLASTICITY HAS LIMITS, THOUGH! IT'S NOT LIKE YOU CAN GROW A THIRD HEMISPHERE!
OF COURSE... BUT DON'T SELL IT SHORT.

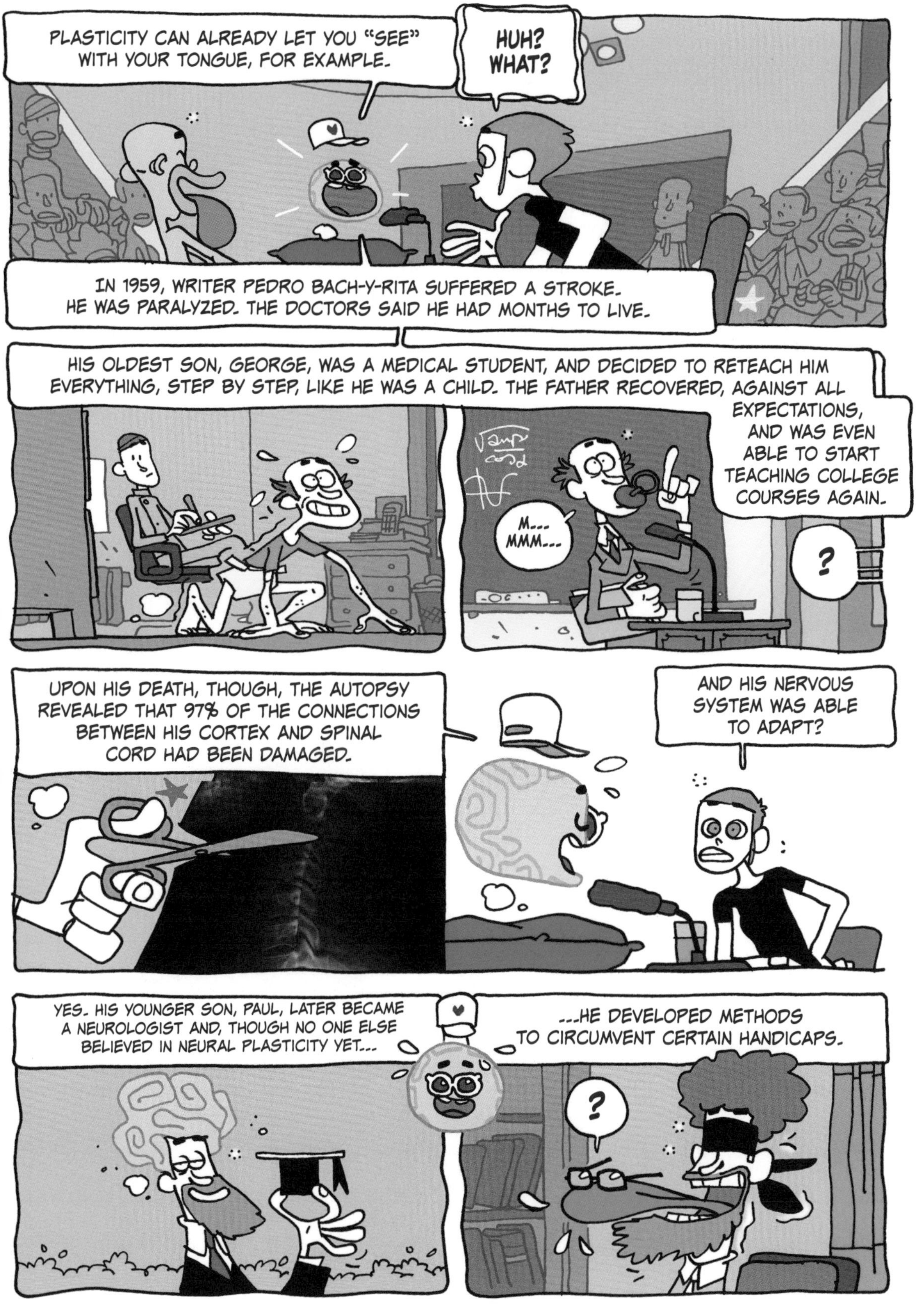
PLASTICITY CAN ALREADY LET YOU "SEE" WITH YOUR TONGUE, FOR EXAMPLE.
HUH? WHAT?
IN 1959, WRITER PEDRO BACH-Y-RITA SUFFERED A STROKE. HE WAS PARALYZED. THE DOCTORS SAID HE HAD MONTHS TO LIVE.
HIS OLDEST SON, GEORGE, WAS A MEDICAL STUDENT, AND DECIDED TO RETEACH HIM EVERYTHING, STEP BY STEP, LIKE HE WAS A CHILD. THE FATHER RECOVERED, AGAINST ALL EXPECTATIONS,
AND WAS EVEN ABLE TO START TEACHING COLLEGE COURSES AGAIN.
M... MMM...
?
UPON HIS DEATH, THOUGH, THE AUTOPSY REVEALED THAT 97% OF THE CONNECTIONS BETWEEN HIS CORTEX AND SPINAL CORD HAD BEEN DAMAGED.
AND HIS NERVOUS SYSTEM WAS ABLE TO ADAPT?
YES. HIS YOUNGER SON, PAUL, LATER BECAME A NEUROLOGIST AND, THOUGH NO ONE ELSE BELIEVED IN NEURAL PLASTICITY YET...
...HE DEVELOPED METHODS TO CIRCUMVENT CERTAIN HANDICAPS.
?

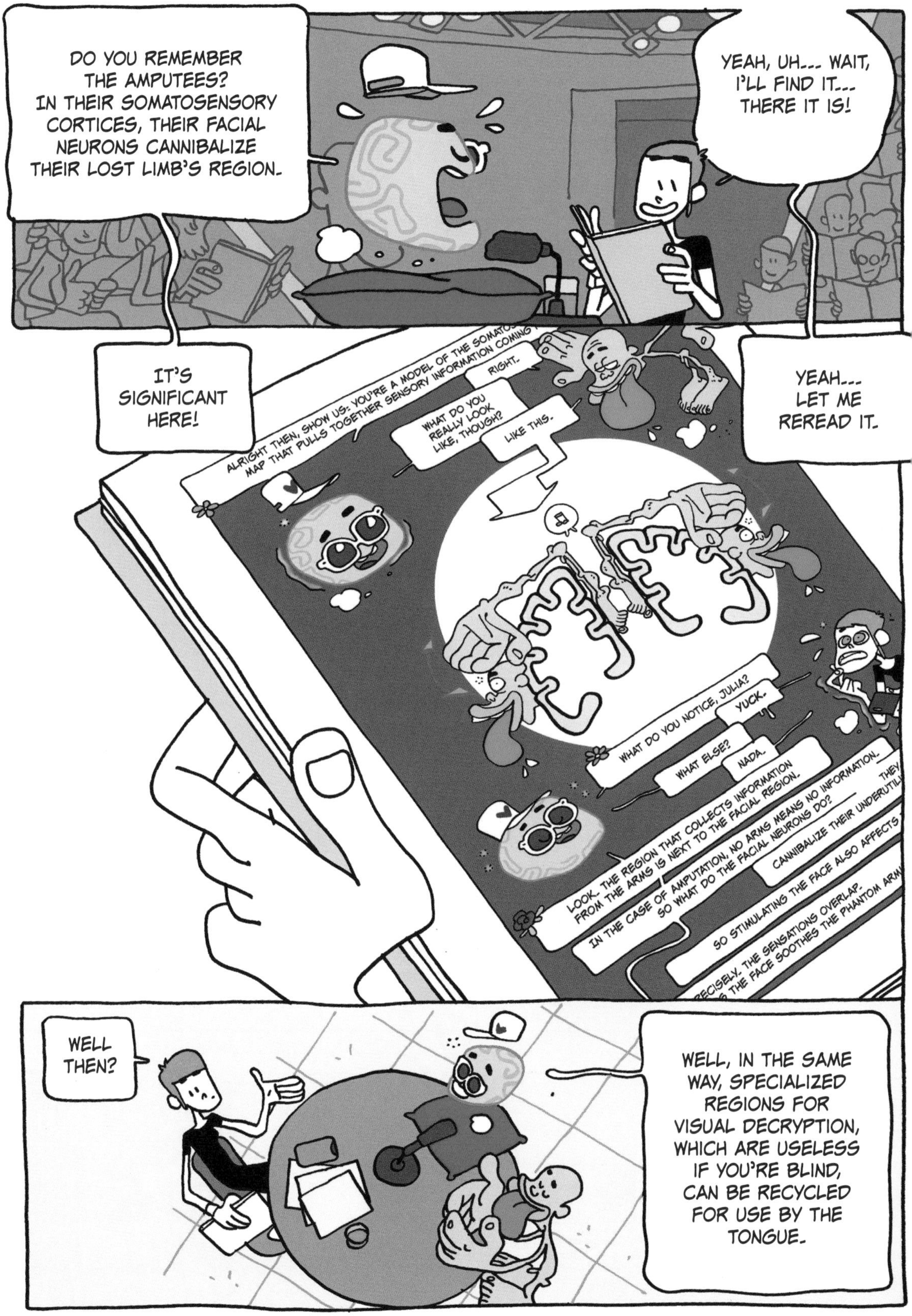

DO YOU REMEMBER THE AMPUTEES? IN THEIR SOMATOSENSORY CORTICES, THEIR FACIAL NEURONS CANNIBALIZE THEIR LOST LIMB'S REGION.
YEAH, UH... WAIT, I'LL FIND IT... THERE IT IS!
IT'S SIGNIFICANT HERE!
YEAH... LET ME REREAD IT.
ALRIGHT THEN, SHOW US: YOU'RE A MODEL OF THE SOMATO
MAP THAT PULLS TOGETHER SENSORY INFORMATION COMING
RIGHT.
WHAT DO YOU REALLY LOOK LIKE, THOUGH?
LIKE THIS.
WHAT DO YOU NOTICE, JULIA?
YUCK.
WHAT ELSE?
NADA.
LOOK. THE REGION THAT COLLECTS INFORMATION FROM THE ARMS IS NEXT TO THE FACIAL REGION.
IN THE CASE OF AMPUTATION, NO ARMS MEANS NO INFORMATION. SO WHAT DO THE FACIAL NEURONS DO?
CANNIBALIZE THEIR UNDERUTIL
SO STIMULATING THE FACE ALSO AFFECTS
THE SENSATIONS OVERLAP.
THE FACE SOOTHES THE PHANTOM ARM
WELL THEN?
WELL, IN THE SAME WAY, SPECIALIZED REGIONS FOR VISUAL DECRYPTION, WHICH ARE USELESS IF YOU'RE BLIND, CAN BE RECYCLED FOR USE BY THE TONGUE.

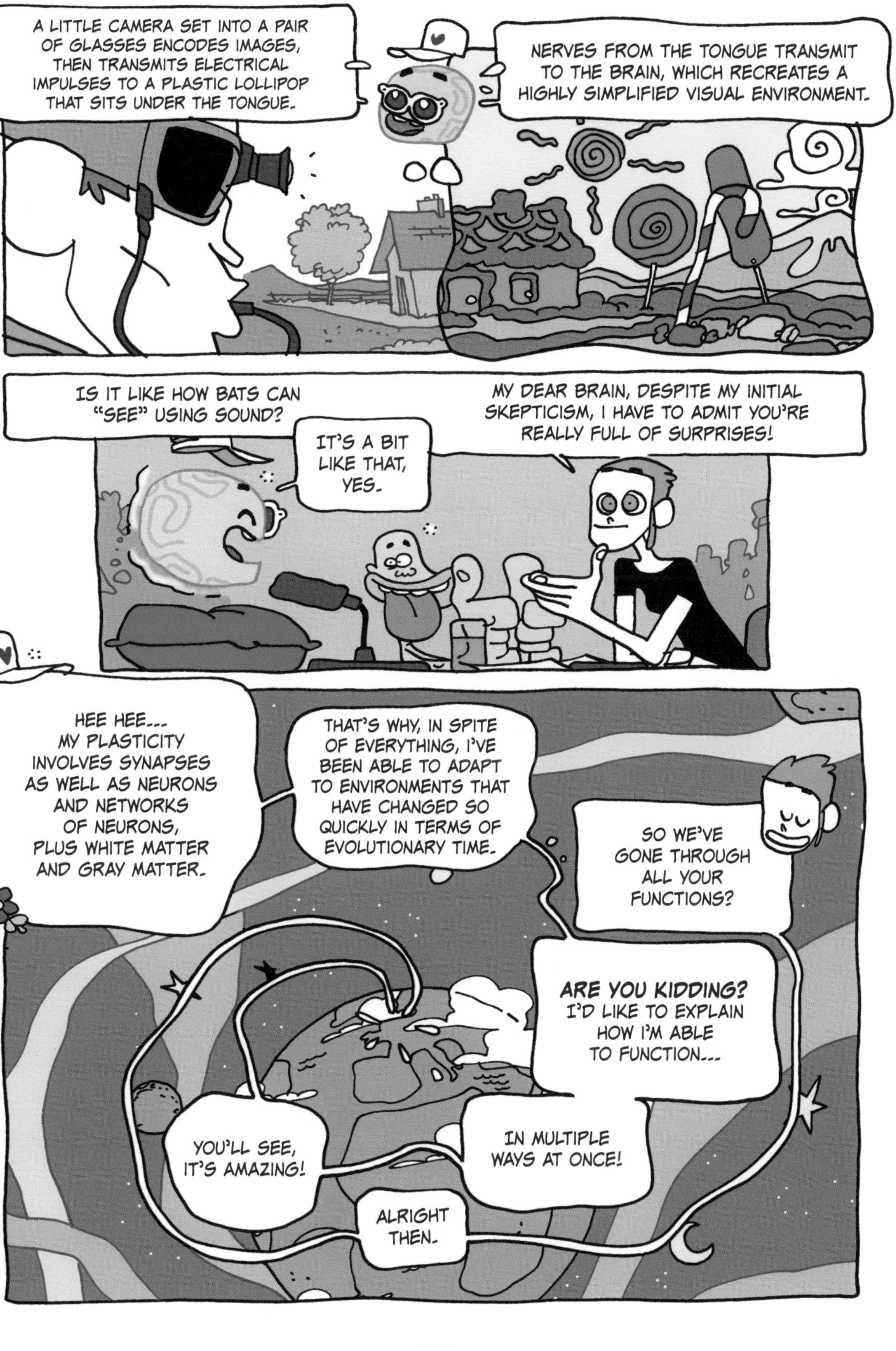
A LITTLE CAMERA SET INTO A PAIR OF GLASSES ENCODES IMAGES, THEN TRANSMITS ELECTRICAL IMPULSES TO A PLASTIC LOLLIPOP THAT SITS UNDER THE TONGUE.
NERVES FROM THE TONGUE TRANSMIT TO THE BRAIN, WHICH RECREATES A HIGHLY SIMPLIFIED VISUAL ENVIRONMENT.
IS IT LIKE HOW BATS CAN "SEE" USING SOUND?
IT'S A BIT LIKE THAT, YES.
MY DEAR BRAIN, DESPITE MY INITIAL SKEPTICISM, I HAVE TO ADMIT YOU'RE REALLY FULL OF SURPRISES!
HEE HEE... MY PLASTICITY INVOLVES SYNAPSES AS WELL AS NEURONS AND NETWORKS OF NEURONS, PLUS WHITE MATTER AND GRAY MATTER.
THAT'S WHY, IN SPITE OF EVERYTHING, I'VE BEEN ABLE TO ADAPT TO ENVIRONMENTS THAT HAVE CHANGED SO QUICKLY IN TERMS OF EVOLUTIONARY TIME.
SO WE'VE GONE THROUGH ALL YOUR FUNCTIONS?
ARE YOU KIDDING? I'D LIKE TO EXPLAIN HOW I'M ABLE TO FUNCTION...
IN MULTIPLE WAYS AT ONCE!
YOU'LL SEE, IT'S AMAZING!
ALRIGHT THEN.

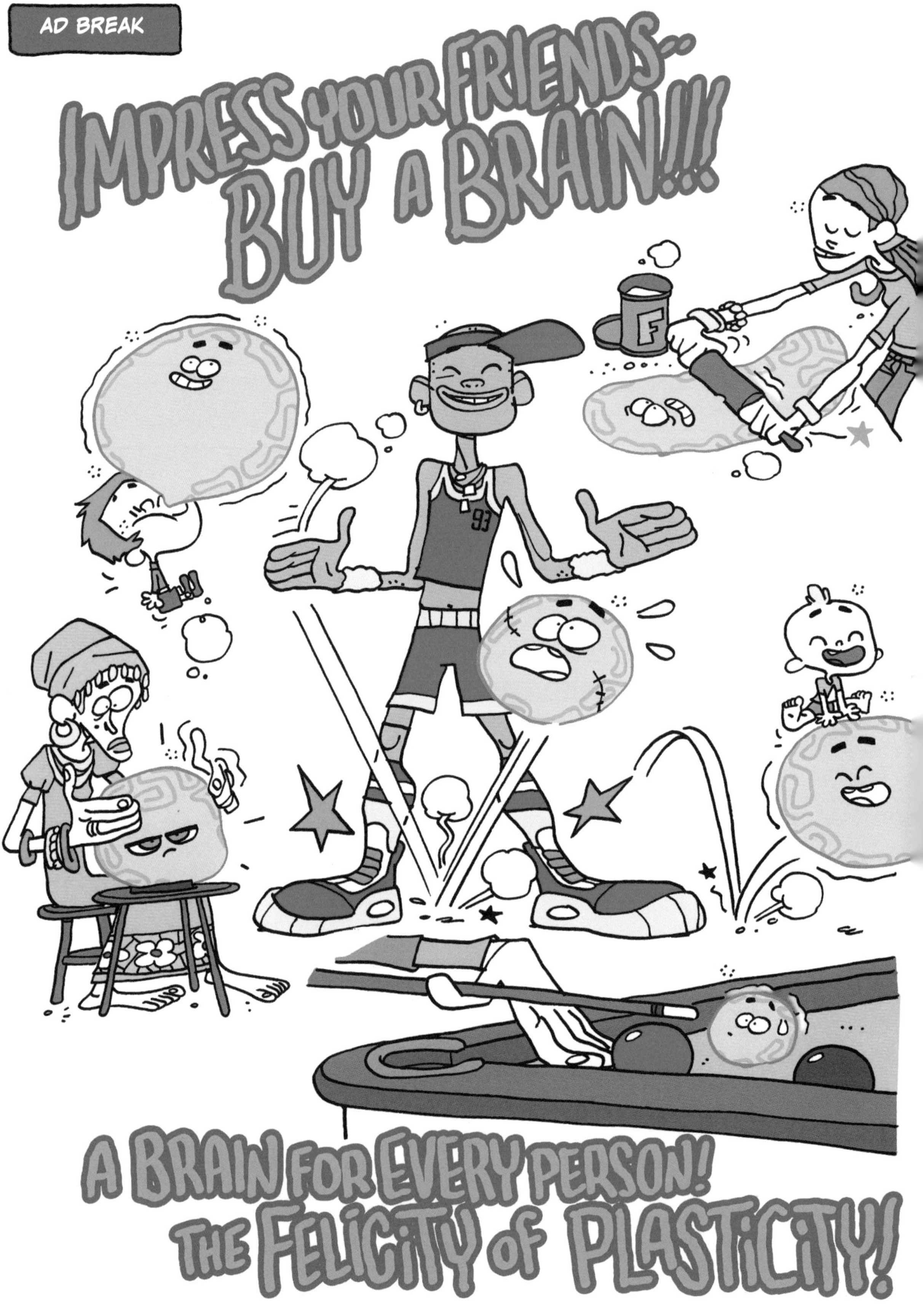
AD BREAK
IMPRESS YOUR FRIENDS--
BUY A BRAIN!!!
F
93
A BRAIN FOR EVERY PERSON!
THE FELICITY of PLASTICITY!

Chapter 8

THE BRAIN HAS TWO SPEEDS

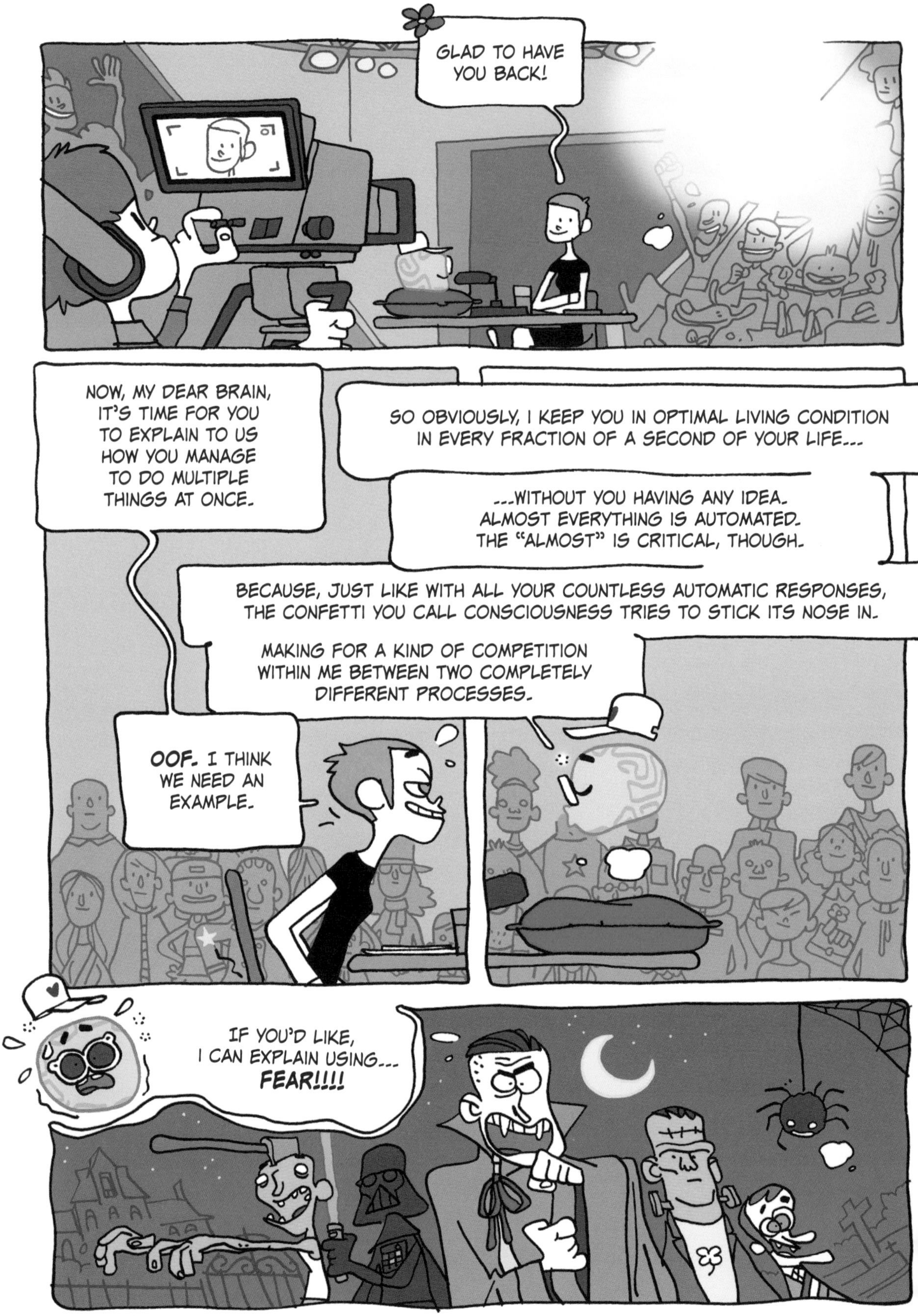
GLAD TO HAVE YOU BACK!
NOW, MY DEAR BRAIN, IT'S TIME FOR YOU TO EXPLAIN TO US HOW YOU MANAGE TO DO MULTIPLE THINGS AT ONCE.
SO OBVIOUSLY, I KEEP YOU IN OPTIMAL LIVING CONDITION IN EVERY FRACTION OF A SECOND OF YOUR LIFE...
...WITHOUT YOU HAVING ANY IDEA. ALMOST EVERYTHING IS AUTOMATED. THE "ALMOST" IS CRITICAL, THOUGH.
BECAUSE, JUST LIKE WITH ALL YOUR COUNTLESS AUTOMATIC RESPONSES, THE CONFETTI YOU CALL CONSCIOUSNESS TRIES TO STICK ITS NOSE IN.
MAKING FOR A KIND OF COMPETITION WITHIN ME BETWEEN TWO COMPLETELY DIFFERENT PROCESSES.
OOF. I THINK WE NEED AN EXAMPLE.
IF YOU'D LIKE, I CAN EXPLAIN USING... FEAR!!!!

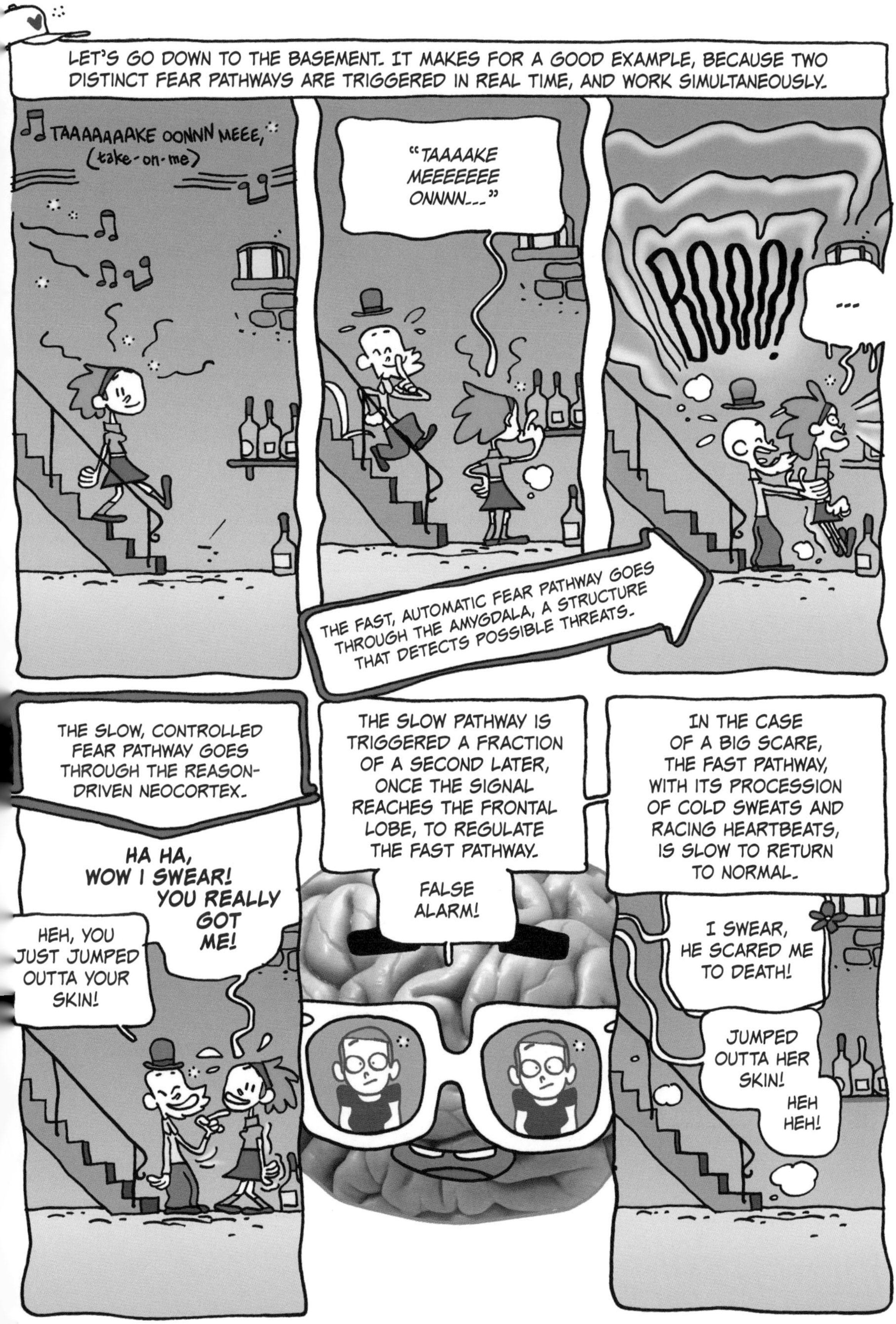
LET'S GO DOWN TO THE BASEMENT. IT MAKES FOR A GOOD EXAMPLE, BECAUSE TWO DISTINCT FEAR PATHWAYS ARE TRIGGERED IN REAL TIME, AND WORK SIMULTANEOUSLY.
TAAAAAAAKE OONNN MEEE, (take-on-me)
"TAAAAKE MEEEEEEE ONNNN..."
BOOO!
...
THE FAST, AUTOMATIC FEAR PATHWAY GOES THROUGH THE AMYGDALA, A STRUCTURE THAT DETECTS POSSIBLE THREATS.
THE SLOW, CONTROLLED FEAR PATHWAY GOES THROUGH THE REASON-DRIVEN NEOCORTEX.
HA HA, WOW I SWEAR! YOU REALLY GOT ME!
HEH, YOU JUST JUMPED OUTTA YOUR SKIN!
THE SLOW PATHWAY IS TRIGGERED A FRACTION OF A SECOND LATER, ONCE THE SIGNAL REACHES THE FRONTAL LOBE, TO REGULATE THE FAST PATHWAY.
FALSE ALARM!
IN THE CASE OF A BIG SCARE, THE FAST PATHWAY, WITH ITS PROCESSION OF COLD SWEATS AND RACING HEARTBEATS, IS SLOW TO RETURN TO NORMAL.
I SWEAR, HE SCARED ME TO DEATH!
JUMPED OUTTA HER SKIN!
HEH HEH!

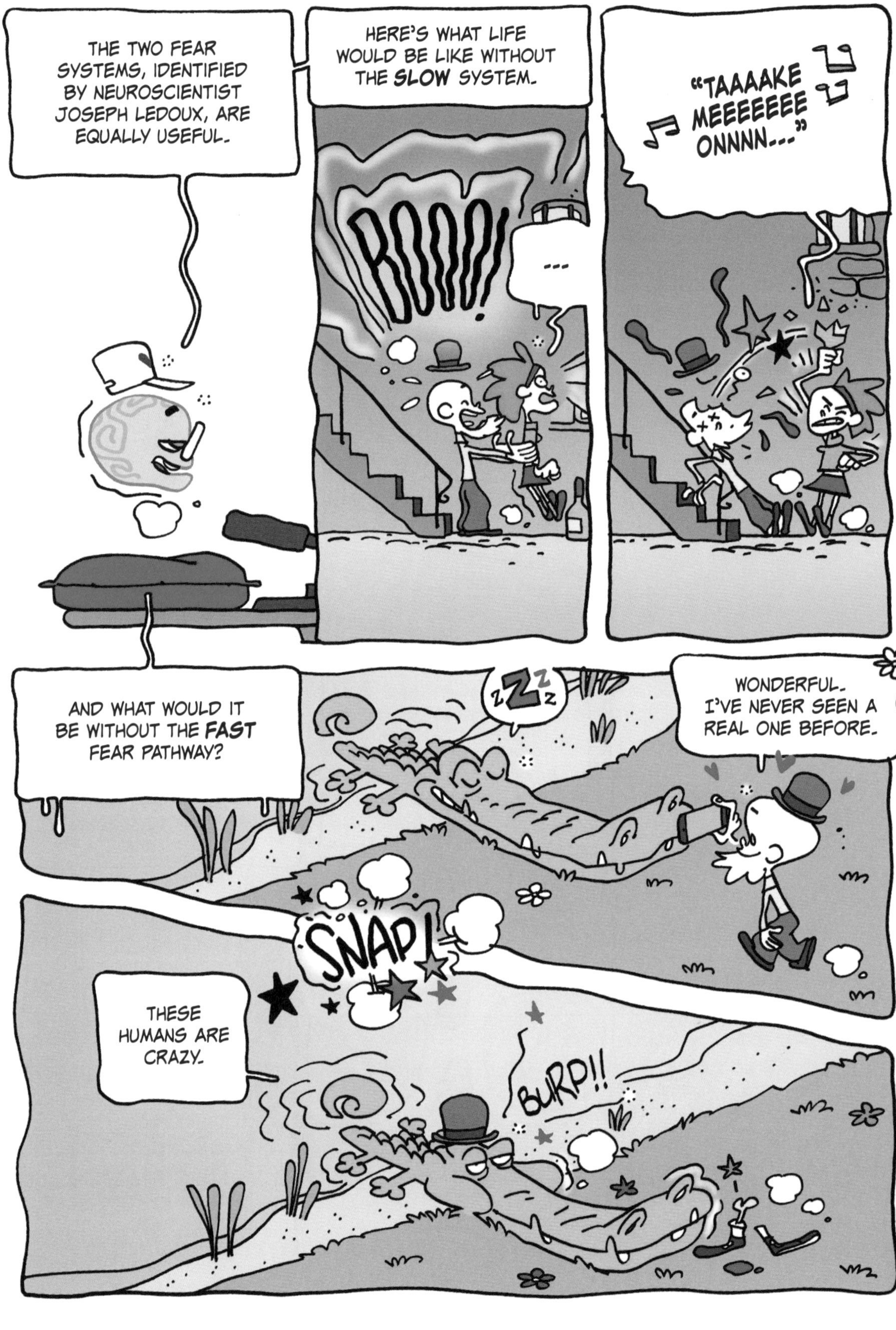
THE TWO FEAR SYSTEMS, IDENTIFIED BY NEUROSCIENTIST JOSEPH LEDOUX, ARE EQUALLY USEFUL.
HERE'S WHAT LIFE WOULD BE LIKE WITHOUT THE SLOW SYSTEM.
BOOO!
...
"TAAAAKE MEEEEEEE ONNNN..."
AND WHAT WOULD IT BE WITHOUT THE FAST FEAR PATHWAY?
ZZZ
WONDERFUL. I'VE NEVER SEEN A REAL ONE BEFORE.
SNAP!
THESE HUMANS ARE CRAZY.
BURP!!

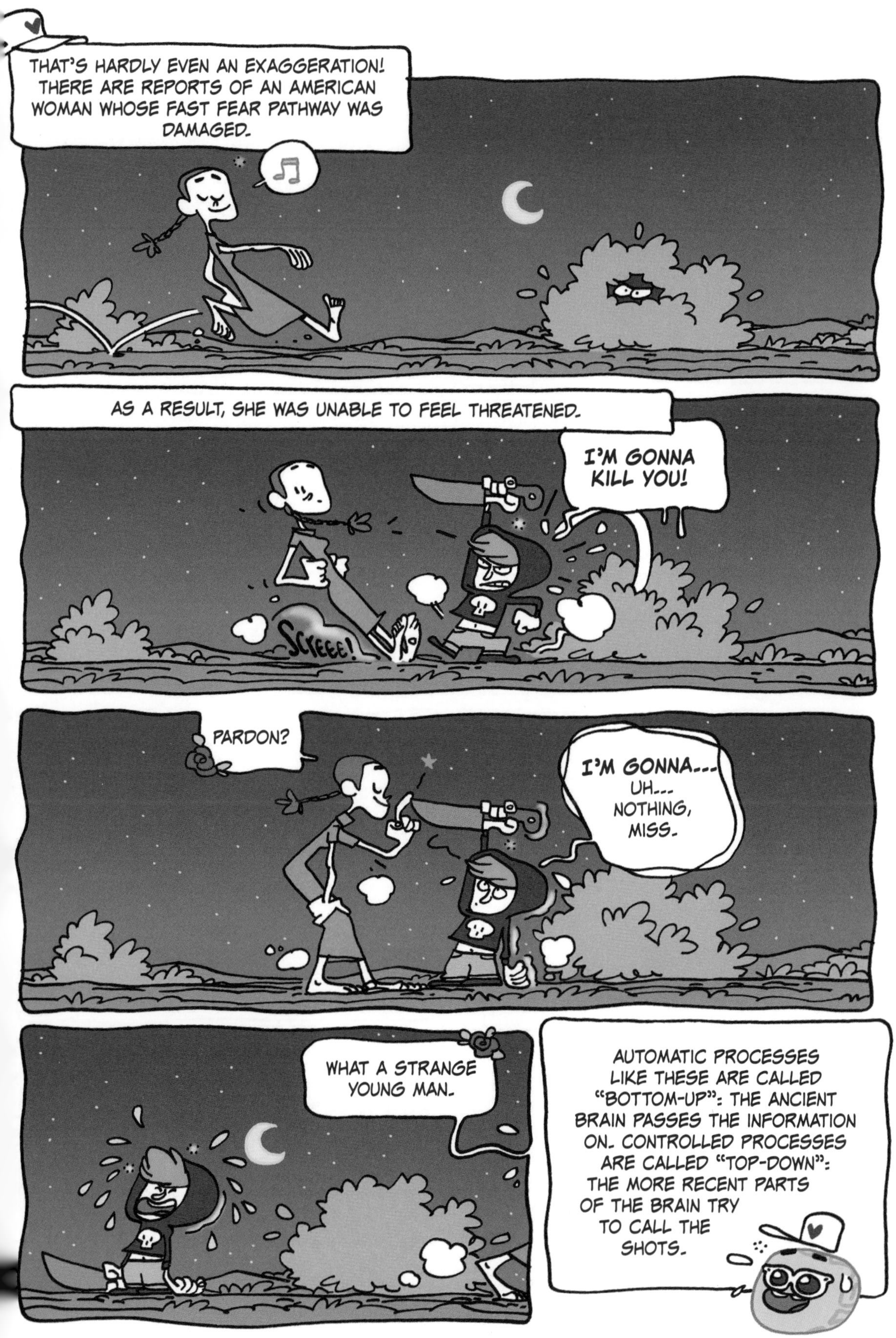
THAT'S HARDLY EVEN AN EXAGGERATION! THERE ARE REPORTS OF AN AMERICAN WOMAN WHOSE FAST FEAR PATHWAY WAS DAMAGED.
AS A RESULT, SHE WAS UNABLE TO FEEL THREATENED.
I'M GONNA KILL YOU!
SCREEEE!
PARDON?
I'M GONNA... UH... NOTHING, MISS.
WHAT A STRANGE YOUNG MAN.
AUTOMATIC PROCESSES LIKE THESE ARE CALLED "BOTTOM-UP": THE ANCIENT BRAIN PASSES THE INFORMATION ON. CONTROLLED PROCESSES ARE CALLED "TOP-DOWN": THE MORE RECENT PARTS OF THE BRAIN TRY TO CALL THE SHOTS.

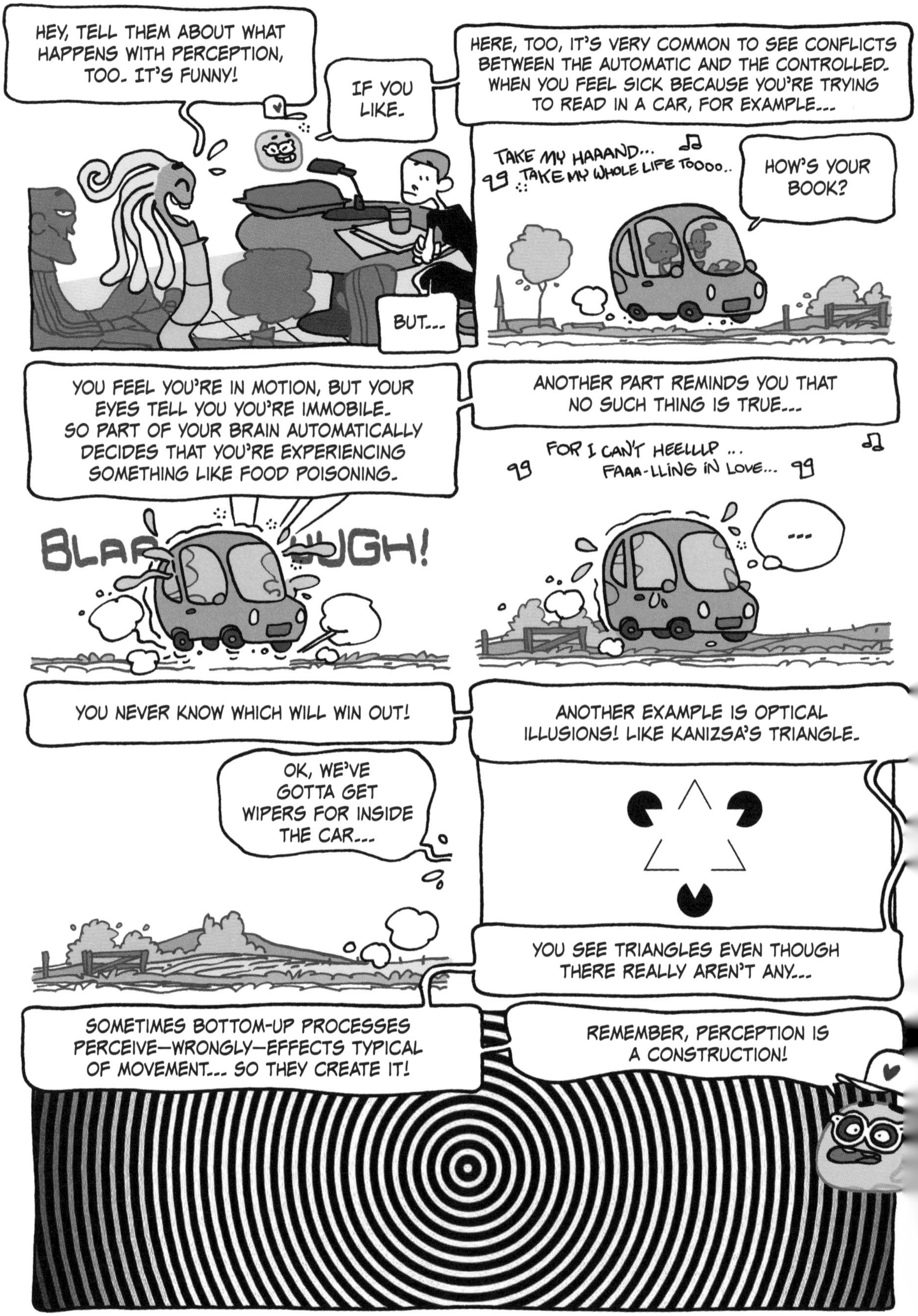
HEY, TELL THEM ABOUT WHAT HAPPENS WITH PERCEPTION, TOO. IT'S FUNNY!
IF YOU LIKE.
BUT...
HERE, TOO, IT'S VERY COMMON TO SEE CONFLICTS BETWEEN THE AUTOMATIC AND THE CONTROLLED. WHEN YOU FEEL SICK BECAUSE YOU'RE TRYING TO READ IN A CAR, FOR EXAMPLE...
TAKE MY HAAAND... TAKE MY WHOLE LIFE TOOOO..
HOW'S YOUR BOOK?
YOU FEEL YOU'RE IN MOTION, BUT YOUR EYES TELL YOU YOU'RE IMMOBILE. SO PART OF YOUR BRAIN AUTOMATICALLY DECIDES THAT YOU'RE EXPERIENCING SOMETHING LIKE FOOD POISONING.
BLAAUGH!
ANOTHER PART REMINDS YOU THAT NO SUCH THING IS TRUE...
FOR I CAN'T HEELLLP ... FAAA-LLING IN LOVE...
...
YOU NEVER KNOW WHICH WILL WIN OUT!
OK, WE'VE GOTTA GET WIPERS FOR INSIDE THE CAR...
ANOTHER EXAMPLE IS OPTICAL ILLUSIONS! LIKE KANIZSA'S TRIANGLE.
YOU SEE TRIANGLES EVEN THOUGH THERE REALLY AREN'T ANY...
SOMETIMES BOTTOM-UP PROCESSES PERCEIVE—WRONGLY—EFFECTS TYPICAL OF MOVEMENT... SO THEY CREATE IT!
REMEMBER, PERCEPTION IS A CONSTRUCTION!

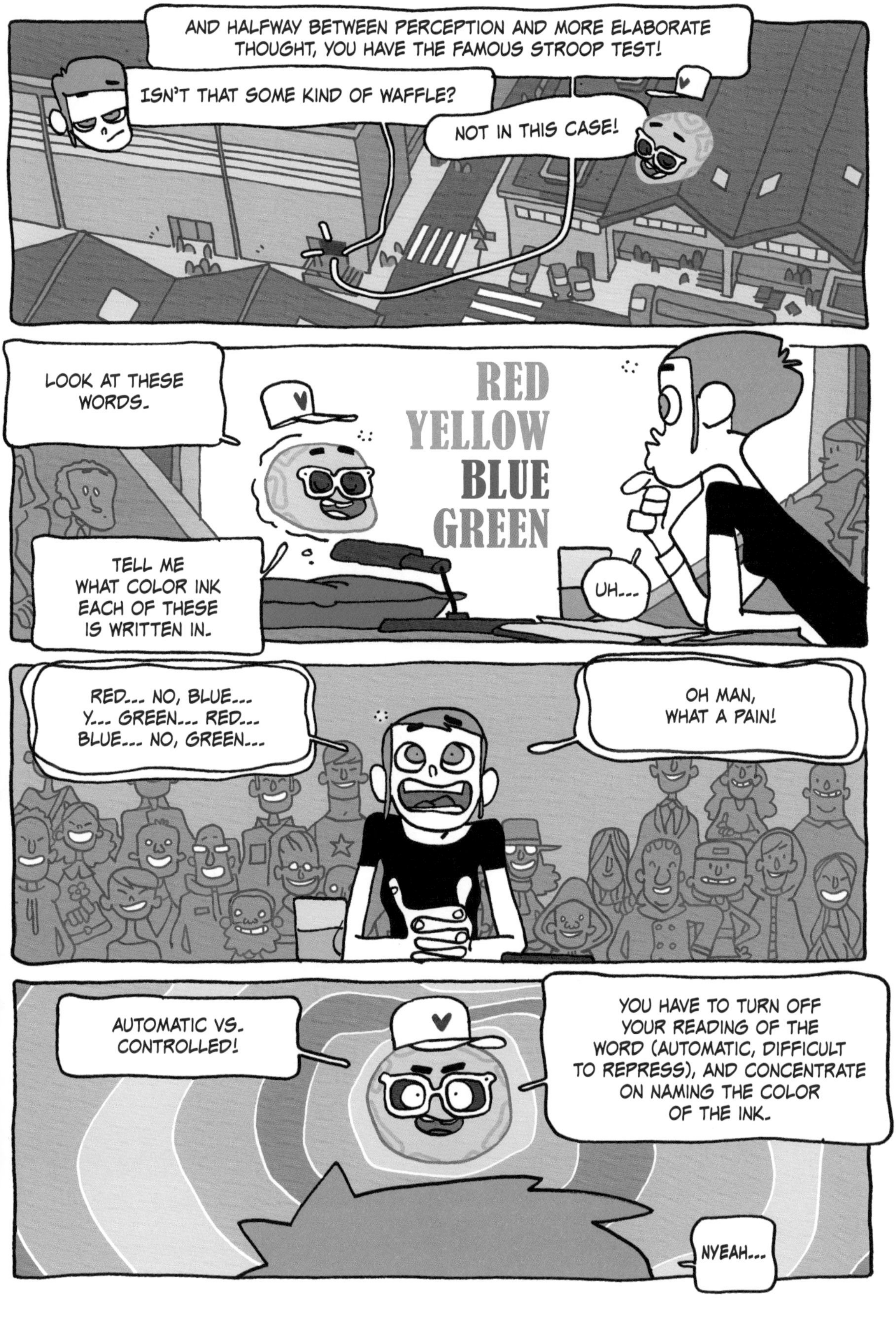
AND HALFWAY BETWEEN PERCEPTION AND MORE ELABORATE THOUGHT, YOU HAVE THE FAMOUS STROOP TEST!
ISN'T THAT SOME KIND OF WAFFLE?
NOT IN THIS CASE!
LOOK AT THESE WORDS.
RED
YELLOW
BLUE
GREEN
TELL ME WHAT COLOR INK EACH OF THESE IS WRITTEN IN.
UH...
RED... NO, BLUE... Y... GREEN... RED... BLUE... NO, GREEN...
OH MAN, WHAT A PAIN!
AUTOMATIC VS. CONTROLLED!
YOU HAVE TO TURN OFF YOUR READING OF THE WORD (AUTOMATIC, DIFFICULT TO REPRESS), AND CONCENTRATE ON NAMING THE COLOR OF THE INK.
NYEAH...

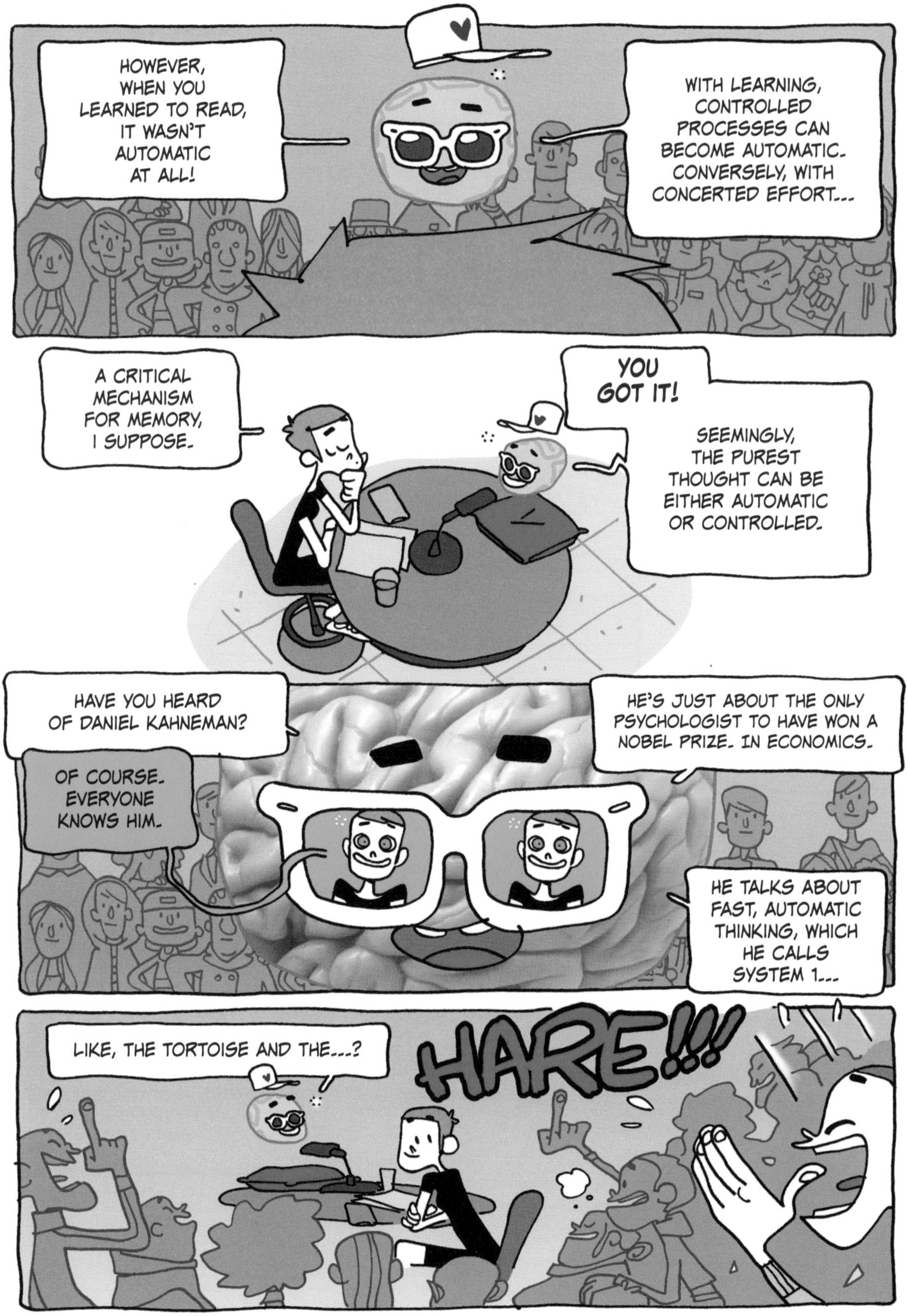
HOWEVER, WHEN YOU LEARNED TO READ, IT WASN'T AUTOMATIC AT ALL!
WITH LEARNING, CONTROLLED PROCESSES CAN BECOME AUTOMATIC. CONVERSELY, WITH CONCERTED EFFORT...
A CRITICAL MECHANISM FOR MEMORY, I SUPPOSE.
YOU GOT IT!
SEEMINGLY, THE PUREST THOUGHT CAN BE EITHER AUTOMATIC OR CONTROLLED.
HAVE YOU HEARD OF DANIEL KAHNEMAN?
OF COURSE. EVERYONE KNOWS HIM.
HE'S JUST ABOUT THE ONLY PSYCHOLOGIST TO HAVE WON A NOBEL PRIZE. IN ECONOMICS.
HE TALKS ABOUT FAST, AUTOMATIC THINKING, WHICH HE CALLS SYSTEM 1...
LIKE, THE TORTOISE AND THE...?
HARE!!!

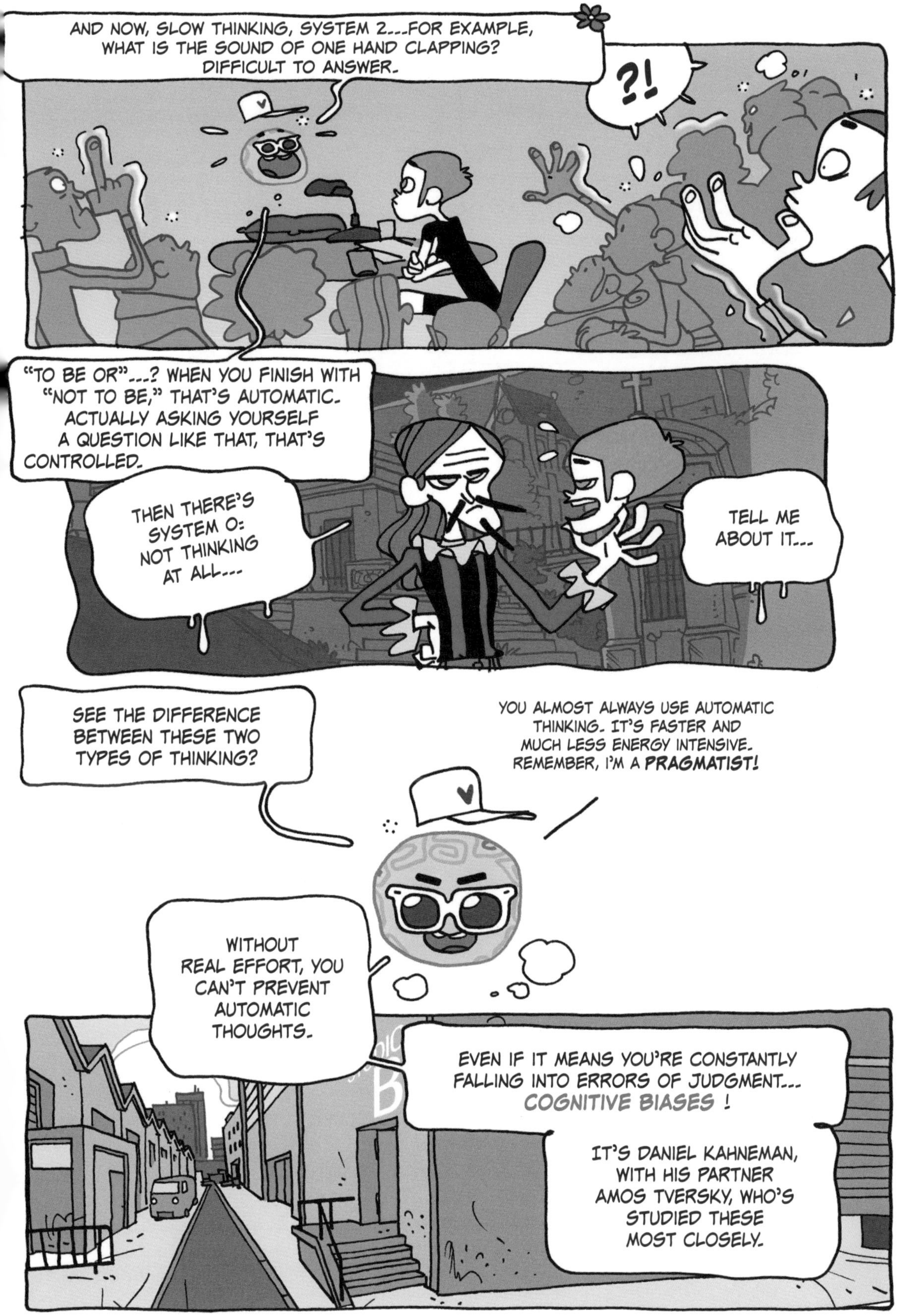
AND NOW, SLOW THINKING, SYSTEM 2...FOR EXAMPLE, WHAT IS THE SOUND OF ONE HAND CLAPPING? DIFFICULT TO ANSWER.
?!
"TO BE OR"...? WHEN YOU FINISH WITH "NOT TO BE," THAT'S AUTOMATIC. ACTUALLY ASKING YOURSELF A QUESTION LIKE THAT, THAT'S CONTROLLED.
THEN THERE'S SYSTEM 0: NOT THINKING AT ALL...
TELL ME ABOUT IT...
SEE THE DIFFERENCE BETWEEN THESE TWO TYPES OF THINKING?
YOU ALMOST ALWAYS USE AUTOMATIC THINKING. IT'S FASTER AND MUCH LESS ENERGY INTENSIVE. REMEMBER, I'M A **PRAGMATIST!**
WITHOUT REAL EFFORT, YOU CAN'T PREVENT AUTOMATIC THOUGHTS.
EVEN IF IT MEANS YOU'RE CONSTANTLY FALLING INTO ERRORS OF JUDGMENT... COGNITIVE BIASES !
IT'S DANIEL KAHNEMAN, WITH HIS PARTNER AMOS TVERSKY, WHO'S STUDIED THESE MOST CLOSELY.

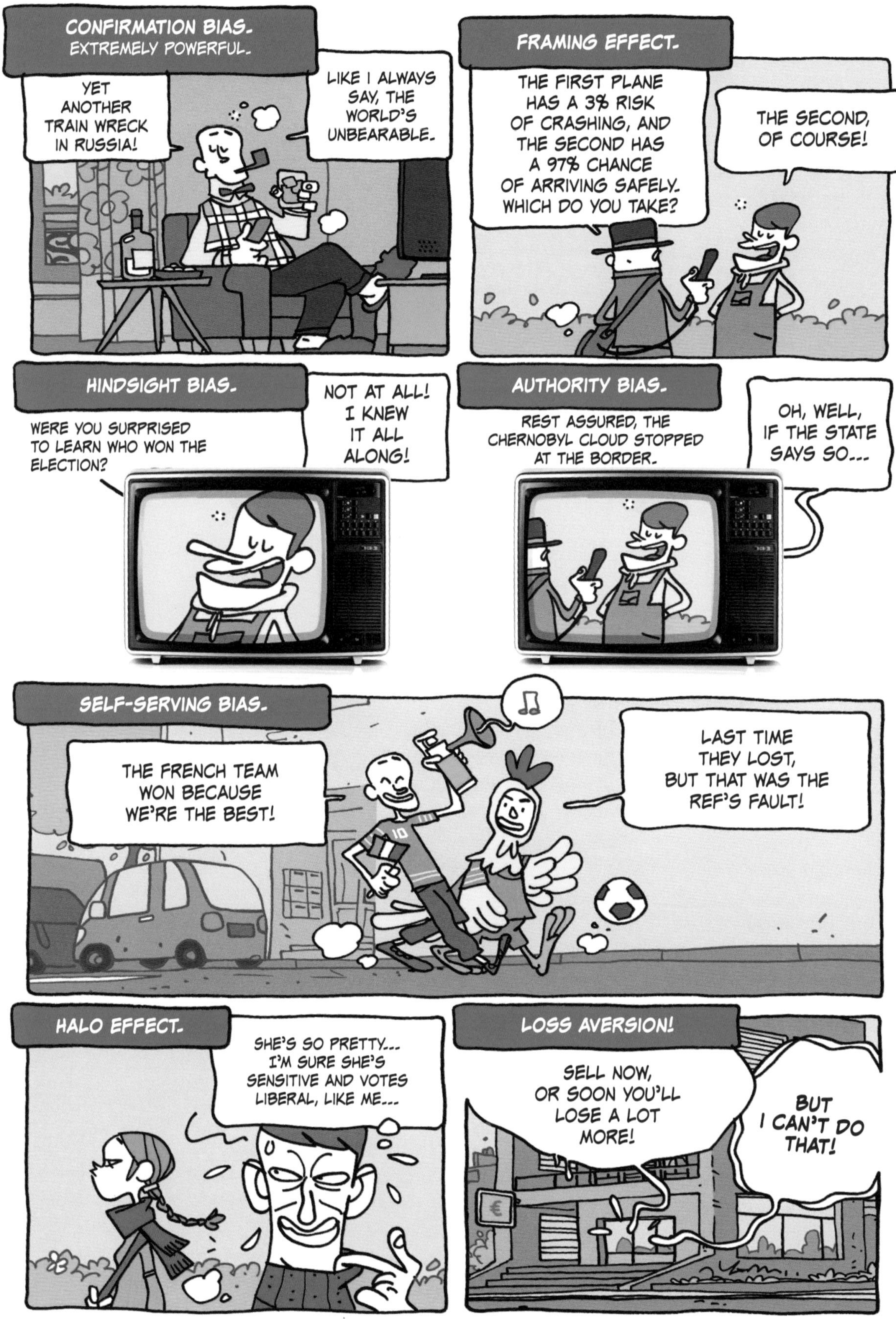
CONFIRMATION BIAS. EXTREMELY POWERFUL.
YET ANOTHER TRAIN WRECK IN RUSSIA!
LIKE I ALWAYS SAY, THE WORLD'S UNBEARABLE.
FRAMING EFFECT.
THE FIRST PLANE HAS A 3% RISK OF CRASHING, AND THE SECOND HAS A 97% CHANCE OF ARRIVING SAFELY. WHICH DO YOU TAKE?
THE SECOND, OF COURSE!
HINDSIGHT BIAS.
WERE YOU SURPRISED TO LEARN WHO WON THE ELECTION?
NOT AT ALL! I KNEW IT ALL ALONG!
AUTHORITY BIAS.
REST ASSURED, THE CHERNOBYL CLOUD STOPPED AT THE BORDER.
OH, WELL, IF THE STATE SAYS SO...
SELF-SERVING BIAS.
THE FRENCH TEAM WON BECAUSE WE'RE THE BEST!
LAST TIME THEY LOST, BUT THAT WAS THE REF'S FAULT!
10
HALO EFFECT.
SHE'S SO PRETTY... I'M SURE SHE'S SENSITIVE AND VOTES LIBERAL, LIKE ME...
LOSS AVERSION!
SELL NOW, OR SOON YOU'LL LOSE A LOT MORE!
BUT I CAN'T DO THAT!
€

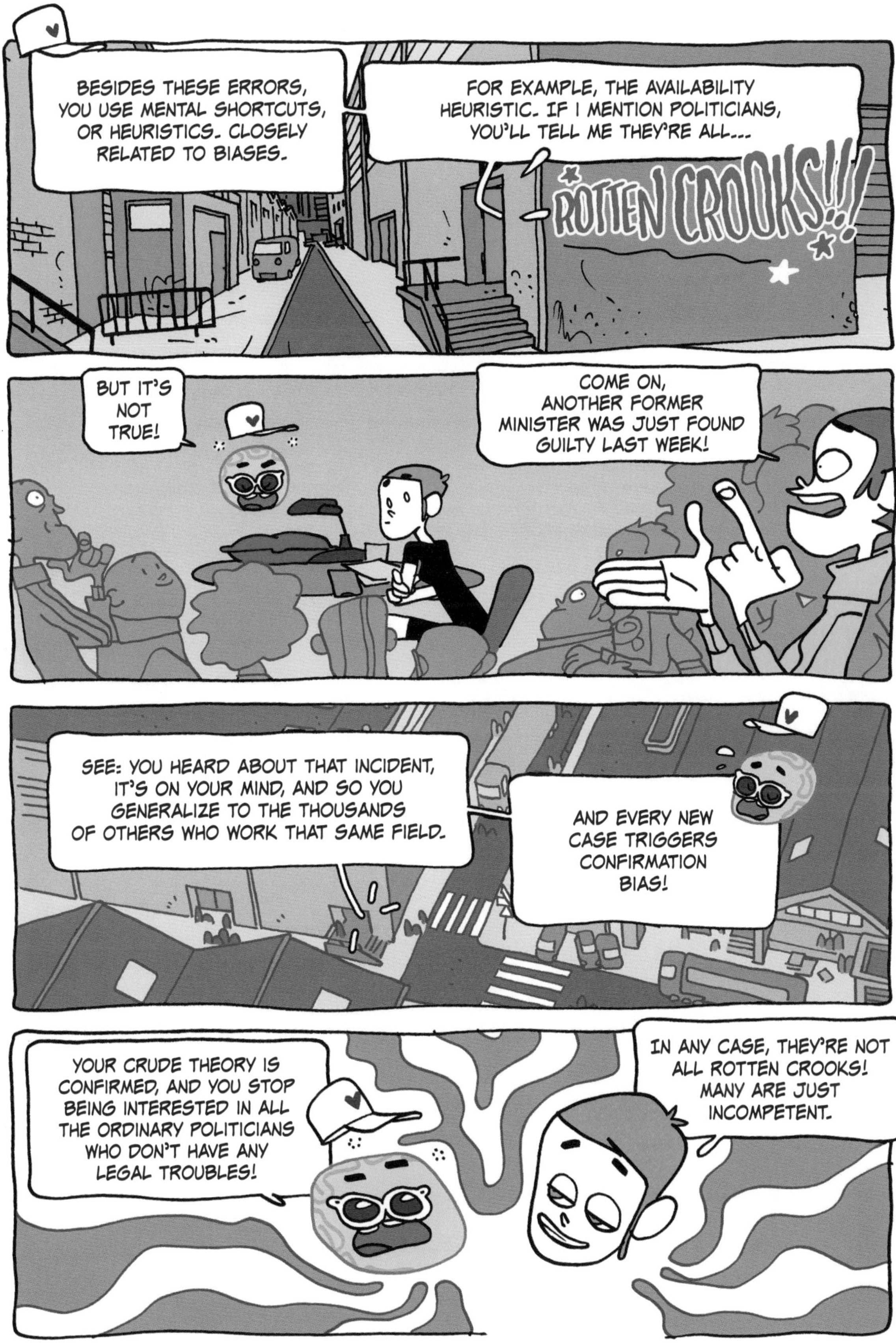
BESIDES THESE ERRORS, YOU USE MENTAL SHORTCUTS, OR HEURISTICS. CLOSELY RELATED TO BIASES.
FOR EXAMPLE, THE AVAILABILITY HEURISTIC. IF I MENTION POLITICIANS, YOU'LL TELL ME THEY'RE ALL...
ROTTEN CROOKS!!!
BUT IT'S NOT TRUE!
COME ON, ANOTHER FORMER MINISTER WAS JUST FOUND GUILTY LAST WEEK!
SEE: YOU HEARD ABOUT THAT INCIDENT, IT'S ON YOUR MIND, AND SO YOU GENERALIZE TO THE THOUSANDS OF OTHERS WHO WORK THAT SAME FIELD.
AND EVERY NEW CASE TRIGGERS CONFIRMATION BIAS!
YOUR CRUDE THEORY IS CONFIRMED, AND YOU STOP BEING INTERESTED IN ALL THE ORDINARY POLITICIANS WHO DON'T HAVE ANY LEGAL TROUBLES!
IN ANY CASE, THEY'RE NOT ALL ROTTEN CROOKS! MANY ARE JUST INCOMPETENT.

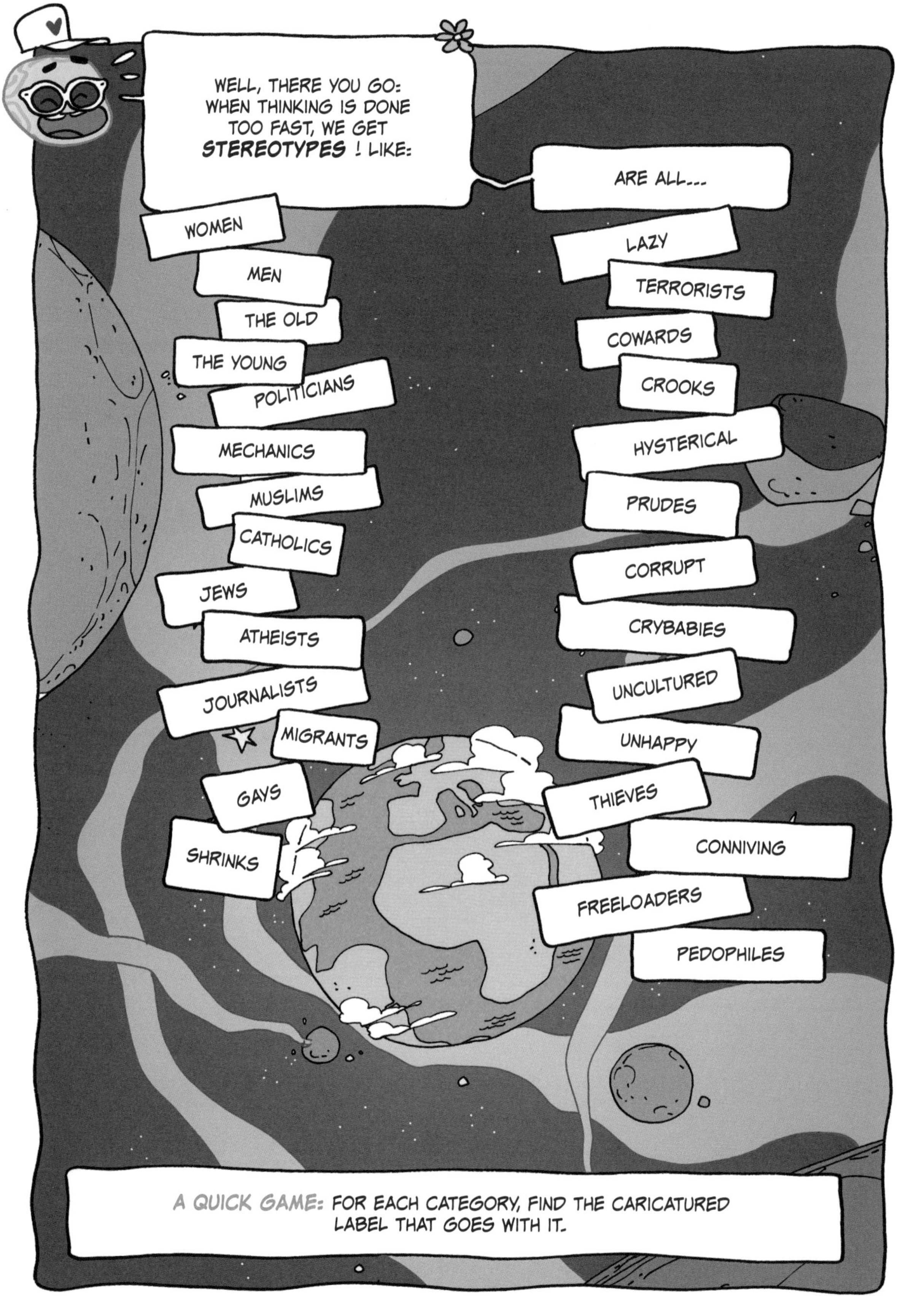

WELL, THERE YOU GO: WHEN THINKING IS DONE TOO FAST, WE GET **STEREOTYPES** ! LIKE:
ARE ALL...
WOMEN
MEN
THE OLD
THE YOUNG
POLITICIANS
MECHANICS
MUSLIMS
CATHOLICS
JEWS
ATHEISTS
JOURNALISTS
MIGRANTS
GAYS
SHRINKS
LAZY
TERRORISTS
COWARDS
CROOKS
HYSTERICAL
PRUDES
CORRUPT
CRYBABIES
UNCULTURED
UNHAPPY
THIEVES
CONNIVING
FREELOADERS
PEDOPHILES
A QUICK GAME: FOR EACH CATEGORY, FIND THE CARICATURED LABEL THAT GOES WITH IT.

EVEN YOUR MOST CLOSELY HELD, MOST THOUGHTFUL RELIGIOUS BELIEFS AND POLITICAL CONVICTIONS BECOME AUTOMATIC AS THEY BECOME ROOTED.

OF COURSE, EVERY THOUGHT YOU HAVE ABOUT YOURSELF OBEYS THE SAME RULES.

I HAVE TO...

I ABSOLUTELY MUST...

OR ELSE I'M WORTHLESS...

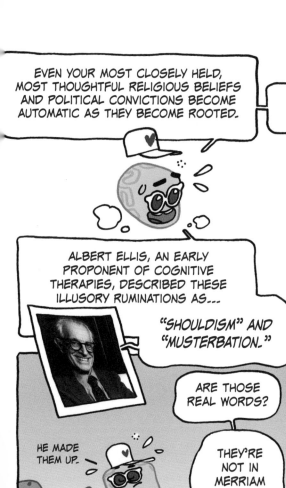

ALBERT ELLIS, AN EARLY PROPONENT OF COGNITIVE THERAPIES, DESCRIBED THESE ILLUSORY RUMINATIONS AS...

"SHOULDISM" AND "MUSTERBATION."

ARE THOSE REAL WORDS?

HE MADE THEM UP.

THEY'RE NOT IN MERRIAM WEBSTER...

THE SYSTEMATIC PESSIMISM OF THE DEPRESSED IS SUSTAINED BY THOUGHTS THAT HAVE BECOME AUTOMATIC.

I'M WORTHLESS.

IN EVERY WAY.

SO'S THE WORLD.

IT WAS BETTER BEFORE.

THINGS WILL NEVER TURN AROUND.

NEVER EVER.

IT'S LIKE WITH STRESS. THE GOOD KIND CAN AUTOMATICALLY SAVE YOUR LIFE.

THE BAD KIND CAN RUIN YOUR LIFE, BLEEDING YOU DRY BY MOBILIZING YOUR DEFENSES AGAINST A VANISHED OR IMAGINARY THREAT.

I HAVEN'T BEEN SAVING ENOUGH FOR RETIREMENT.

WHAT IF I LOSE MY JOB?

OR MY HUSBAND CHEATS?

OR MY SON STARTS DOING DRUGS?

OR I START DRINKING?

WHAT WOULD I DO?

OR...?

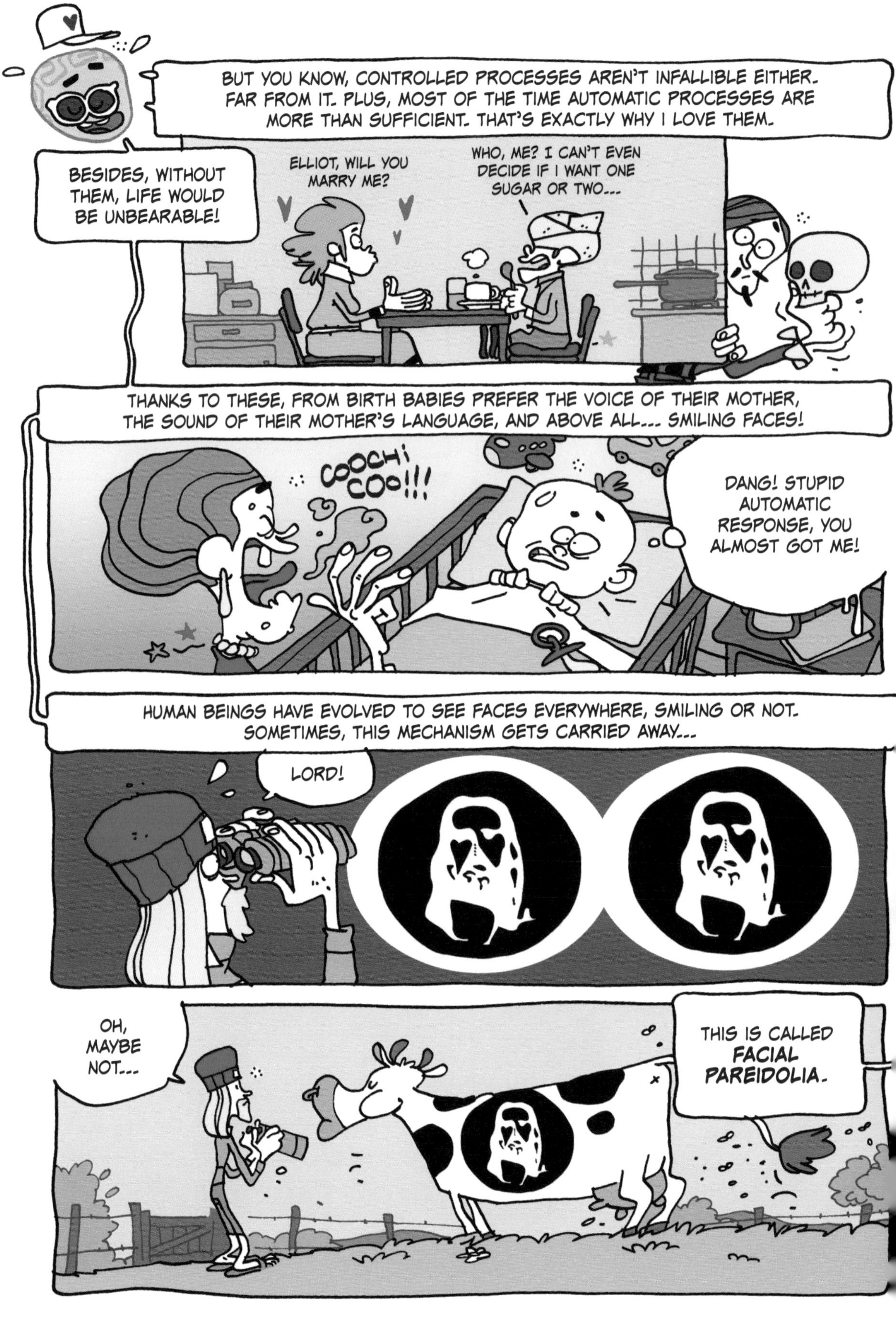

BUT YOU KNOW, CONTROLLED PROCESSES AREN'T INFALLIBLE EITHER. FAR FROM IT. PLUS, MOST OF THE TIME AUTOMATIC PROCESSES ARE MORE THAN SUFFICIENT. THAT'S EXACTLY WHY I LOVE THEM.
BESIDES, WITHOUT THEM, LIFE WOULD BE UNBEARABLE!
ELLIOT, WILL YOU MARRY ME?
WHO, ME? I CAN'T EVEN DECIDE IF I WANT ONE SUGAR OR TWO...
THANKS TO THESE, FROM BIRTH BABIES PREFER THE VOICE OF THEIR MOTHER, THE SOUND OF THEIR MOTHER'S LANGUAGE, AND ABOVE ALL... SMILING FACES!
COOCHI COO!!!
DANG! STUPID AUTOMATIC RESPONSE, YOU ALMOST GOT ME!
HUMAN BEINGS HAVE EVOLVED TO SEE FACES EVERYWHERE, SMILING OR NOT. SOMETIMES, THIS MECHANISM GETS CARRIED AWAY...
LORD!
OH, MAYBE NOT...
THIS IS CALLED FACIAL PAREIDOLIA.

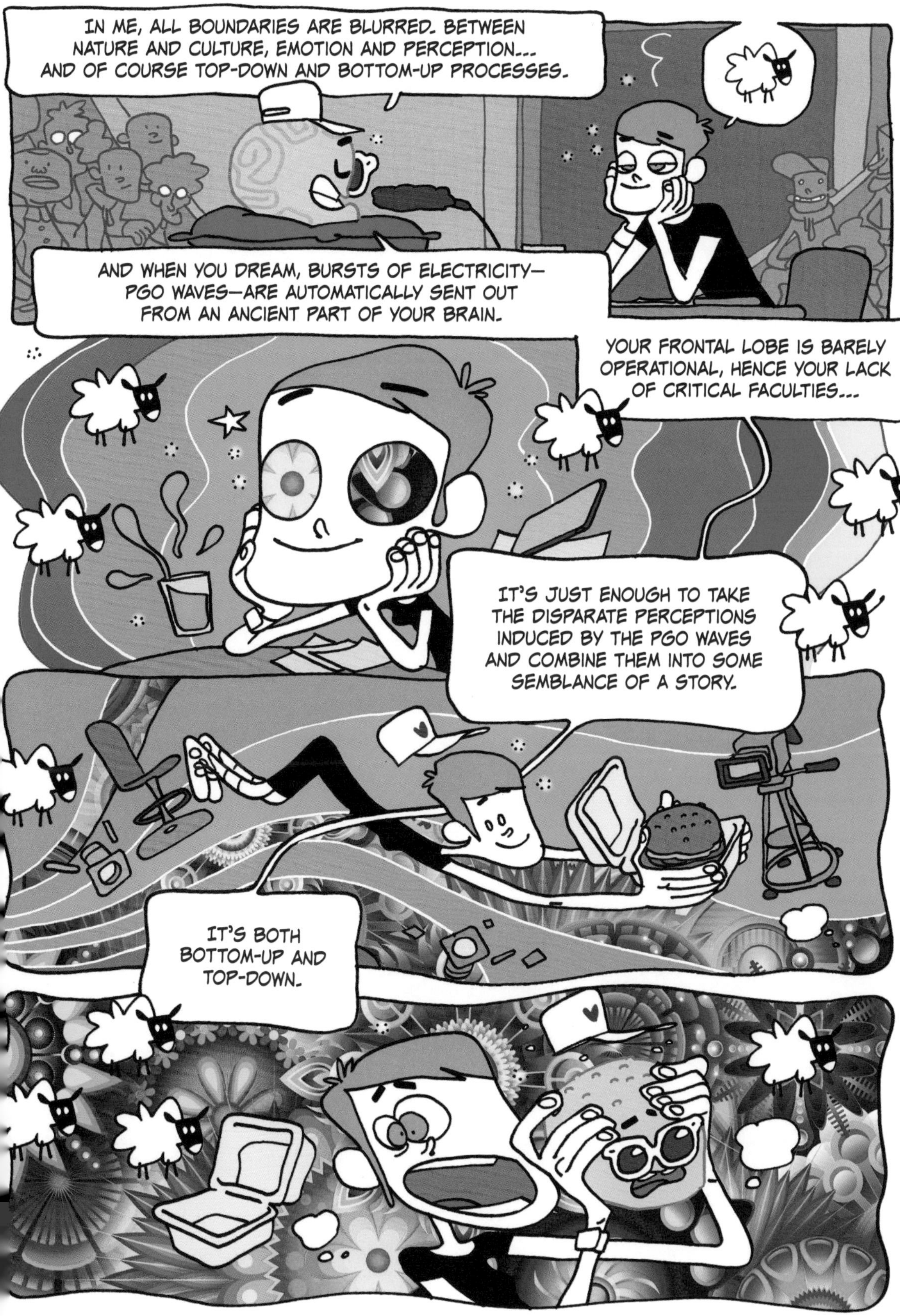
IN ME, ALL BOUNDARIES ARE BLURRED. BETWEEN NATURE AND CULTURE, EMOTION AND PERCEPTION... AND OF COURSE TOP-DOWN AND BOTTOM-UP PROCESSES.
AND WHEN YOU DREAM, BURSTS OF ELECTRICITY—PGO WAVES—ARE AUTOMATICALLY SENT OUT FROM AN ANCIENT PART OF YOUR BRAIN.
YOUR FRONTAL LOBE IS BARELY OPERATIONAL, HENCE YOUR LACK OF CRITICAL FACULTIES...
IT'S JUST ENOUGH TO TAKE THE DISPARATE PERCEPTIONS INDUCED BY THE PGO WAVES AND COMBINE THEM INTO SOME SEMBLANCE OF A STORY.
IT'S BOTH BOTTOM-UP AND TOP-DOWN.

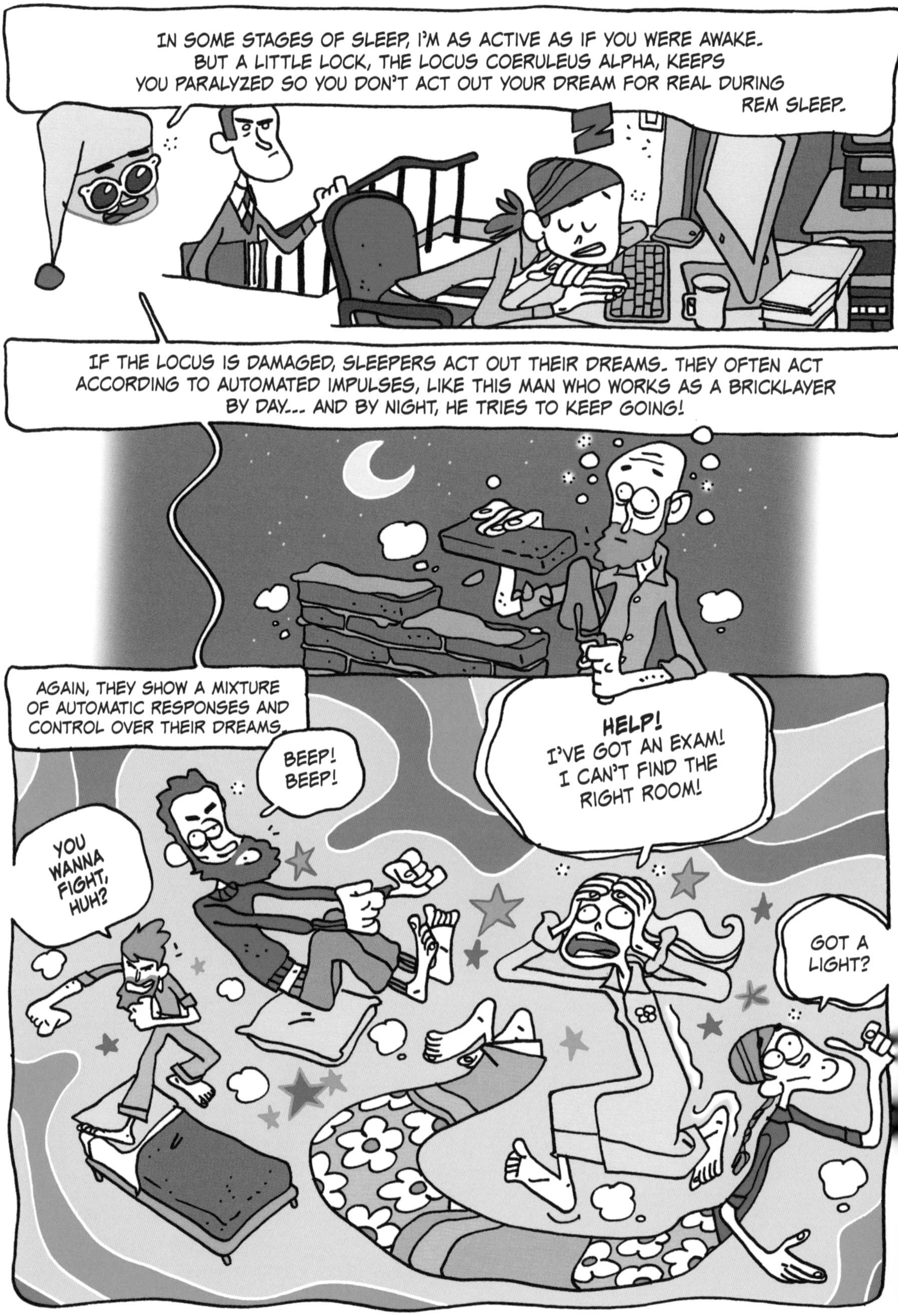

IN SOME STAGES OF SLEEP, I'M AS ACTIVE AS IF YOU WERE AWAKE. BUT A LITTLE LOCK, THE LOCUS COERULEUS ALPHA, KEEPS YOU PARALYZED SO YOU DON'T ACT OUT YOUR DREAM FOR REAL DURING REM SLEEP.
Z
IF THE LOCUS IS DAMAGED, SLEEPERS ACT OUT THEIR DREAMS. THEY OFTEN ACT ACCORDING TO AUTOMATED IMPULSES, LIKE THIS MAN WHO WORKS AS A BRICKLAYER BY DAY... AND BY NIGHT, HE TRIES TO KEEP GOING!
AGAIN, THEY SHOW A MIXTURE OF AUTOMATIC RESPONSES AND CONTROL OVER THEIR DREAMS.
BEEP! BEEP!
HELP! I'VE GOT AN EXAM! I CAN'T FIND THE RIGHT ROOM!
YOU WANNA FIGHT, HUH?
GOT A LIGHT?

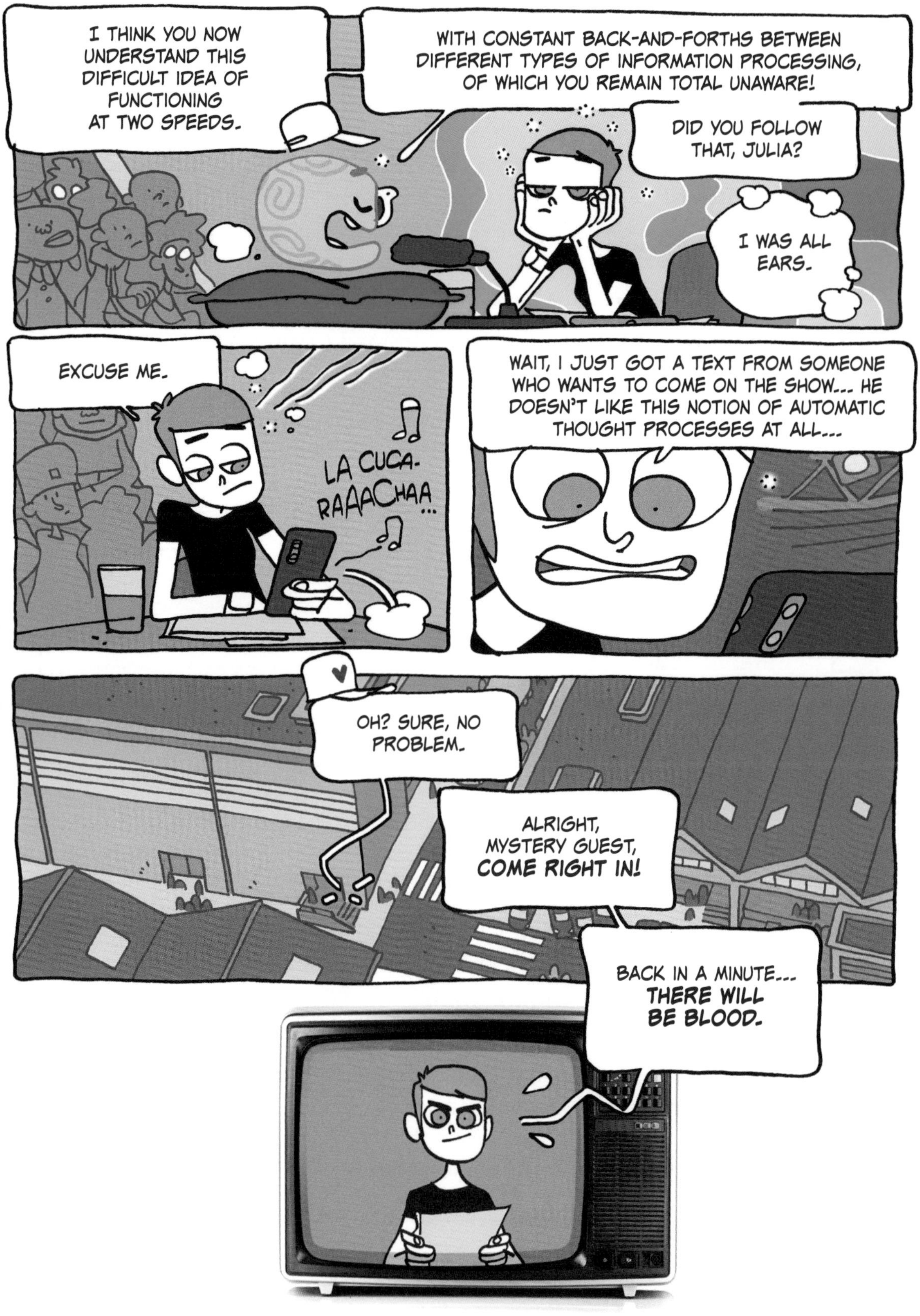
I THINK YOU NOW UNDERSTAND THIS DIFFICULT IDEA OF FUNCTIONING AT TWO SPEEDS.
WITH CONSTANT BACK-AND-FORTHS BETWEEN DIFFERENT TYPES OF INFORMATION PROCESSING, OF WHICH YOU REMAIN TOTAL UNAWARE!
DID YOU FOLLOW THAT, JULIA?
I WAS ALL EARS.
EXCUSE ME.
LA CUCA-RAAACHAA...
WAIT, I JUST GOT A TEXT FROM SOMEONE WHO WANTS TO COME ON THE SHOW... HE DOESN'T LIKE THIS NOTION OF AUTOMATIC THOUGHT PROCESSES AT ALL...
OH? SURE, NO PROBLEM.
ALRIGHT, MYSTERY GUEST, **COME RIGHT IN!**
BACK IN A MINUTE... **THERE WILL BE BLOOD.**

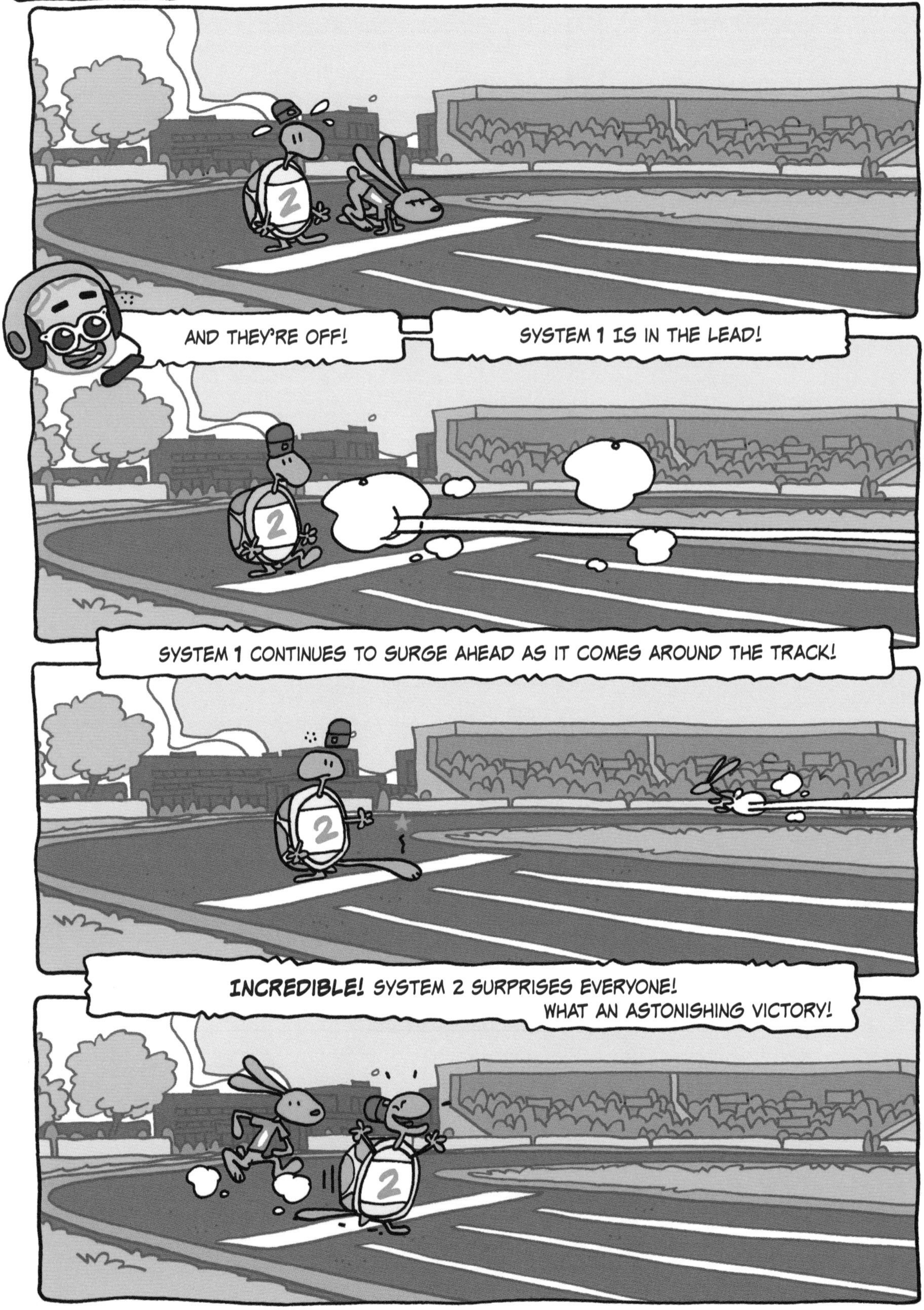
AND THEY'RE OFF!
SYSTEM 1 IS IN THE LEAD!
SYSTEM 1 CONTINUES TO SURGE AHEAD AS IT COMES AROUND THE TRACK!
INCREDIBLE! SYSTEM 2 SURPRISES EVERYONE! WHAT AN ASTONISHING VICTORY!

Chapter 9

THE GREAT INTERPRETER

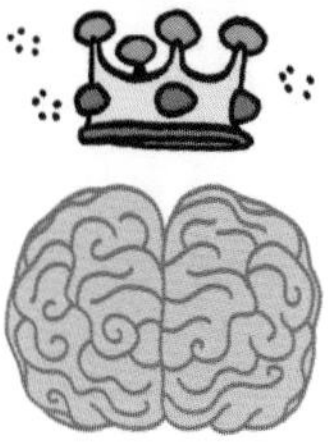

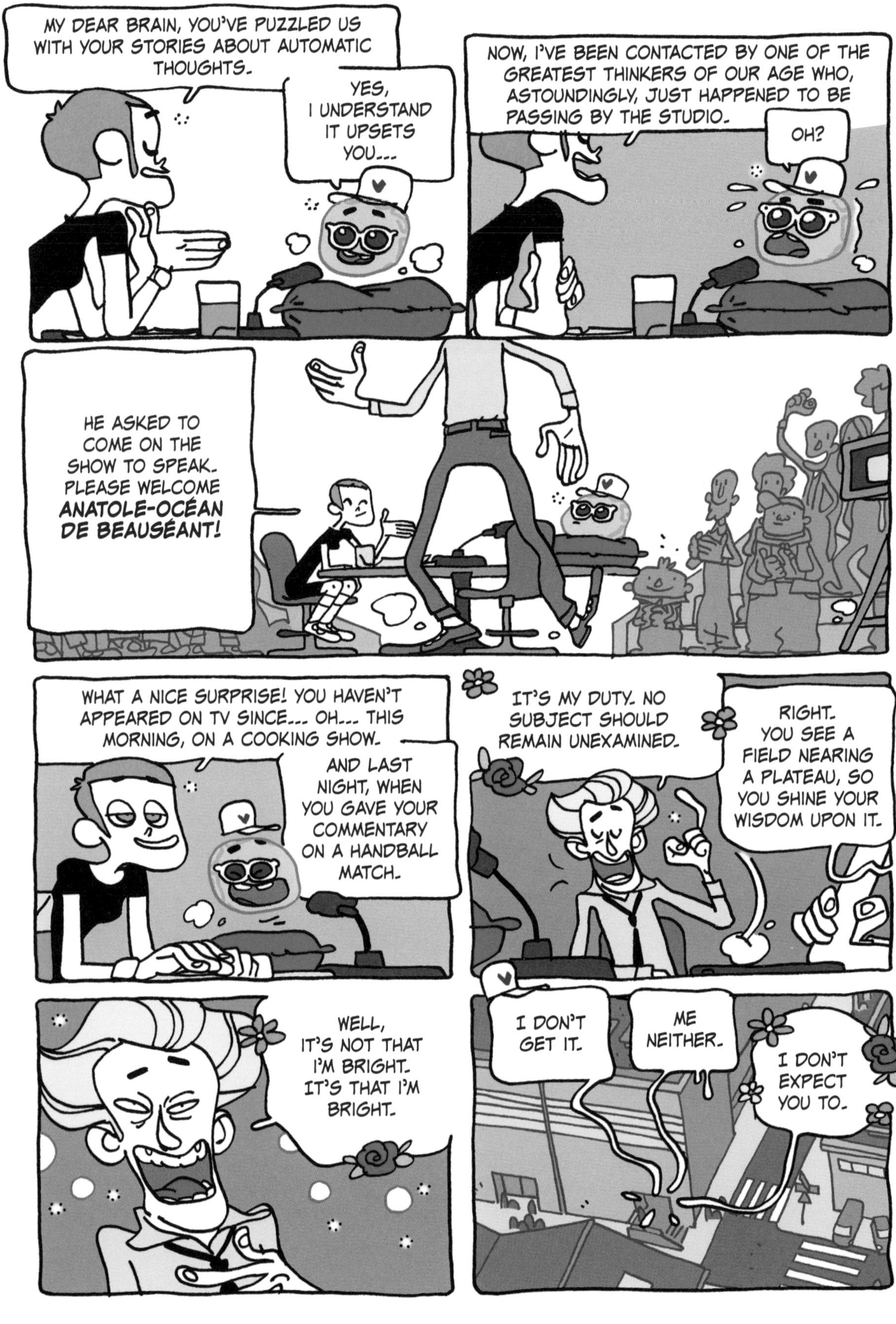
MY DEAR BRAIN, YOU'VE PUZZLED US WITH YOUR STORIES ABOUT AUTOMATIC THOUGHTS.
YES, I UNDERSTAND IT UPSETS YOU...
NOW, I'VE BEEN CONTACTED BY ONE OF THE GREATEST THINKERS OF OUR AGE WHO, ASTOUNDINGLY, JUST HAPPENED TO BE PASSING BY THE STUDIO.
OH?
HE ASKED TO COME ON THE SHOW TO SPEAK. PLEASE WELCOME **ANATOLE-OCÉAN DE BEAUSÉANT!**
WHAT A NICE SURPRISE! YOU HAVEN'T APPEARED ON TV SINCE... OH... THIS MORNING, ON A COOKING SHOW.
AND LAST NIGHT, WHEN YOU GAVE YOUR COMMENTARY ON A HANDBALL MATCH.
IT'S MY DUTY. NO SUBJECT SHOULD REMAIN UNEXAMINED.
RIGHT. YOU SEE A FIELD NEARING A PLATEAU, SO YOU SHINE YOUR WISDOM UPON IT.
WELL, IT'S NOT THAT I'M BRIGHT. IT'S THAT I'M BRIGHT.
I DON'T GET IT.
ME NEITHER.
I DON'T EXPECT YOU TO.

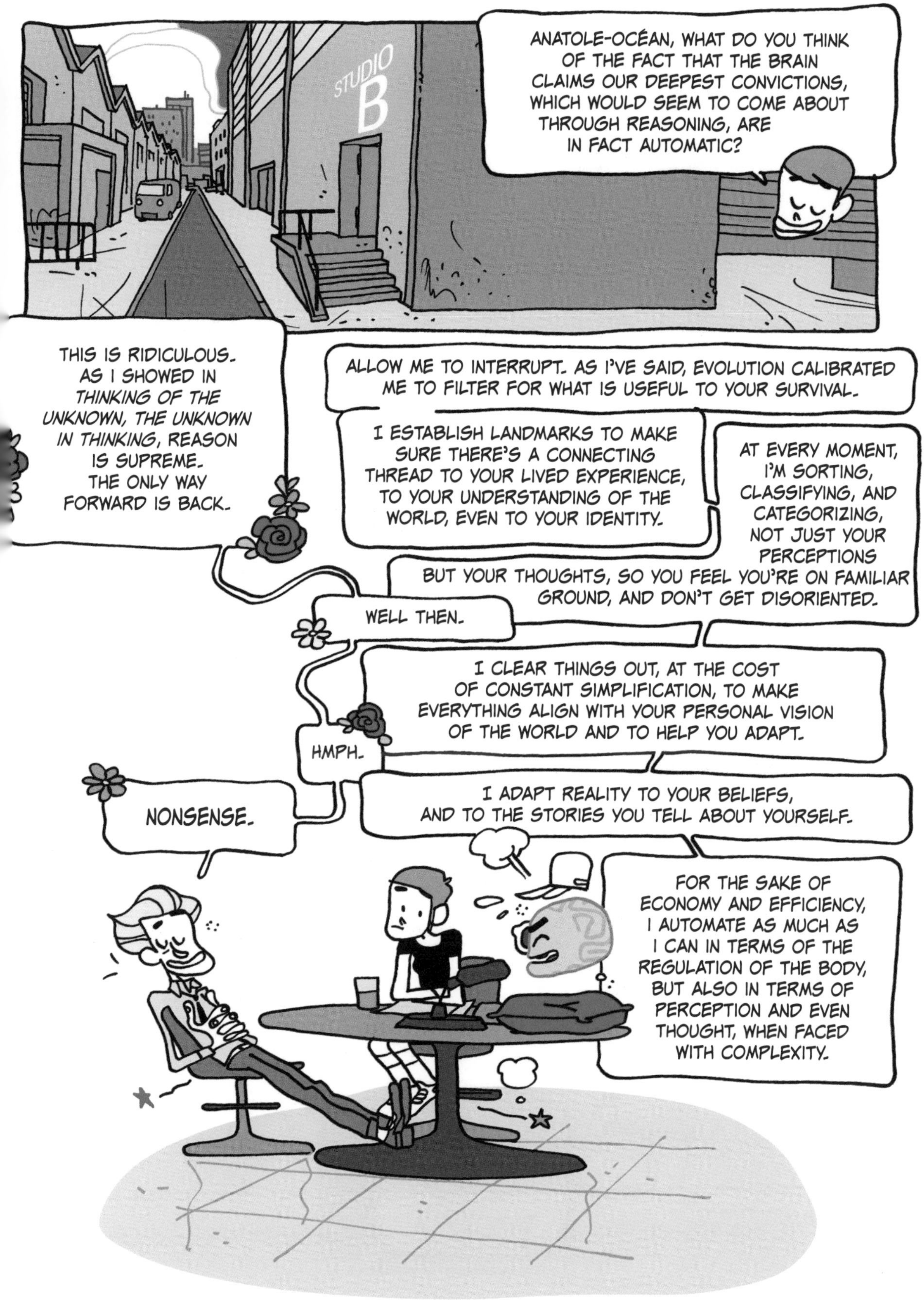
STUDIO B
ANATOLE-OCÉAN, WHAT DO YOU THINK OF THE FACT THAT THE BRAIN CLAIMS OUR DEEPEST CONVICTIONS, WHICH WOULD SEEM TO COME ABOUT THROUGH REASONING, ARE IN FACT AUTOMATIC?
THIS IS RIDICULOUS. AS I SHOWED IN *THINKING OF THE UNKNOWN, THE UNKNOWN IN THINKING*, REASON IS SUPREME. THE ONLY WAY FORWARD IS BACK.
ALLOW ME TO INTERRUPT. AS I'VE SAID, EVOLUTION CALIBRATED ME TO FILTER FOR WHAT IS USEFUL TO YOUR SURVIVAL.
I ESTABLISH LANDMARKS TO MAKE SURE THERE'S A CONNECTING THREAD TO YOUR LIVED EXPERIENCE, TO YOUR UNDERSTANDING OF THE WORLD, EVEN TO YOUR IDENTITY.
AT EVERY MOMENT, I'M SORTING, CLASSIFYING, AND CATEGORIZING, NOT JUST YOUR PERCEPTIONS
BUT YOUR THOUGHTS, SO YOU FEEL YOU'RE ON FAMILIAR GROUND, AND DON'T GET DISORIENTED.
WELL THEN.
I CLEAR THINGS OUT, AT THE COST OF CONSTANT SIMPLIFICATION, TO MAKE EVERYTHING ALIGN WITH YOUR PERSONAL VISION OF THE WORLD AND TO HELP YOU ADAPT.
HMPH.
NONSENSE.
I ADAPT REALITY TO YOUR BELIEFS, AND TO THE STORIES YOU TELL ABOUT YOURSELF.
FOR THE SAKE OF ECONOMY AND EFFICIENCY, I AUTOMATE AS MUCH AS I CAN IN TERMS OF THE REGULATION OF THE BODY, BUT ALSO IN TERMS OF PERCEPTION AND EVEN THOUGHT, WHEN FACED WITH COMPLEXITY.

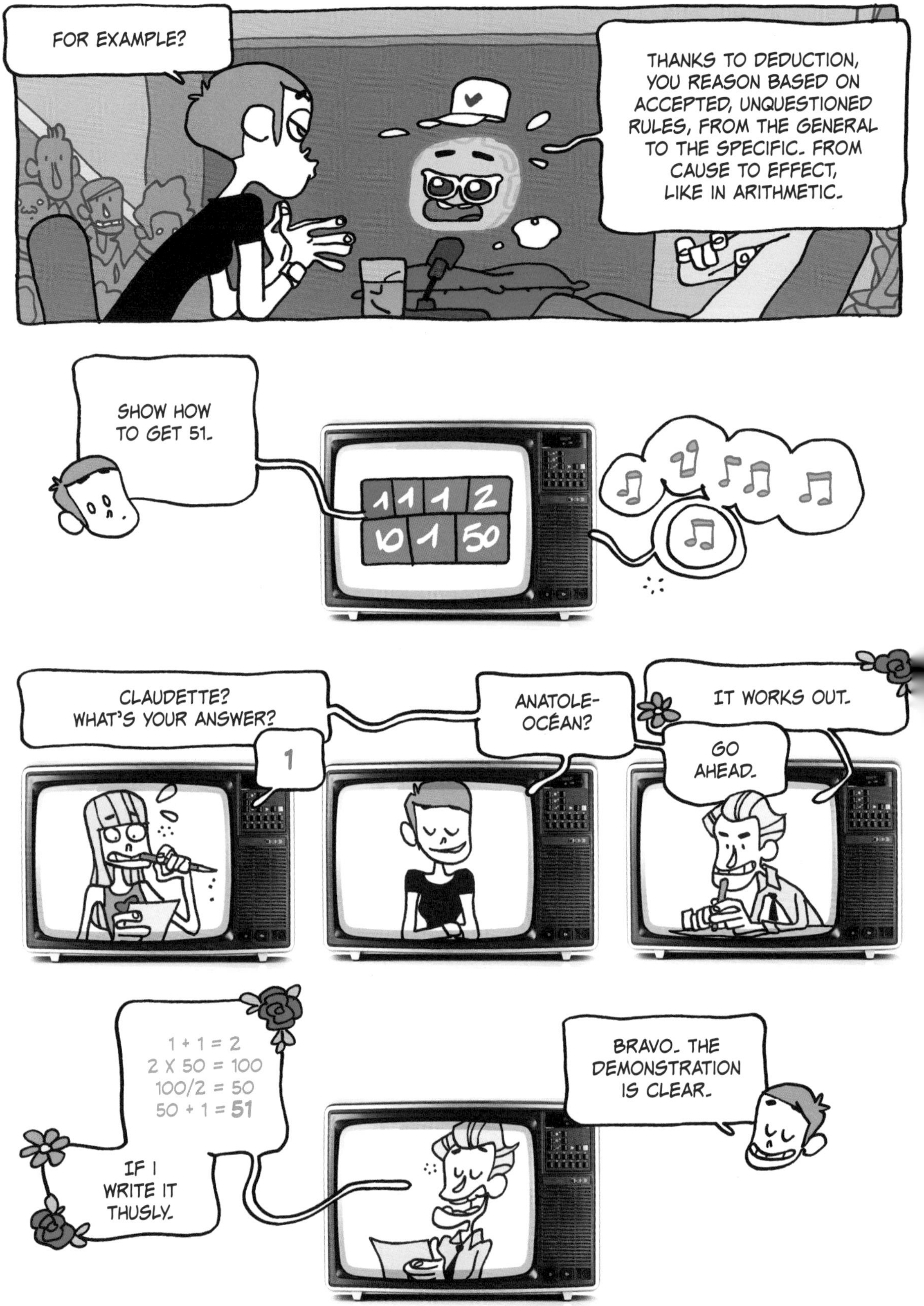
FOR EXAMPLE?
THANKS TO DEDUCTION, YOU REASON BASED ON ACCEPTED, UNQUESTIONED RULES, FROM THE GENERAL TO THE SPECIFIC. FROM CAUSE TO EFFECT, LIKE IN ARITHMETIC.
SHOW HOW TO GET 51.
1 1 1 2
10 1 50
CLAUDETTE? WHAT'S YOUR ANSWER?
1
ANATOLE-OCÉAN?
IT WORKS OUT.
GO AHEAD.
1 + 1 = 2
2 X 50 = 100
100/2 = 50
50 + 1 = 51
IF I WRITE IT THUSLY.
BRAVO. THE DEMONSTRATION IS CLEAR.

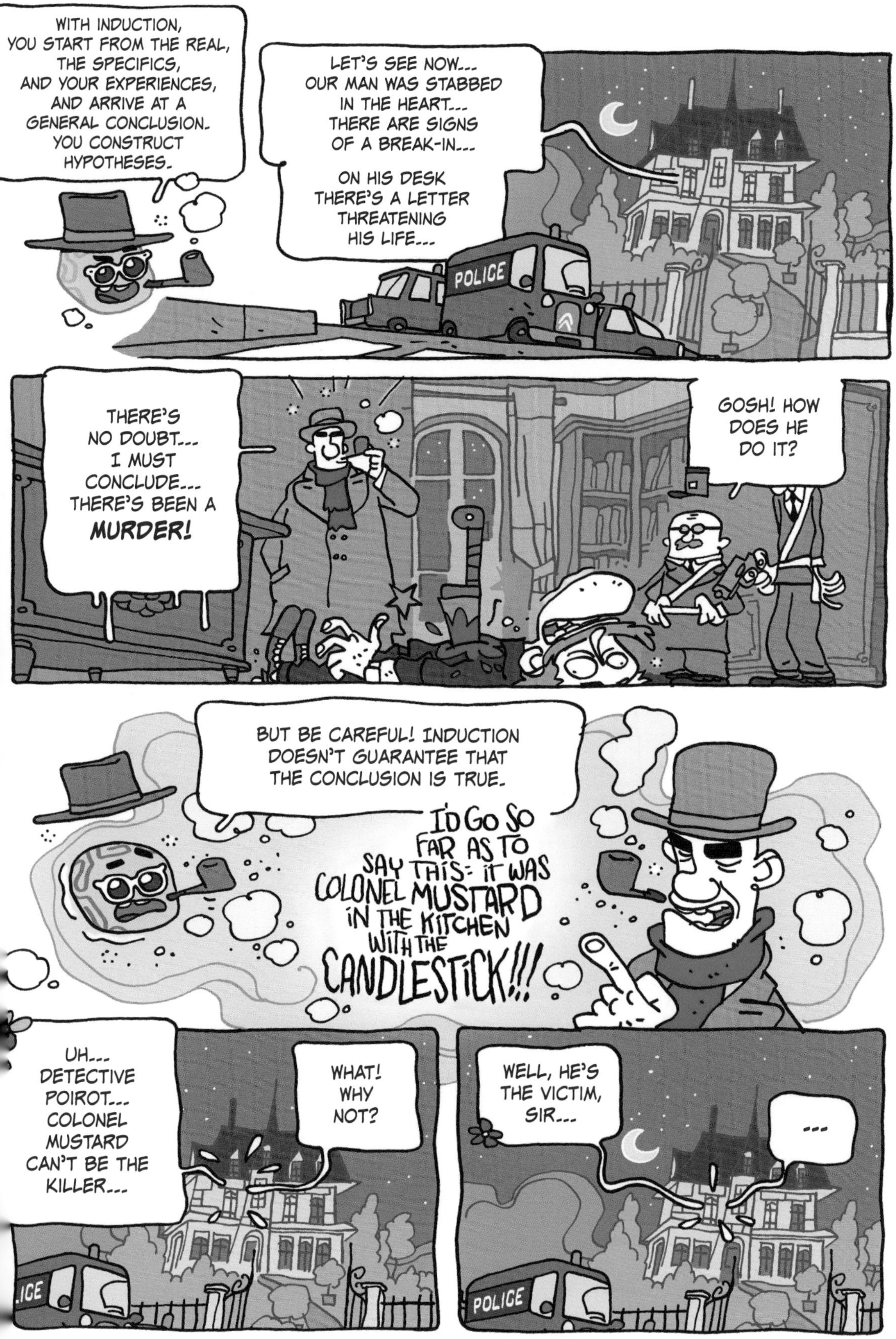
WITH INDUCTION, YOU START FROM THE REAL, THE SPECIFICS, AND YOUR EXPERIENCES, AND ARRIVE AT A GENERAL CONCLUSION. YOU CONSTRUCT HYPOTHESES.
LET'S SEE NOW... OUR MAN WAS STABBED IN THE HEART... THERE ARE SIGNS OF A BREAK-IN...
ON HIS DESK THERE'S A LETTER THREATENING HIS LIFE...
POLICE
THERE'S NO DOUBT... I MUST CONCLUDE... THERE'S BEEN A MURDER!
GOSH! HOW DOES HE DO IT?
BUT BE CAREFUL! INDUCTION DOESN'T GUARANTEE THAT THE CONCLUSION IS TRUE.
I'D GO SO FAR AS TO SAY THIS: IT WAS COLONEL MUSTARD IN THE KITCHEN WITH THE CANDLESTICK!!!
UH... DETECTIVE POIROT... COLONEL MUSTARD CAN'T BE THE KILLER...
WHAT! WHY NOT?
WELL, HE'S THE VICTIM, SIR...
...
POLICE

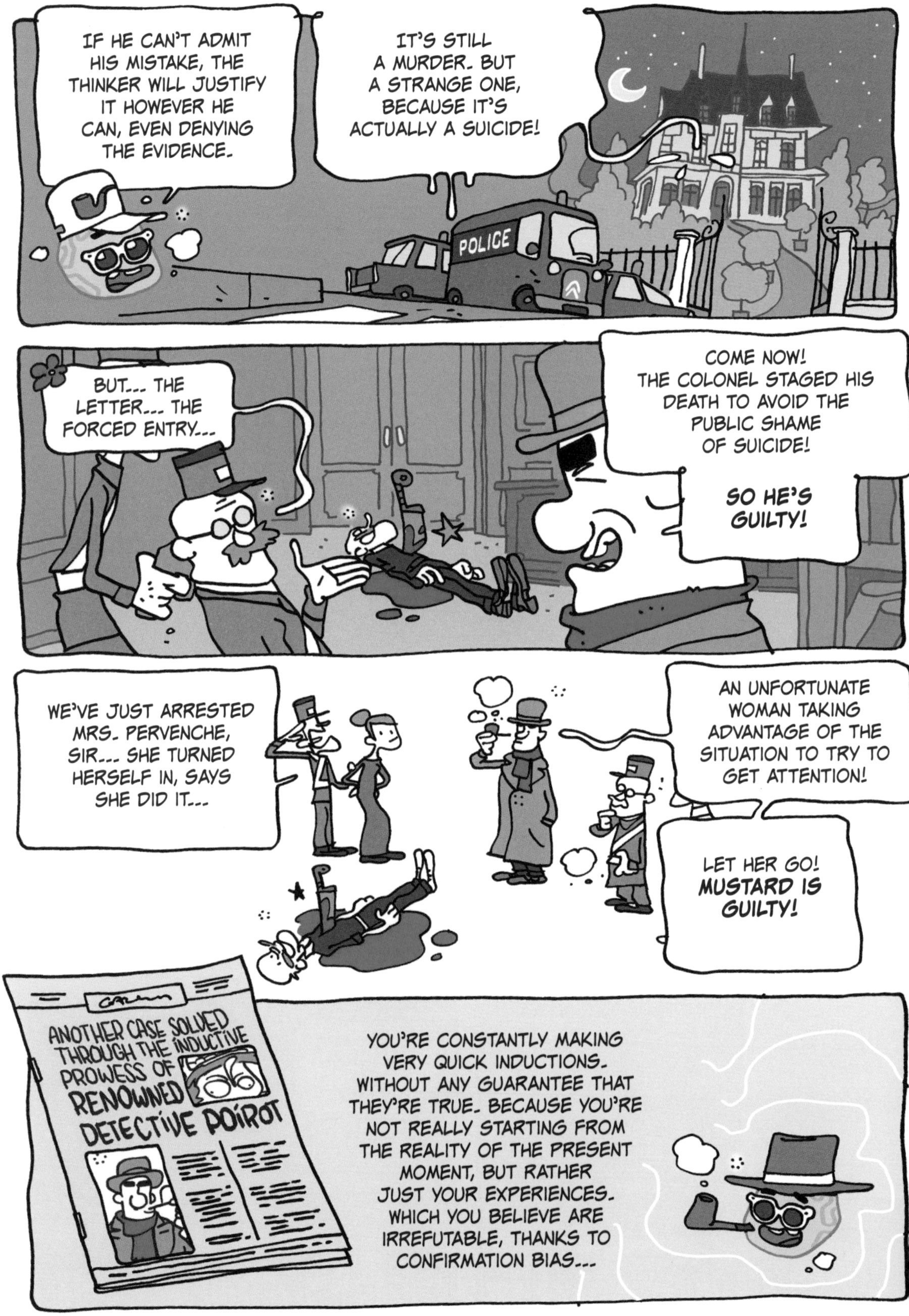
IF HE CAN'T ADMIT HIS MISTAKE, THE THINKER WILL JUSTIFY IT HOWEVER HE CAN, EVEN DENYING THE EVIDENCE.
IT'S STILL A MURDER. BUT A STRANGE ONE, BECAUSE IT'S ACTUALLY A SUICIDE!
POLICE
BUT... THE LETTER... THE FORCED ENTRY...
COME NOW! THE COLONEL STAGED HIS DEATH TO AVOID THE PUBLIC SHAME OF SUICIDE!
SO HE'S GUILTY!
WE'VE JUST ARRESTED MRS. PERVENCHE, SIR... SHE TURNED HERSELF IN, SAYS SHE DID IT...
AN UNFORTUNATE WOMAN TAKING ADVANTAGE OF THE SITUATION TO TRY TO GET ATTENTION!
LET HER GO! **MUSTARD IS GUILTY!**
ANOTHER CASE SOLVED THROUGH THE INDUCTIVE PROWESS OF RENOWNED DETECTIVE POIROT
YOU'RE CONSTANTLY MAKING VERY QUICK INDUCTIONS. WITHOUT ANY GUARANTEE THAT THEY'RE TRUE. BECAUSE YOU'RE NOT REALLY STARTING FROM THE REALITY OF THE PRESENT MOMENT, BUT RATHER JUST YOUR EXPERIENCES. WHICH YOU BELIEVE ARE IRREFUTABLE, THANKS TO CONFIRMATION BIAS...

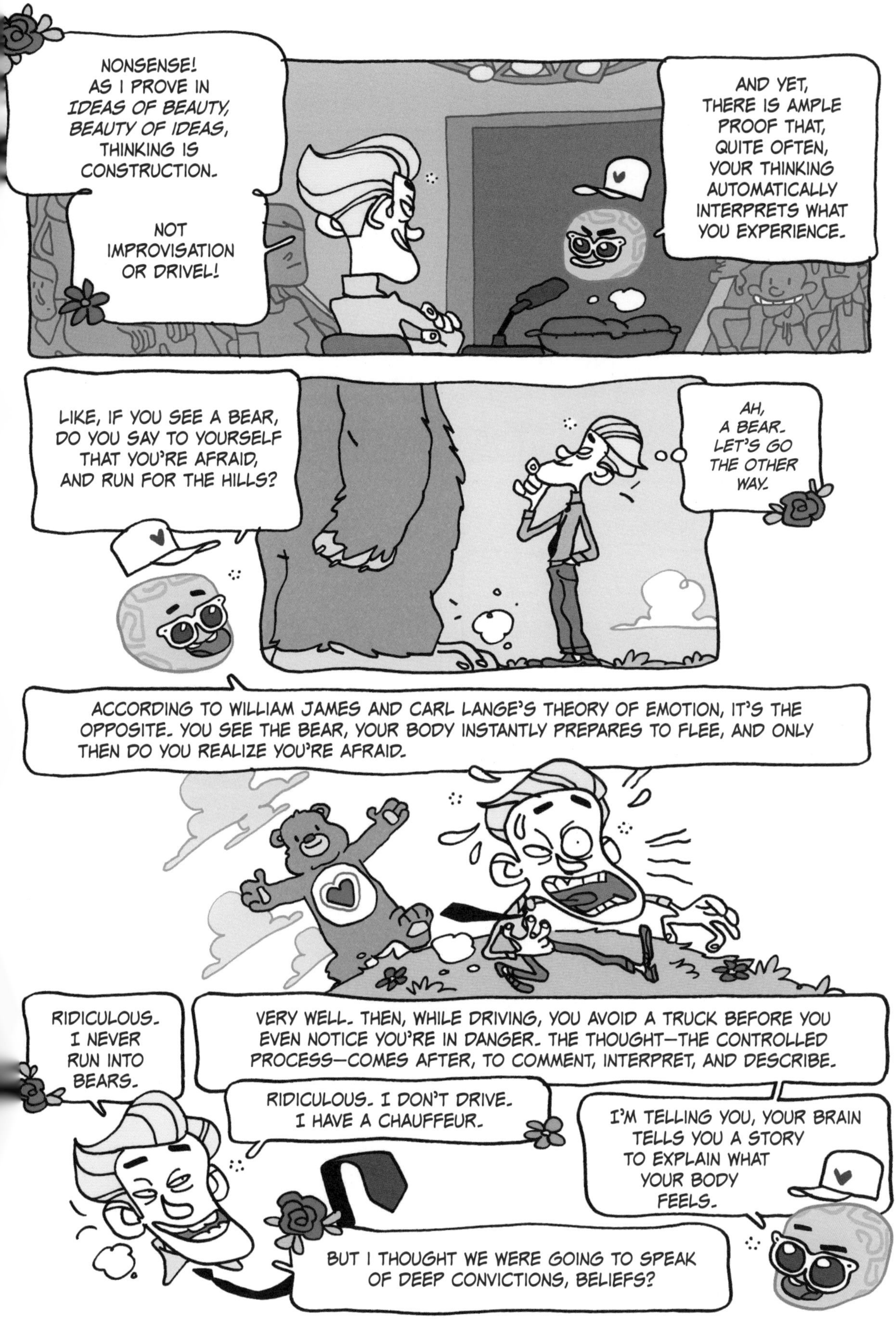
NONSENSE! AS I PROVE IN *IDEAS OF BEAUTY, BEAUTY OF IDEAS,* THINKING IS CONSTRUCTION.
NOT IMPROVISATION OR DRIVEL!
AND YET, THERE IS AMPLE PROOF THAT, QUITE OFTEN, YOUR THINKING AUTOMATICALLY INTERPRETS WHAT YOU EXPERIENCE.
LIKE, IF YOU SEE A BEAR, DO YOU SAY TO YOURSELF THAT YOU'RE AFRAID, AND RUN FOR THE HILLS?
AH, A BEAR. LET'S GO THE OTHER WAY.
ACCORDING TO WILLIAM JAMES AND CARL LANGE'S THEORY OF EMOTION, IT'S THE OPPOSITE. YOU SEE THE BEAR, YOUR BODY INSTANTLY PREPARES TO FLEE, AND ONLY THEN DO YOU REALIZE YOU'RE AFRAID.
RIDICULOUS. I NEVER RUN INTO BEARS.
VERY WELL. THEN, WHILE DRIVING, YOU AVOID A TRUCK BEFORE YOU EVEN NOTICE YOU'RE IN DANGER. THE THOUGHT—THE CONTROLLED PROCESS—COMES AFTER, TO COMMENT, INTERPRET, AND DESCRIBE.
RIDICULOUS. I DON'T DRIVE. I HAVE A CHAUFFEUR.
I'M TELLING YOU, YOUR BRAIN TELLS YOU A STORY TO EXPLAIN WHAT YOUR BODY FEELS.
BUT I THOUGHT WE WERE GOING TO SPEAK OF DEEP CONVICTIONS, BELIEFS?

IT'S THE SAME PRINCIPLE! YOU CREATE BIG, SIMPLISTIC CATEGORIES TO HELP YOU INTERPRET THE COMPLEX WORLD...

...YOUR THINKING BECOMES AUTOMATIC, AND YOUR COGNITIVE BIASES TAKE IT FROM THERE! RELENTLESSLY CONFIRMING YOUR VIEW OF THE WORLD!

TO MY MIND, NOTHING HOLDS A CANDLE TO CENTRISM! BESIDES, THE *WALL STREET JOURNAL* SUPPORTS IT!

I DON'T READ THAT RAG! I PREFER THE *POST*! BESIDES, EVERY TIME I BUY IT, THEY SAY THE LEFT HAS THE RIGHT OF IT!

ANOTHER VICTIM OF CAPITALISM!

JUST LOOK WHERE HANDOUTS GET YOU!

YOU CAN'T LIKEN THINKING TO A REFLEX! NOT MINE, AT LEAST.

OH, BUT THE MOST COMPLEX, COHERENT THOUGHT, AS WELL AS THE MOST OFF-THE-MARK, CAN BE JUST AS QUICK AS A REFLEX!

LIKE DO YOU KNOW WHAT PSYCHOLOGY STUDENT MICHAEL GAZZANIGA DISCOVERED IN THE EARLY 1960S?

...OF COURSE. BUT I'LL LET YOU EXPLAIN, SINCE YOU'RE CLEARLY DYING TO.

ALRIGHT. YOU REMEMBER THAT MY TWO HEMISPHERES ARE CONNECTED BY THE CORPUS CALLOSUM, A MASSIVE BUNDLE OF WHITE MATTER?

GAZZANIGA DISCOVERED WHAT HAPPENS IF WE CUT THE CORPUS CALLOSUM, FOR EXAMPLE TO PREVENT AN EPILEPTIC FIT FROM SPREADING FROM ONE HEMISPHERE TO THE OTHER.

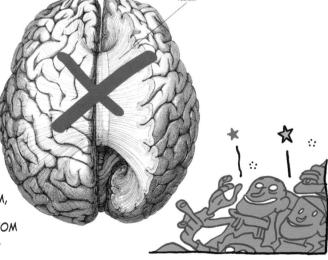

Corpus callosum

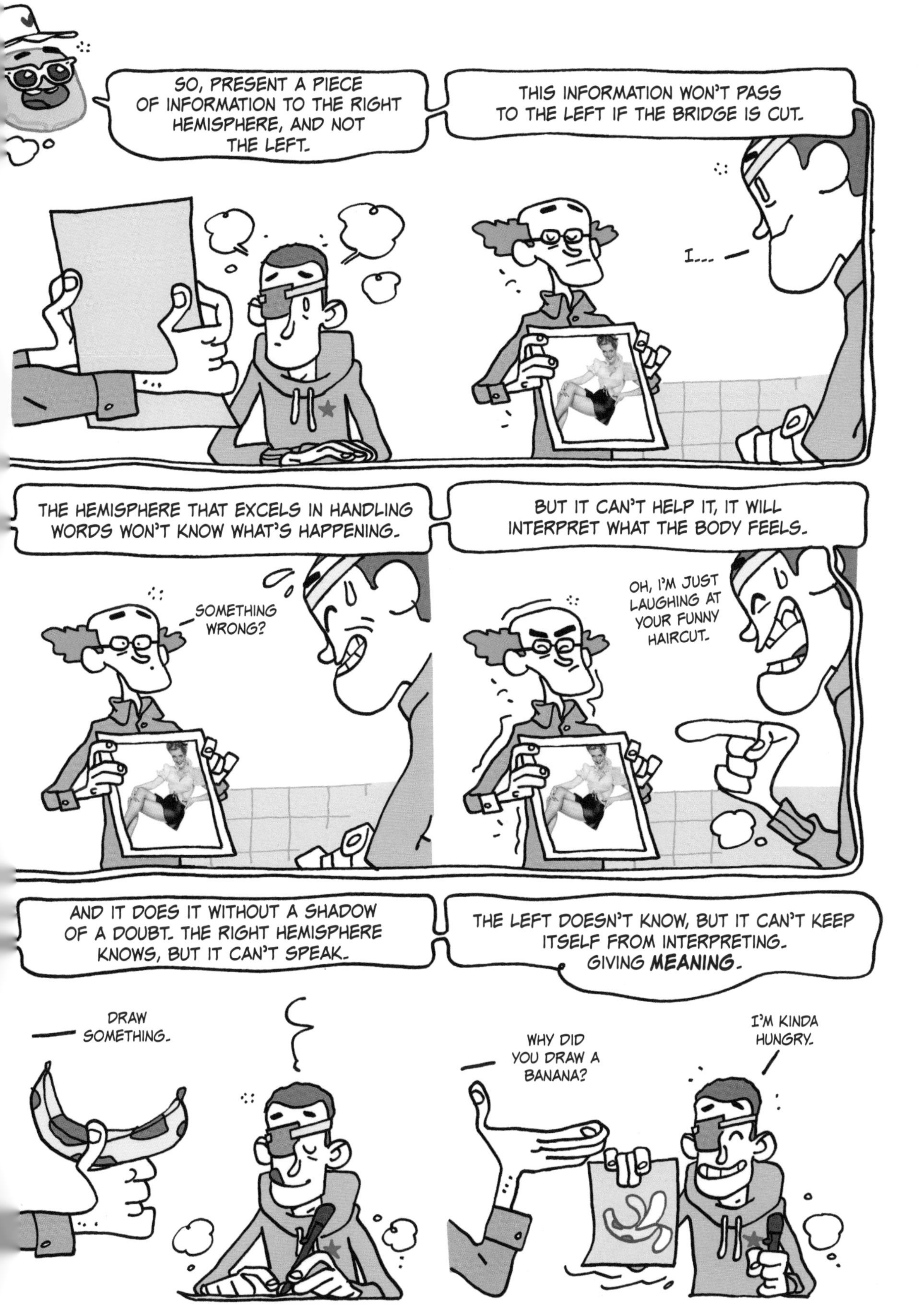

SO, PRESENT A PIECE OF INFORMATION TO THE RIGHT HEMISPHERE, AND NOT THE LEFT.
THIS INFORMATION WON'T PASS TO THE LEFT IF THE BRIDGE IS CUT.
I...
THE HEMISPHERE THAT EXCELS IN HANDLING WORDS WON'T KNOW WHAT'S HAPPENING.
BUT IT CAN'T HELP IT, IT WILL INTERPRET WHAT THE BODY FEELS.
SOMETHING WRONG?
OH, I'M JUST LAUGHING AT YOUR FUNNY HAIRCUT.
AND IT DOES IT WITHOUT A SHADOW OF A DOUBT. THE RIGHT HEMISPHERE KNOWS, BUT IT CAN'T SPEAK.
THE LEFT DOESN'T KNOW, BUT IT CAN'T KEEP ITSELF FROM INTERPRETING. GIVING **MEANING**.
DRAW SOMETHING.
WHY DID YOU DRAW A BANANA?
I'M KINDA HUNGRY.

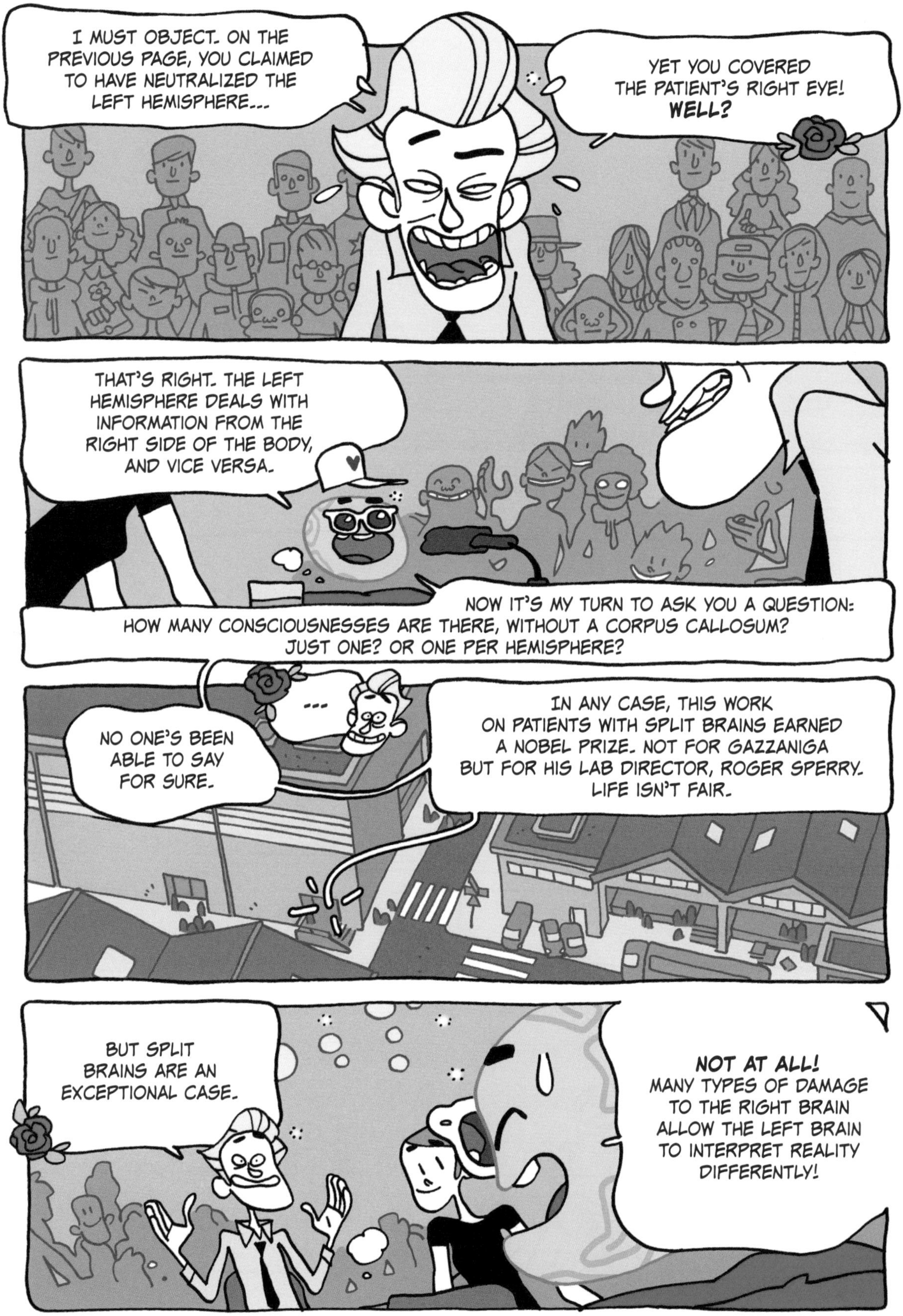
I MUST OBJECT. ON THE PREVIOUS PAGE, YOU CLAIMED TO HAVE NEUTRALIZED THE LEFT HEMISPHERE...
YET YOU COVERED THE PATIENT'S RIGHT EYE! **WELL?**
THAT'S RIGHT. THE LEFT HEMISPHERE DEALS WITH INFORMATION FROM THE RIGHT SIDE OF THE BODY, AND VICE VERSA.
NOW IT'S MY TURN TO ASK YOU A QUESTION: HOW MANY CONSCIOUSNESSES ARE THERE, WITHOUT A CORPUS CALLOSUM? JUST ONE? OR ONE PER HEMISPHERE?
...
NO ONE'S BEEN ABLE TO SAY FOR SURE.
IN ANY CASE, THIS WORK ON PATIENTS WITH SPLIT BRAINS EARNED A NOBEL PRIZE. NOT FOR GAZZANIGA BUT FOR HIS LAB DIRECTOR, ROGER SPERRY. LIFE ISN'T FAIR.
BUT SPLIT BRAINS ARE AN EXCEPTIONAL CASE.
NOT AT ALL! MANY TYPES OF DAMAGE TO THE RIGHT BRAIN ALLOW THE LEFT BRAIN TO INTERPRET REALITY DIFFERENTLY!

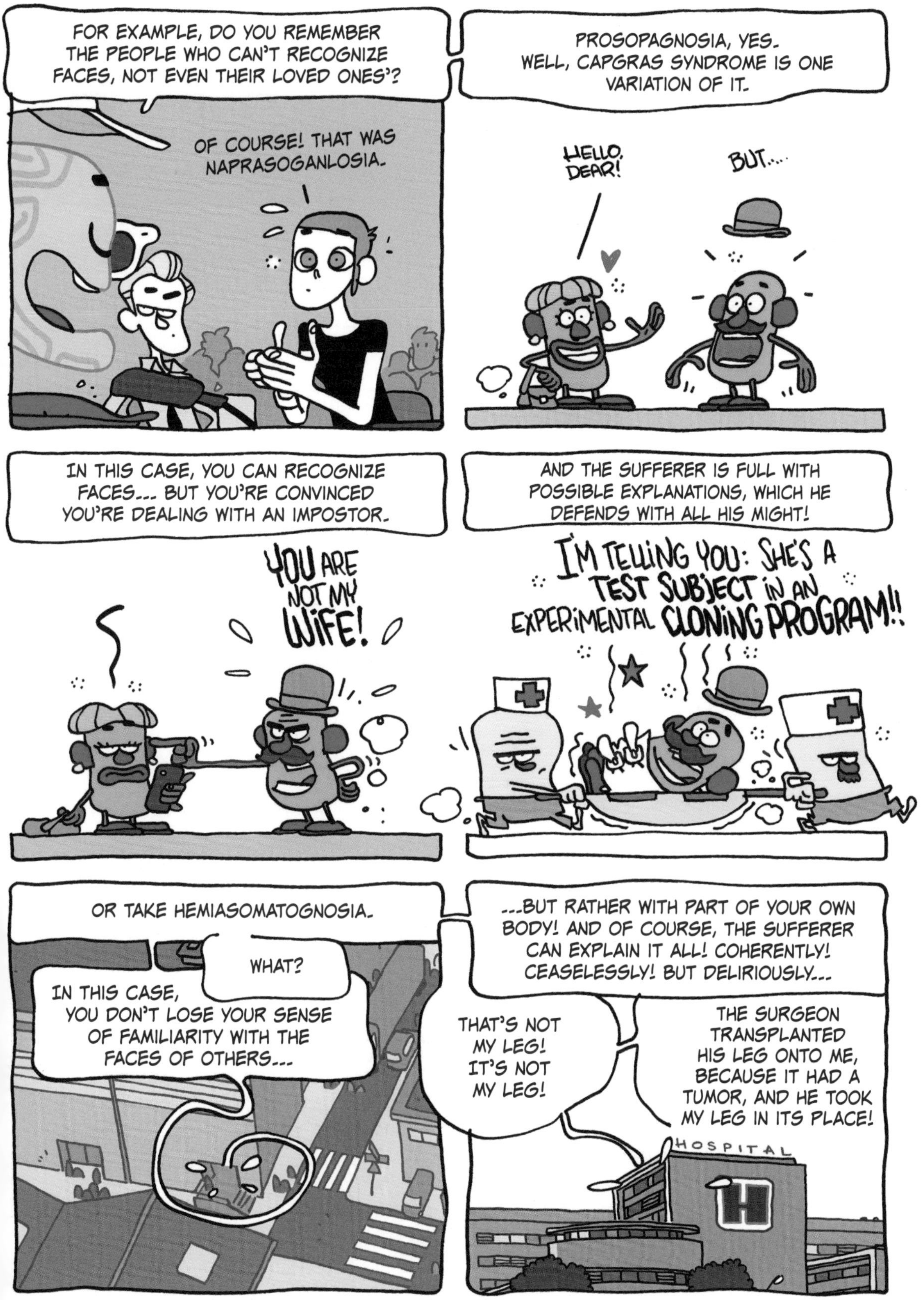
FOR EXAMPLE, DO YOU REMEMBER THE PEOPLE WHO CAN'T RECOGNIZE FACES, NOT EVEN THEIR LOVED ONES'?
OF COURSE! THAT WAS NAPRASOGANLOSIA.
PROSOPAGNOSIA, YES. WELL, CAPGRAS SYNDROME IS ONE VARIATION OF IT.
HELLO, DEAR!
BUT....
IN THIS CASE, YOU CAN RECOGNIZE FACES... BUT YOU'RE CONVINCED YOU'RE DEALING WITH AN IMPOSTOR.
YOU ARE NOT MY WIFE!
AND THE SUFFERER IS FULL WITH POSSIBLE EXPLANATIONS, WHICH HE DEFENDS WITH ALL HIS MIGHT!
I'M TELLING YOU: SHE'S A TEST SUBJECT IN AN EXPERIMENTAL CLONING PROGRAM!!
OR TAKE HEMIASOMATOGNOSIA.
WHAT?
IN THIS CASE, YOU DON'T LOSE YOUR SENSE OF FAMILIARITY WITH THE FACES OF OTHERS...
...BUT RATHER WITH PART OF YOUR OWN BODY! AND OF COURSE, THE SUFFERER CAN EXPLAIN IT ALL! COHERENTLY! CEASELESSLY! BUT DELIRIOUSLY...
THAT'S NOT MY LEG! IT'S NOT MY LEG!
THE SURGEON TRANSPLANTED HIS LEG ONTO ME, BECAUSE IT HAD A TUMOR, AND HE TOOK MY LEG IN ITS PLACE!
HOSPITAL

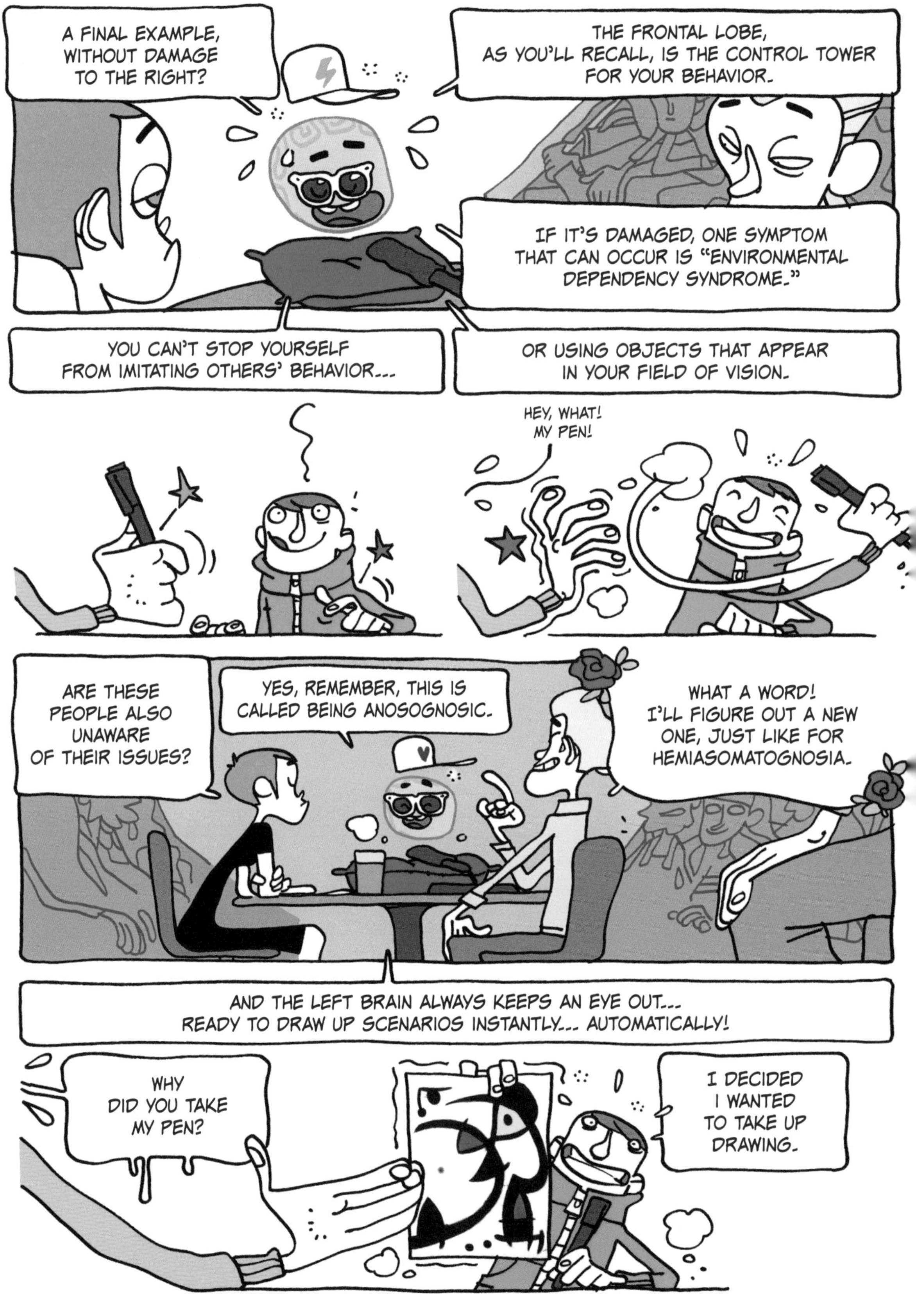
A FINAL EXAMPLE, WITHOUT DAMAGE TO THE RIGHT?
THE FRONTAL LOBE, AS YOU'LL RECALL, IS THE CONTROL TOWER FOR YOUR BEHAVIOR.
IF IT'S DAMAGED, ONE SYMPTOM THAT CAN OCCUR IS "ENVIRONMENTAL DEPENDENCY SYNDROME."
YOU CAN'T STOP YOURSELF FROM IMITATING OTHERS' BEHAVIOR...
OR USING OBJECTS THAT APPEAR IN YOUR FIELD OF VISION.
HEY, WHAT! MY PEN!
ARE THESE PEOPLE ALSO UNAWARE OF THEIR ISSUES?
YES, REMEMBER, THIS IS CALLED BEING ANOSOGNOSIC.
WHAT A WORD! I'LL FIGURE OUT A NEW ONE, JUST LIKE FOR HEMIASOMATOGNOSIA.
AND THE LEFT BRAIN ALWAYS KEEPS AN EYE OUT... READY TO DRAW UP SCENARIOS INSTANTLY... AUTOMATICALLY!
WHY DID YOU TAKE MY PEN?
I DECIDED I WANTED TO TAKE UP DRAWING.

LIKE WITH THE BLIND SPOT THAT SHOULD OBSCURE HALF YOUR FIELD OF VISION, BUT WHICH I CONSTANTLY FILL IN... I FILL INTELLECTUAL HOLES WHEN YOUR PERCEPTIONS DON'T MAKE SENSE.
I FILTER AND WRITE EVERYTHING, INCLUDING THE MEANING YOU GIVE TO YOUR EXPERIENCES.
I'M A MACHINE DESIGNED TO GIVE SENSE TO THINGS, AS COHERENTLY AND RELIABLY AS POSSIBLE...
BECAUSE ABOVE ALL I'M DESIGNED FOR SURVIVAL.
AND MEANING IS ESSENTIAL FOR SURVIVAL.
WHAT ARE THE LIMITS OF INTERPRETATION? OF ITS FRENZY OF SENSE, IN ALL ITS SENSELESSNESS?
WE CAN AT LEAST AGREE ON THE IMPORTANCE OF MEANING. WHICH I HAVE WRITTEN A BOOK ON: *MEANING OF LIFE, LIFE OF MEANING.*
SORRY, I HAVEN'T READ IT.
THAT DOESN'T BOTHER ME. I DON'T WRITE SO PEOPLE WILL READ IT, BUT SO PEOPLE WILL **TALK** ABOUT IT.
OR RATHER, SO I CAN TALK ABOUT IT. WHO ELSE WOULD BE CAPABLE OF IT? IT REQUIRES HEIGHTENED SENSIBILITIES, AND FREE THINKING.

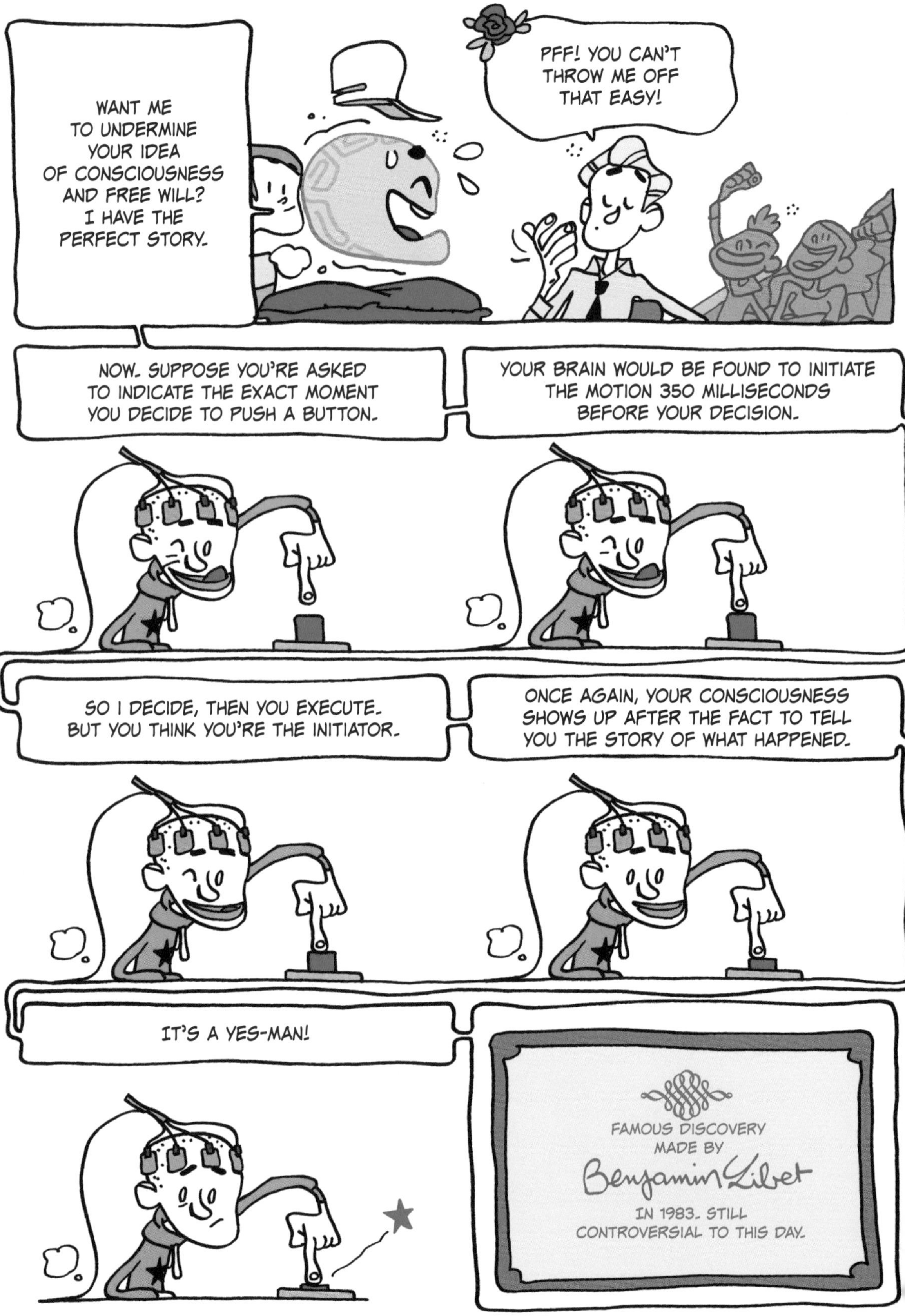
WANT ME TO UNDERMINE YOUR IDEA OF CONSCIOUSNESS AND FREE WILL? I HAVE THE PERFECT STORY.
PFF! YOU CAN'T THROW ME OFF THAT EASY!
NOW. SUPPOSE YOU'RE ASKED TO INDICATE THE EXACT MOMENT YOU DECIDE TO PUSH A BUTTON.
YOUR BRAIN WOULD BE FOUND TO INITIATE THE MOTION 350 MILLISECONDS BEFORE YOUR DECISION.
SO I DECIDE, THEN YOU EXECUTE. BUT YOU THINK YOU'RE THE INITIATOR.
ONCE AGAIN, YOUR CONSCIOUSNESS SHOWS UP AFTER THE FACT TO TELL YOU THE STORY OF WHAT HAPPENED.
IT'S A YES-MAN!
FAMOUS DISCOVERY MADE BY Benjamin Libet IN 1983. STILL CONTROVERSIAL TO THIS DAY.

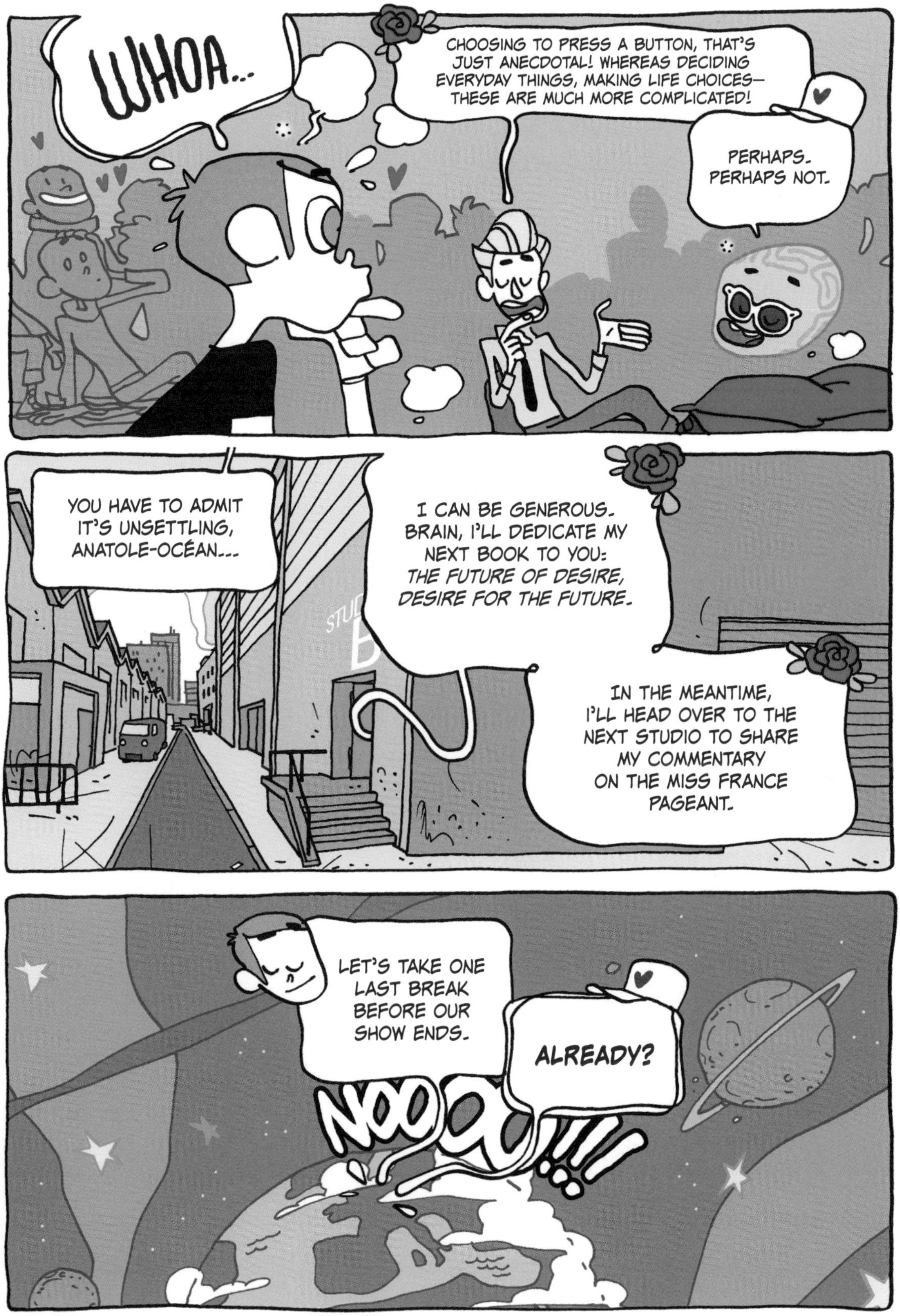
WHOA...
CHOOSING TO PRESS A BUTTON, THAT'S JUST ANECDOTAL! WHEREAS DECIDING EVERYDAY THINGS, MAKING LIFE CHOICES— THESE ARE MUCH MORE COMPLICATED!
PERHAPS. PERHAPS NOT.
YOU HAVE TO ADMIT IT'S UNSETTLING, ANATOLE-OCÉAN...
I CAN BE GENEROUS. BRAIN, I'LL DEDICATE MY NEXT BOOK TO YOU: *THE FUTURE OF DESIRE, DESIRE FOR THE FUTURE.*
IN THE MEANTIME, I'LL HEAD OVER TO THE NEXT STUDIO TO SHARE MY COMMENTARY ON THE MISS FRANCE PAGEANT.
LET'S TAKE ONE LAST BREAK BEFORE OUR SHOW ENDS.
ALREADY?
NOOOO!!!...

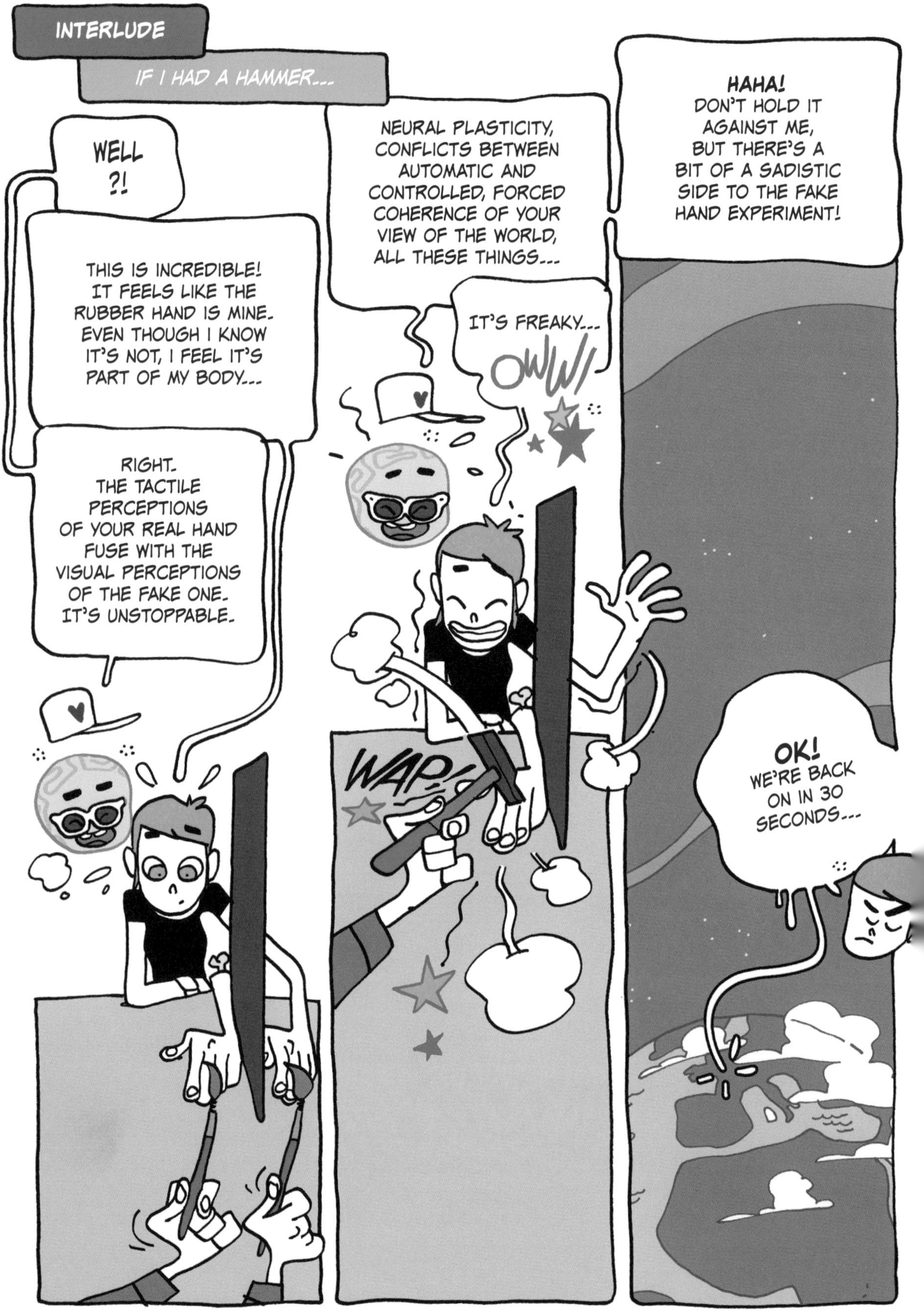
INTERLUDE
IF I HAD A HAMMER...
WELL ?!
THIS IS INCREDIBLE! IT FEELS LIKE THE RUBBER HAND IS MINE. EVEN THOUGH I KNOW IT'S NOT, I FEEL IT'S PART OF MY BODY...
RIGHT. THE TACTILE PERCEPTIONS OF YOUR REAL HAND FUSE WITH THE VISUAL PERCEPTIONS OF THE FAKE ONE. IT'S UNSTOPPABLE.
NEURAL PLASTICITY, CONFLICTS BETWEEN AUTOMATIC AND CONTROLLED, FORCED COHERENCE OF YOUR VIEW OF THE WORLD, ALL THESE THINGS...
IT'S FREAKY...
OWW!
WAP!
HAHA! DON'T HOLD IT AGAINST ME, BUT THERE'S A BIT OF A SADISTIC SIDE TO THE FAKE HAND EXPERIMENT!
OK! WE'RE BACK ON IN 30 SECONDS...

Chapter 10

THE CURIOSITY CABINET

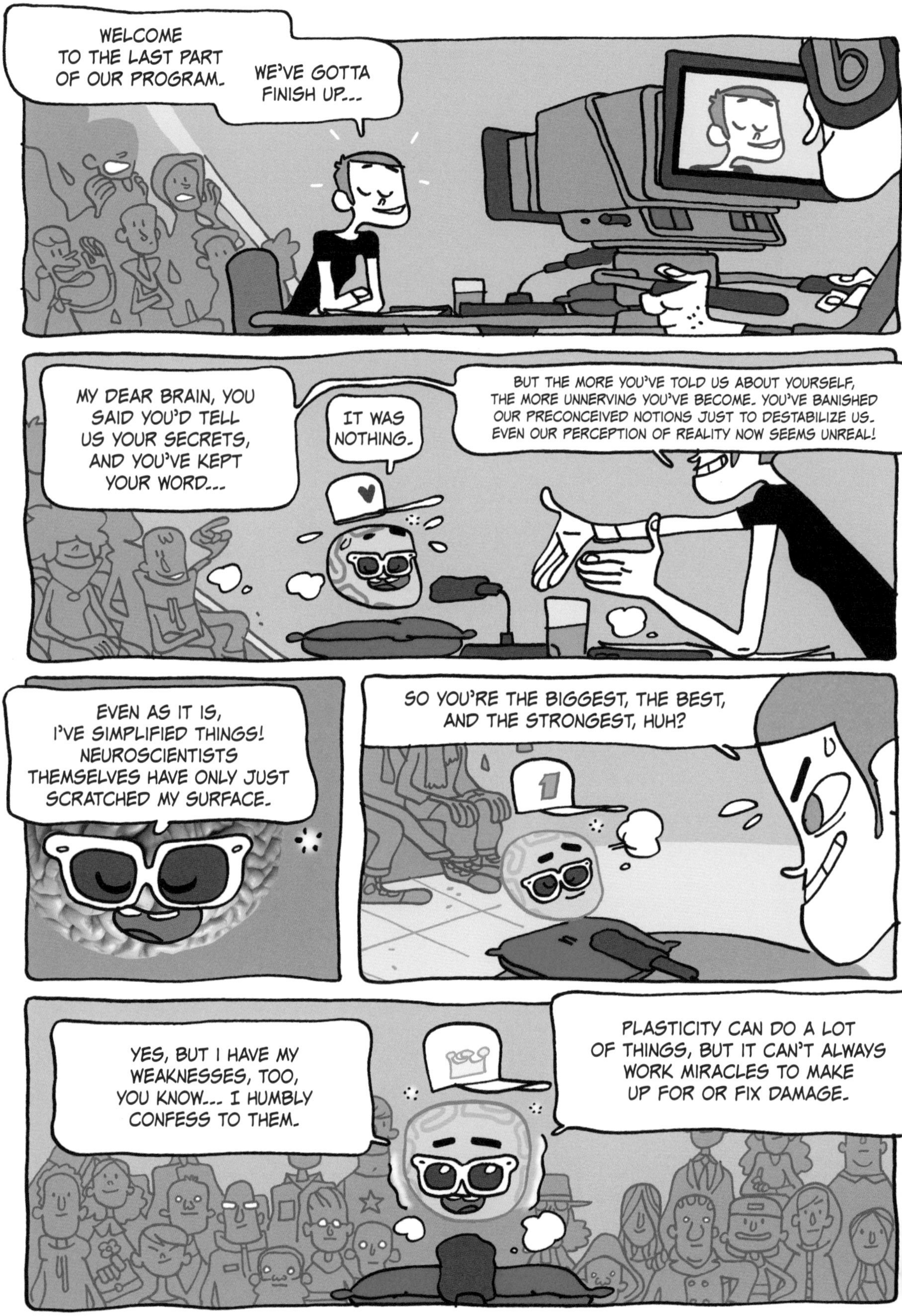
WELCOME TO THE LAST PART OF OUR PROGRAM.
WE'VE GOTTA FINISH UP...
MY DEAR BRAIN, YOU SAID YOU'D TELL US YOUR SECRETS, AND YOU'VE KEPT YOUR WORD...
IT WAS NOTHING.
BUT THE MORE YOU'VE TOLD US ABOUT YOURSELF, THE MORE UNNERVING YOU'VE BECOME. YOU'VE BANISHED OUR PRECONCEIVED NOTIONS JUST TO DESTABILIZE US. EVEN OUR PERCEPTION OF REALITY NOW SEEMS UNREAL!
EVEN AS IT IS, I'VE SIMPLIFIED THINGS! NEUROSCIENTISTS THEMSELVES HAVE ONLY JUST SCRATCHED MY SURFACE.
SO YOU'RE THE BIGGEST, THE BEST, AND THE STRONGEST, HUH?
YES, BUT I HAVE MY WEAKNESSES, TOO, YOU KNOW... I HUMBLY CONFESS TO THEM.
PLASTICITY CAN DO A LOT OF THINGS, BUT IT CAN'T ALWAYS WORK MIRACLES TO MAKE UP FOR OR FIX DAMAGE.

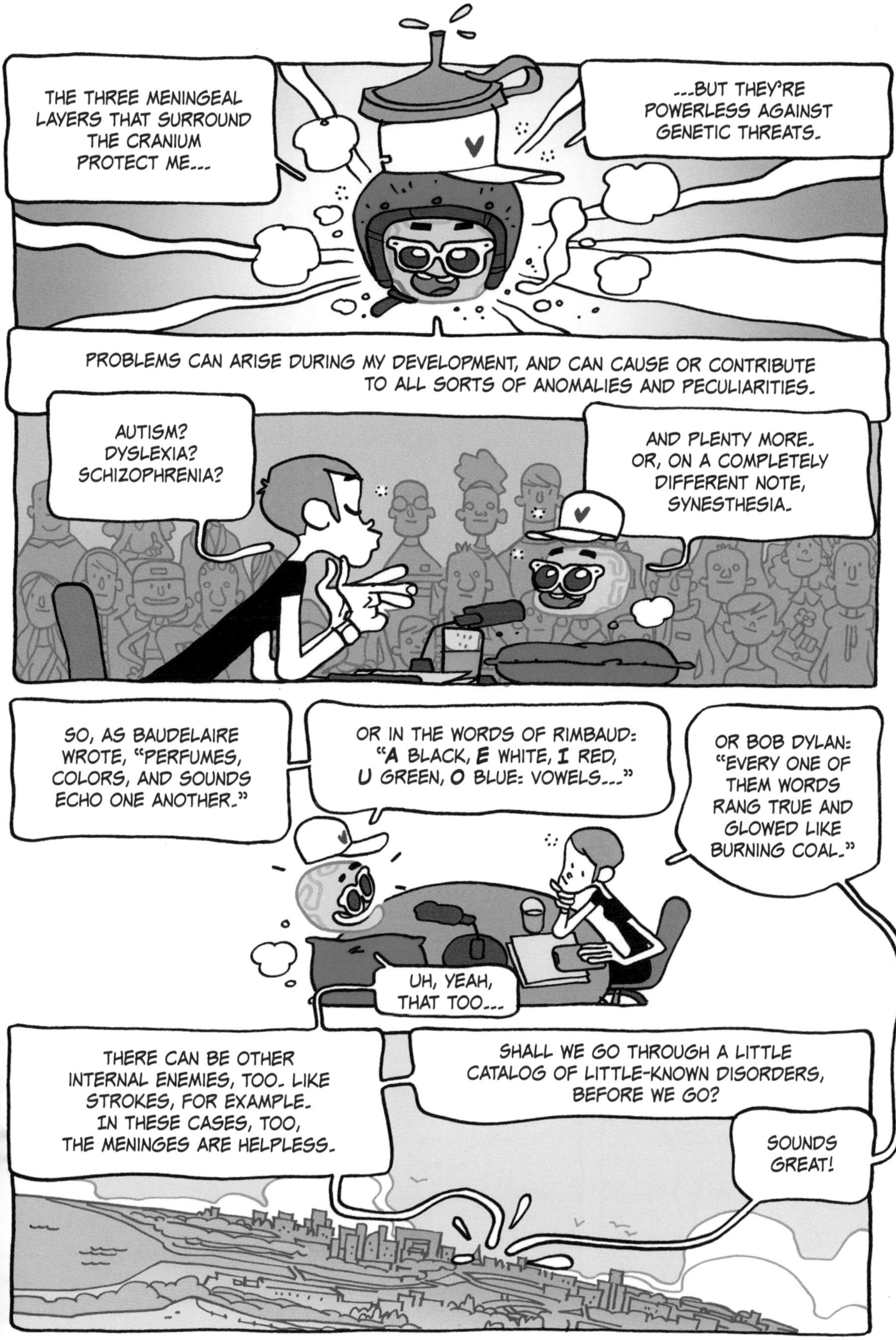
THE THREE MENINGEAL LAYERS THAT SURROUND THE CRANIUM PROTECT ME...
...BUT THEY'RE POWERLESS AGAINST GENETIC THREATS.
PROBLEMS CAN ARISE DURING MY DEVELOPMENT, AND CAN CAUSE OR CONTRIBUTE TO ALL SORTS OF ANOMALIES AND PECULIARITIES.
AUTISM? DYSLEXIA? SCHIZOPHRENIA?
AND PLENTY MORE. OR, ON A COMPLETELY DIFFERENT NOTE, SYNESTHESIA.
SO, AS BAUDELAIRE WROTE, "PERFUMES, COLORS, AND SOUNDS ECHO ONE ANOTHER."
OR IN THE WORDS OF RIMBAUD: "A BLACK, E WHITE, I RED, U GREEN, O BLUE: VOWELS..."
OR BOB DYLAN: "EVERY ONE OF THEM WORDS RANG TRUE AND GLOWED LIKE BURNING COAL."
UH, YEAH, THAT TOO...
THERE CAN BE OTHER INTERNAL ENEMIES, TOO. LIKE STROKES, FOR EXAMPLE. IN THESE CASES, TOO, THE MENINGES ARE HELPLESS.
SHALL WE GO THROUGH A LITTLE CATALOG OF LITTLE-KNOWN DISORDERS, BEFORE WE GO?
SOUNDS GREAT!

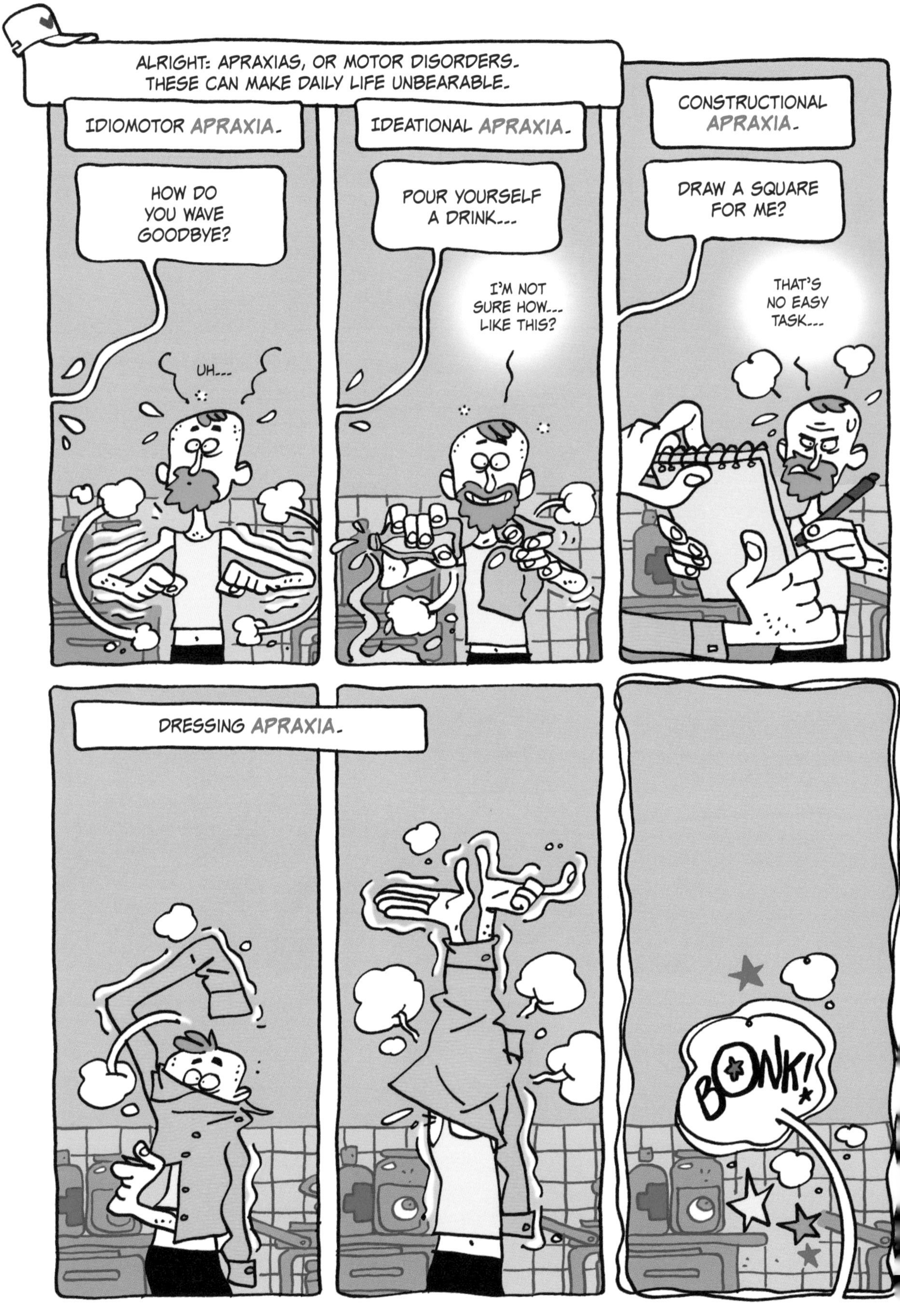
ALRIGHT: APRAXIAS, OR MOTOR DISORDERS. THESE CAN MAKE DAILY LIFE UNBEARABLE.
IDIOMOTOR APRAXIA.
HOW DO YOU WAVE GOODBYE?
UH...
IDEATIONAL APRAXIA.
POUR YOURSELF A DRINK...
I'M NOT SURE HOW... LIKE THIS?
CONSTRUCTIONAL APRAXIA.
DRAW A SQUARE FOR ME?
THAT'S NO EASY TASK...
DRESSING APRAXIA.
BONK!

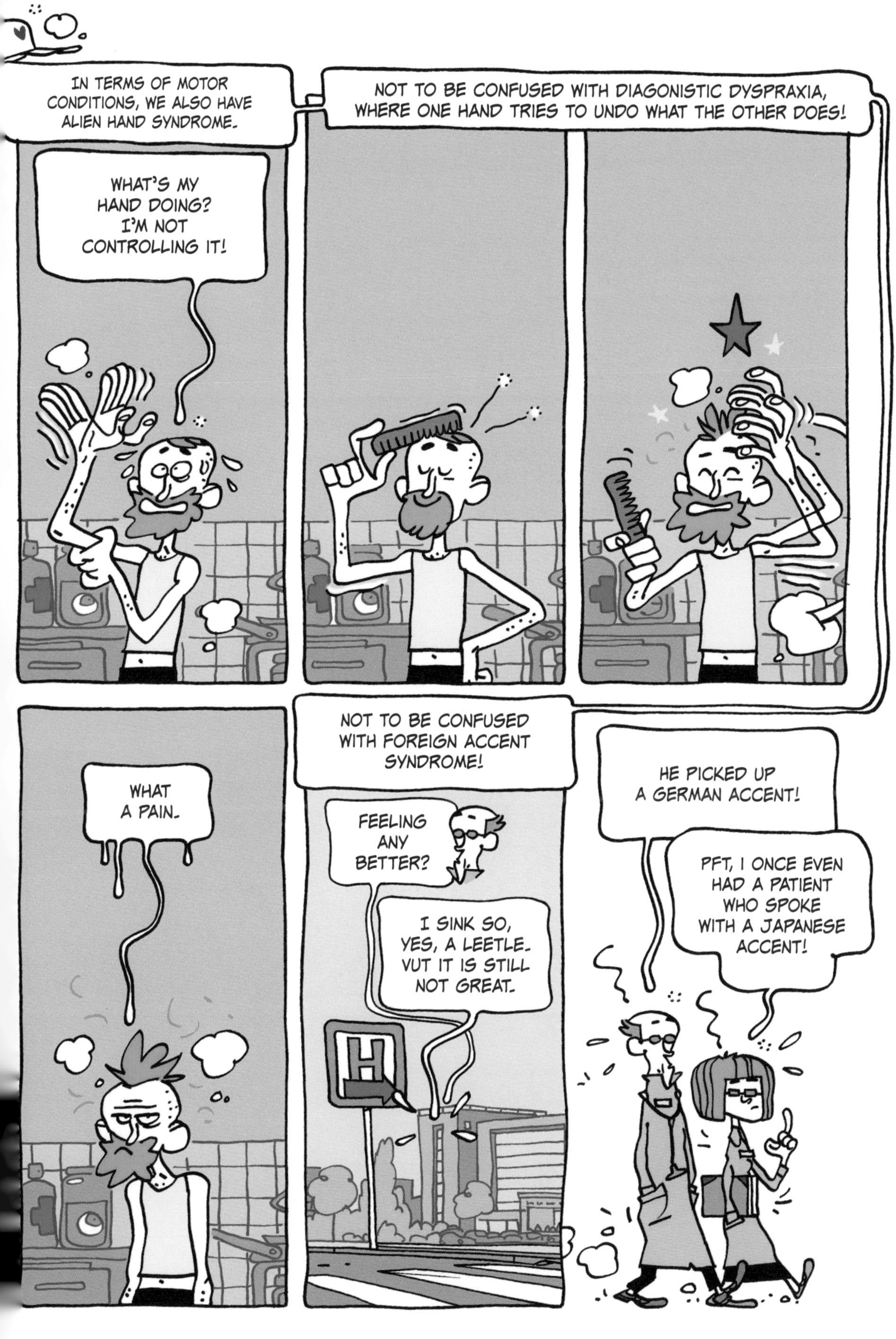

IN TERMS OF MOTOR CONDITIONS, WE ALSO HAVE ALIEN HAND SYNDROME.
WHAT'S MY HAND DOING? I'M NOT CONTROLLING IT!
NOT TO BE CONFUSED WITH DIAGONISTIC DYSPRAXIA, WHERE ONE HAND TRIES TO UNDO WHAT THE OTHER DOES!
WHAT A PAIN.
NOT TO BE CONFUSED WITH FOREIGN ACCENT SYNDROME!
FEELING ANY BETTER?
I SINK SO, YES, A LEETLE. VUT IT IS STILL NOT GREAT.
HE PICKED UP A GERMAN ACCENT!
PFT, I ONCE EVEN HAD A PATIENT WHO SPOKE WITH A JAPANESE ACCENT!

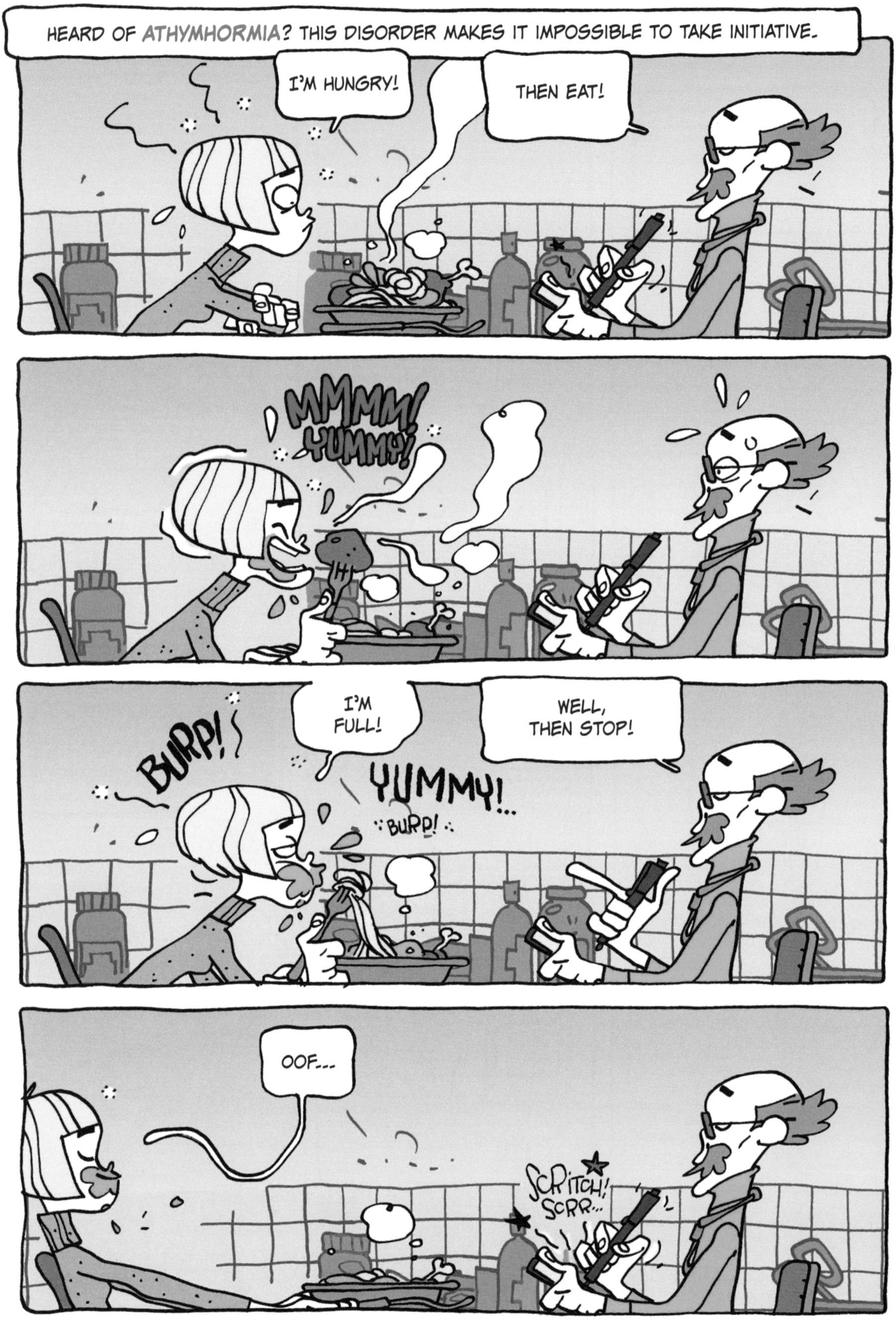
HEARD OF ATHYMHORMIA? THIS DISORDER MAKES IT IMPOSSIBLE TO TAKE INITIATIVE.
I'M HUNGRY!
THEN EAT!
MMMM!
YUMMY!
BURP!
I'M FULL!
YUMMY!...
BURP!
WELL, THEN STOP!
OOF...
SCRITCH! SCRR...

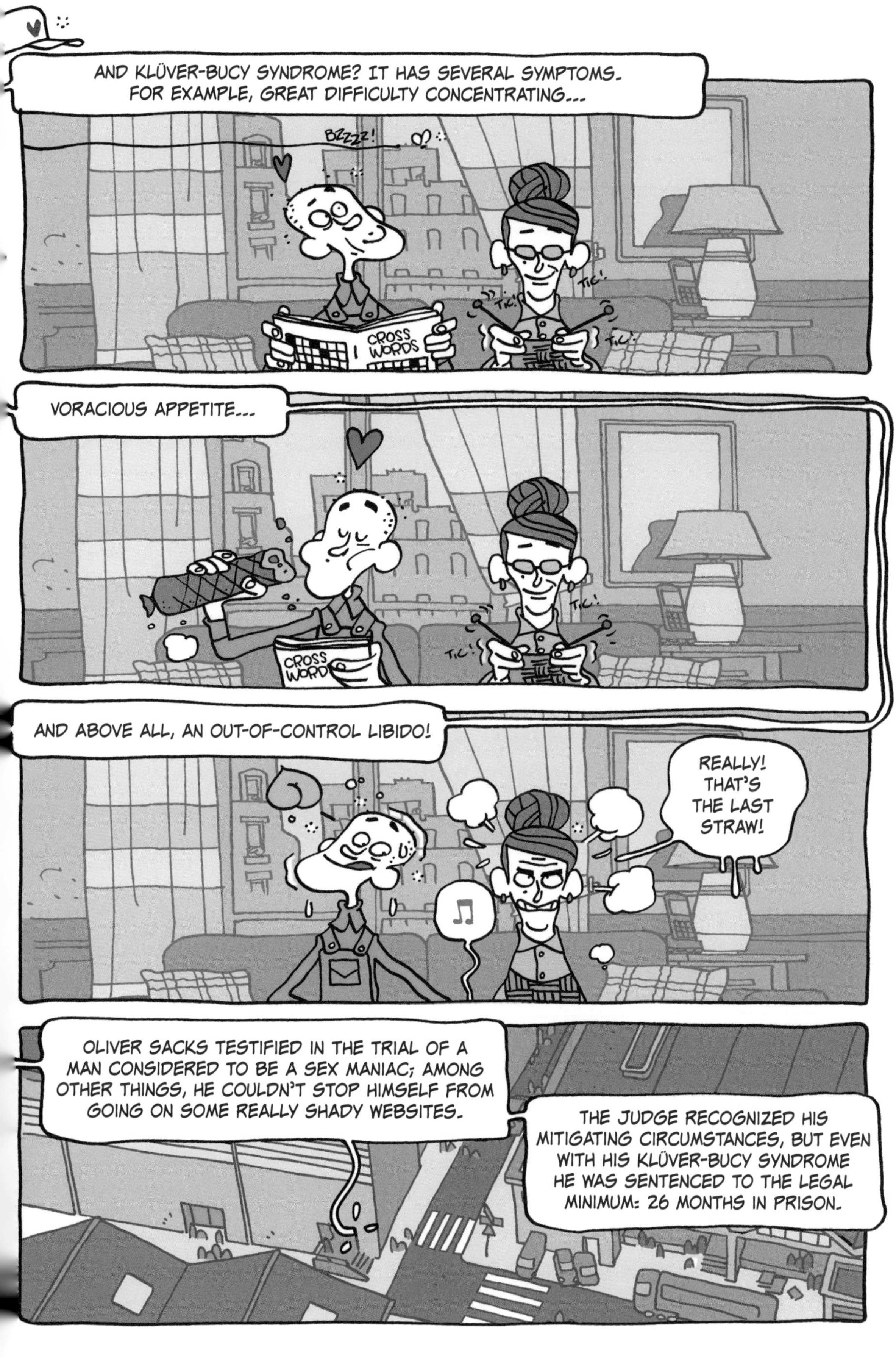
AND KLÜVER-BUCY SYNDROME? IT HAS SEVERAL SYMPTOMS. FOR EXAMPLE, GREAT DIFFICULTY CONCENTRATING...
BZZZZ!
CROSS WORDS
TIC!
TIC!
TIC!
VORACIOUS APPETITE...
CROSS WORDS
TIC!
TIC!
AND ABOVE ALL, AN OUT-OF-CONTROL LIBIDO!
REALLY! THAT'S THE LAST STRAW!
OLIVER SACKS TESTIFIED IN THE TRIAL OF A MAN CONSIDERED TO BE A SEX MANIAC; AMONG OTHER THINGS, HE COULDN'T STOP HIMSELF FROM GOING ON SOME REALLY SHADY WEBSITES.
THE JUDGE RECOGNIZED HIS MITIGATING CIRCUMSTANCES, BUT EVEN WITH HIS KLÜVER-BUCY SYNDROME HE WAS SENTENCED TO THE LEGAL MINIMUM: 26 MONTHS IN PRISON.

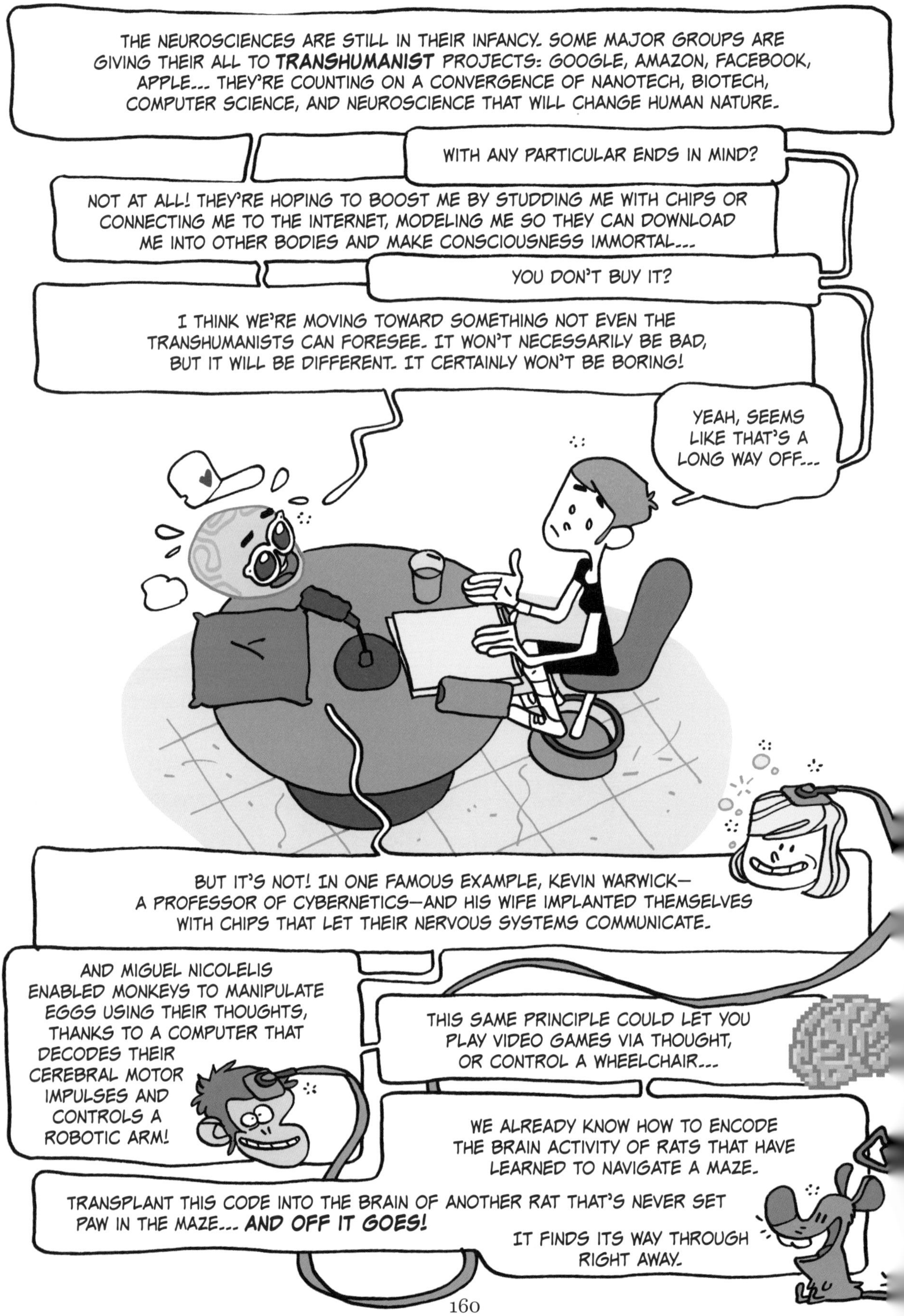
THE NEUROSCIENCES ARE STILL IN THEIR INFANCY. SOME MAJOR GROUPS ARE GIVING THEIR ALL TO **TRANSHUMANIST** PROJECTS: GOOGLE, AMAZON, FACEBOOK, APPLE... THEY'RE COUNTING ON A CONVERGENCE OF NANOTECH, BIOTECH, COMPUTER SCIENCE, AND NEUROSCIENCE THAT WILL CHANGE HUMAN NATURE.
WITH ANY PARTICULAR ENDS IN MIND?
NOT AT ALL! THEY'RE HOPING TO BOOST ME BY STUDDING ME WITH CHIPS OR CONNECTING ME TO THE INTERNET, MODELING ME SO THEY CAN DOWNLOAD ME INTO OTHER BODIES AND MAKE CONSCIOUSNESS IMMORTAL...
YOU DON'T BUY IT?
I THINK WE'RE MOVING TOWARD SOMETHING NOT EVEN THE TRANSHUMANISTS CAN FORESEE. IT WON'T NECESSARILY BE BAD, BUT IT WILL BE DIFFERENT. IT CERTAINLY WON'T BE BORING!
YEAH, SEEMS LIKE THAT'S A LONG WAY OFF...
BUT IT'S NOT! IN ONE FAMOUS EXAMPLE, KEVIN WARWICK—A PROFESSOR OF CYBERNETICS—AND HIS WIFE IMPLANTED THEMSELVES WITH CHIPS THAT LET THEIR NERVOUS SYSTEMS COMMUNICATE.
AND MIGUEL NICOLELIS ENABLED MONKEYS TO MANIPULATE EGGS USING THEIR THOUGHTS, THANKS TO A COMPUTER THAT DECODES THEIR CEREBRAL MOTOR IMPULSES AND CONTROLS A ROBOTIC ARM!
THIS SAME PRINCIPLE COULD LET YOU PLAY VIDEO GAMES VIA THOUGHT, OR CONTROL A WHEELCHAIR...
WE ALREADY KNOW HOW TO ENCODE THE BRAIN ACTIVITY OF RATS THAT HAVE LEARNED TO NAVIGATE A MAZE.
TRANSPLANT THIS CODE INTO THE BRAIN OF ANOTHER RAT THAT'S NEVER SET PAW IN THE MAZE... **AND OFF IT GOES!**
IT FINDS ITS WAY THROUGH RIGHT AWAY.

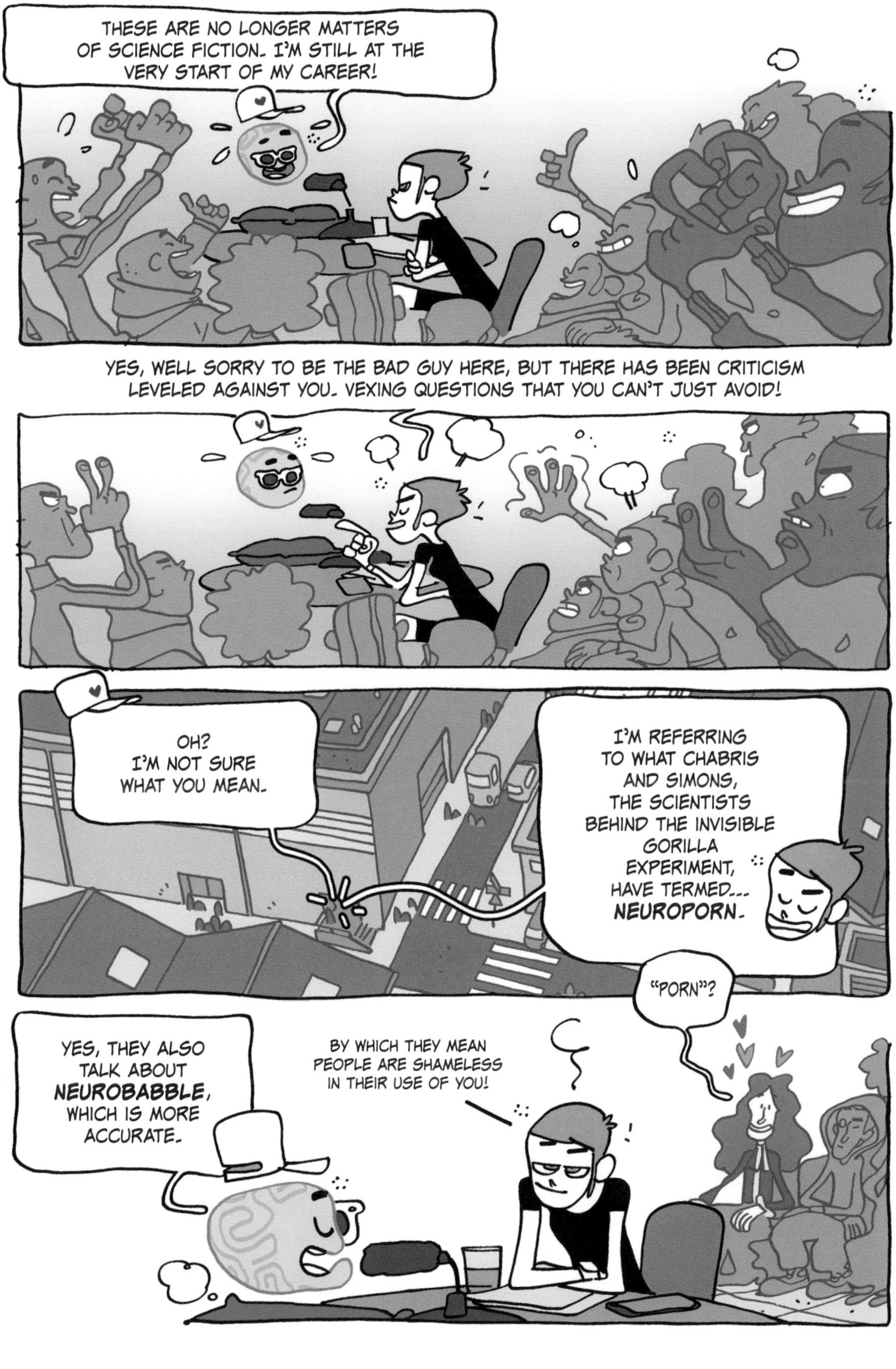
THESE ARE NO LONGER MATTERS OF SCIENCE FICTION. I'M STILL AT THE VERY START OF MY CAREER!
YES, WELL SORRY TO BE THE BAD GUY HERE, BUT THERE HAS BEEN CRITICISM LEVELED AGAINST YOU. VEXING QUESTIONS THAT YOU CAN'T JUST AVOID!
OH? I'M NOT SURE WHAT YOU MEAN.
I'M REFERRING TO WHAT CHABRIS AND SIMONS, THE SCIENTISTS BEHIND THE INVISIBLE GORILLA EXPERIMENT, HAVE TERMED... **NEUROPORN**.
"PORN"?
YES, THEY ALSO TALK ABOUT **NEUROBABBLE**, WHICH IS MORE ACCURATE.
BY WHICH THEY MEAN PEOPLE ARE SHAMELESS IN THEIR USE OF YOU!

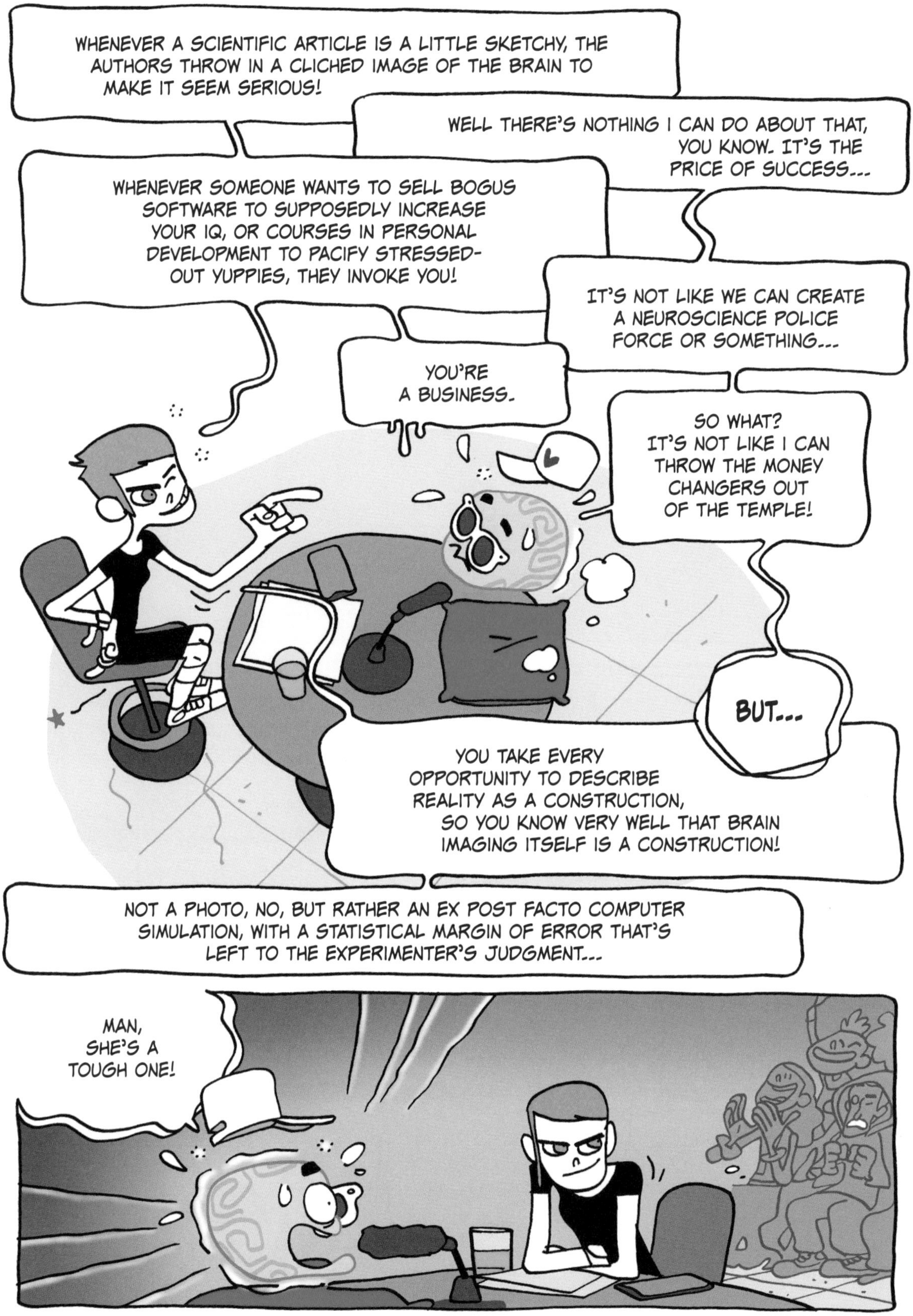
WHENEVER A SCIENTIFIC ARTICLE IS A LITTLE SKETCHY, THE AUTHORS THROW IN A CLICHED IMAGE OF THE BRAIN TO MAKE IT SEEM SERIOUS!
WELL THERE'S NOTHING I CAN DO ABOUT THAT, YOU KNOW. IT'S THE PRICE OF SUCCESS...
WHENEVER SOMEONE WANTS TO SELL BOGUS SOFTWARE TO SUPPOSEDLY INCREASE YOUR IQ, OR COURSES IN PERSONAL DEVELOPMENT TO PACIFY STRESSED-OUT YUPPIES, THEY INVOKE YOU!
IT'S NOT LIKE WE CAN CREATE A NEUROSCIENCE POLICE FORCE OR SOMETHING...
YOU'RE A BUSINESS.
SO WHAT? IT'S NOT LIKE I CAN THROW THE MONEY CHANGERS OUT OF THE TEMPLE!
BUT...
YOU TAKE EVERY OPPORTUNITY TO DESCRIBE REALITY AS A CONSTRUCTION, SO YOU KNOW VERY WELL THAT BRAIN IMAGING ITSELF IS A CONSTRUCTION!
NOT A PHOTO, NO, BUT RATHER AN EX POST FACTO COMPUTER SIMULATION, WITH A STATISTICAL MARGIN OF ERROR THAT'S LEFT TO THE EXPERIMENTER'S JUDGMENT...
MAN, SHE'S A TOUGH ONE!

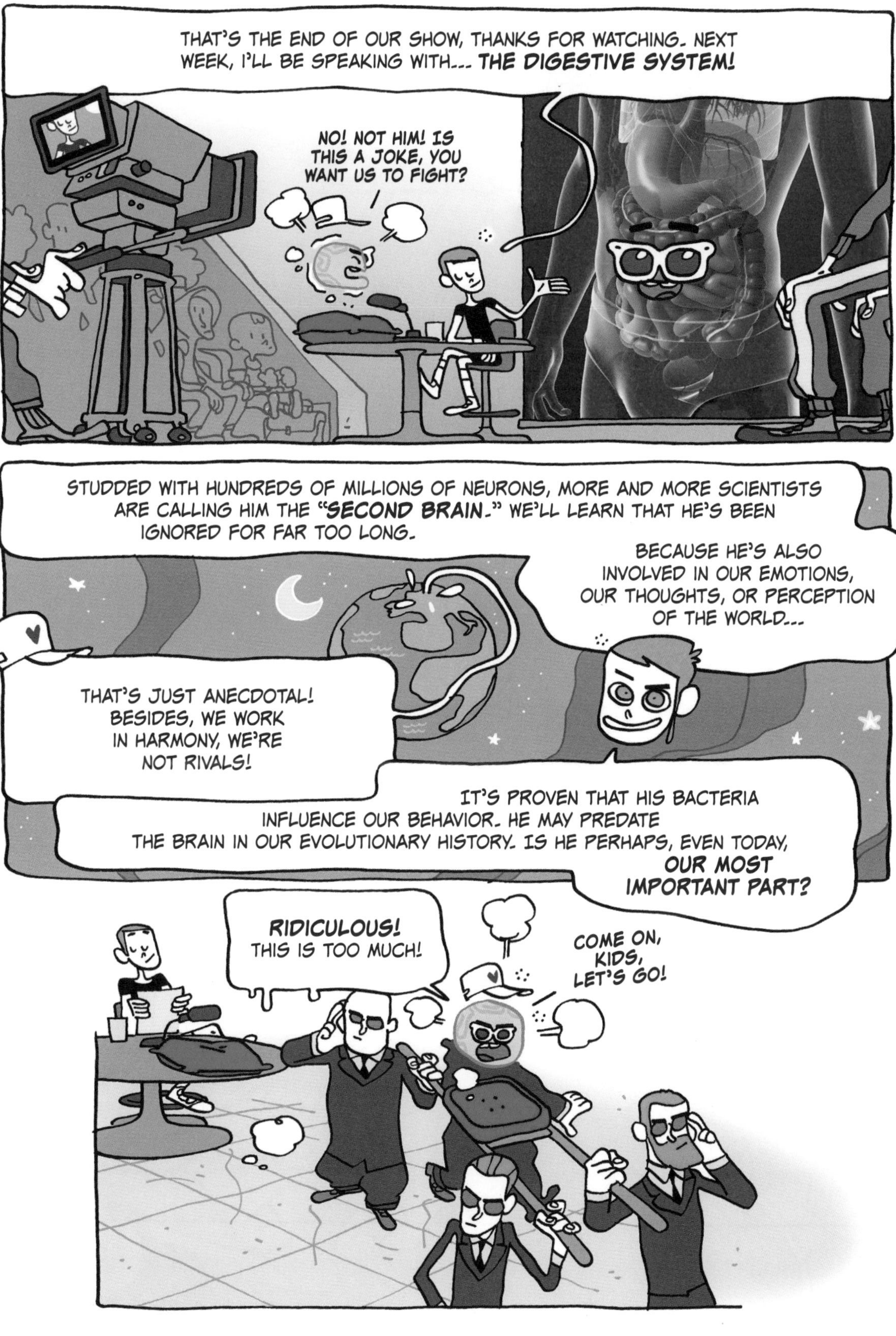
THAT'S THE END OF OUR SHOW, THANKS FOR WATCHING. NEXT WEEK, I'LL BE SPEAKING WITH... **THE DIGESTIVE SYSTEM!**
NO! NOT HIM! IS THIS A JOKE, YOU WANT US TO FIGHT?
STUDDED WITH HUNDREDS OF MILLIONS OF NEURONS, MORE AND MORE SCIENTISTS ARE CALLING HIM THE **"SECOND BRAIN."** WE'LL LEARN THAT HE'S BEEN IGNORED FOR FAR TOO LONG.
BECAUSE HE'S ALSO INVOLVED IN OUR EMOTIONS, OUR THOUGHTS, OR PERCEPTION OF THE WORLD...
THAT'S JUST ANECDOTAL! BESIDES, WE WORK IN HARMONY, WE'RE NOT RIVALS!
IT'S PROVEN THAT HIS BACTERIA INFLUENCE OUR BEHAVIOR. HE MAY PREDATE THE BRAIN IN OUR EVOLUTIONARY HISTORY. IS HE PERHAPS, EVEN TODAY, **OUR MOST IMPORTANT PART?**
RIDICULOUS! THIS IS TOO MUCH!
COME ON, KIDS, LET'S GO!

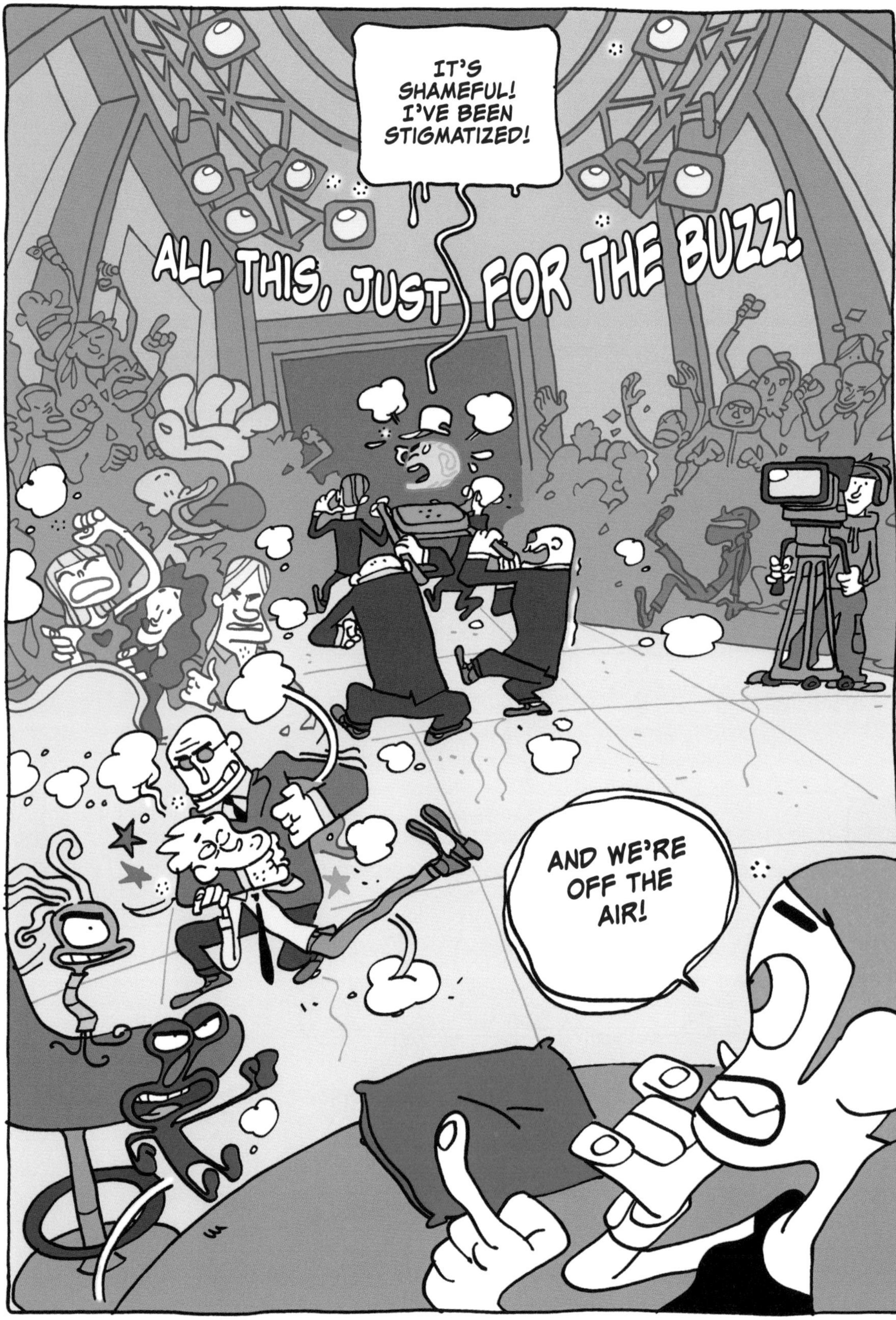
IT'S SHAMEFUL! I'VE BEEN STIGMATIZED!
ALL THIS, JUST FOR THE BUZZ!
AND WE'RE OFF THE AIR!

Library of Congress Cataloging-in-Publication Data

Names: Marmion, Jean-François, author. | Monsieur B, 1966– illustrator.
Title: Braincomix / Jean-François Marmion and Monsieur B
Other titles: Cervocomix. English
Description: University Park, Pennsylvania : The Pennsylvania State University Press/Graphic Mundi, [2021] | Originally published as Cervocomix: Le cerveau expliqué en BD by Jean-François Marmion and Monsieur B Paris : Les Arènes, 2019; translated by Hannah Chute.
Summary: "An exploration of the complexities of the human brain in graphic novel format"—Provided by publisher.
Identifiers: LCCN 2021010042 | ISBN 9781637790021 (hardback ; alk. paper)
Subjects: MESH: Brain—physiology | Mental Processes—physiology | Graphic Novel
Classification: LCC QP376 | NLM WL 17 | DDC 612.8/2—dc23
LC record available at https://lccn.loc.gov/2021010042

Printed in Lithuania
Published by The Pennsylvania State University Press, University Park, PA 16802-1003

graphic mundi
drawing our worlds together

Graphic Mundi is an imprint of The Pennsylvania State University Press.

Translated by Hannah Chute
Supplemental lettering by Indigo Kelleigh

Originally published as *Cervocomix: Le cerveau expliqué en BD* by Jean-François Marmion and Monsieur B

The Pennsylvania State University Press is a member of the Association of University Presses.

It is the policy of The Pennsylvania State University Press to use acid-free paper. Publications on uncoated stock satisfy the minimum requirements of American National Standard for Information Sciences—Permanence of Paper for Printed Library Material, ANSI Z39.48–1992.

The script writer would like to thank Monsieur B for breathing life into his crazy ideas, and Manolo for colorizing them. Thanks as well to Laurent Beccaria, Catherine Meyer, and Laurent Muller for giving us such wonderful freedom.

A special thanks to Gérard Depardieu, who lent his likeness to the historical reenactments throughout this book, and who would surely have given us his blessing to use it if we had thought to ask him....

To you, dear reader!
Monsieur B

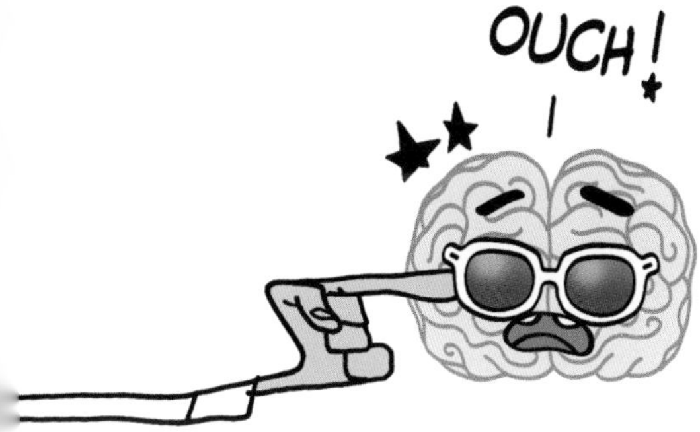
OUCH!